PLANT INDICATORS OF SOILS, ROCKS, AND SUBSURFACE WATERS

RASTITEL'NYE INDIKATORY POCHV, GORNYKH POROD I PODZEMNYKH VOD

РАСТИТЕЛЬНЫЕ ИНДИКАТОРЫ ПОЧВ, ГОРНЫХ ПОРОД И ПОДЗЕМНЫХ ВОД

# PLANT INDICATORS
## OF SOILS, ROCKS,
## AND
## SUBSURFACE WATERS

Edited by
A. G. Chikishev

*Authorized translation from the Russian*

SPRINGER SCIENCE+BUSINESS MEDIA, LLC
1965

The Russian text, the Proceedings of the Conference on Indicational Geobotany held on February 3-6, 1961, was published by Nauka in Moscow in 1964 as the Transactions (Trudy) of the Moscow Society of Naturalists, Volume VIII, 1964, Geological-Geographical Series, Geography Division.

Library of Congress Catalog Card Number 65-15596

ISBN 978-1-4899-4916-5     ISBN 978-1-4899-4914-1 (eBook)
DOI 10.1007/978-1-4899-4914-1

# PREFACE TO THE AMERICAN EDITION

The problem of the possible use of vegetation for determining the properties of the substrate on which the vegetation grows was a matter of interest even in ancient times. We find in the works of Vitruvius, Virgil, Pausanius, and Palladius references to plants which can be used in searching for subsurface waters or evaluating soil quality.

With the progressive development of botanical science, knowledge concerning plant indicators gradually developed into a special branch of science. Its rudiments should clearly be sought in the works of Linnaeus himself, as has been demonstrated conclusively by E. Du-Rietr (1957), in an article describing Linnaeus as a phytogeographer. A number of important concepts relating to the close association of plants with the substrate, which opened up the possibility of using plants as indicators of soils and rocks, are found in the works of Humboldt, Griesebach, Unger, Turmann, Warming, Schimper, and other leading botanists.

In Russia, the earliest investigations involving the use of indicators were those of P. S. Pallas (1786), É. A. Eversman (1825), and A. Yagmin (1845), which showed the possibility of prospecting for water in sands, and for evaluating the fertility of soils in steppe and desert regions, on the basis of the character of the vegetation cover. The geologists A. M. Karpinskii (1841), P. A. Ososkov (1886, 1909-1912), and N. K. Vysotskii (1904) should, however, be regarded as the founders of specialized indicator research in Russia, as they were the first to deal with the problem of the indicator role of vegetation with rocks and soils and to make wide use of geobotanical observations in practical geological surveys. Later, the works of B. A. Keller (1912) and L. V. Larin (1926) appeared, representing attempts to compile special handbooks for determining lithological and soil conditions on the basis of geobotanical data. Finally, the theoretical premises of indicator research were dealt with in the work of D. N. Kashkarov (1933), who made wide use in his research of many of the views put forward in the well-known work of F. Clements (1920). Problems of indicator research in subsurface water were surveyed in papers by V. A. Priklonskii (1935), and the same topic was dealt with in the studies of O. Meintzer (1927) in the USA, which were mainly concerned with phreatophytes.

Indicator research reached its greatest scope in the Soviet Union after 1945, when, as a result of the introduction of aerial methods into geology and hydrogeology, several organizations, notably the All-Union Aerogeological Trust, established special geobotanical units in geological expeditions. Geobotanical research methods became widely used as an aid in geological mapping, prospecting for subsurface waters, mapping of salinated substrates, and prospecting for certain useful fossils. These studies yielded rich factual material, which was reviewed for the first time by S. V. Viktorov in his monograph, "The Use of the Geobotanical Method in Geological and Hydrogeological Research," published by the USSR Academy of Sciences Publishing House in 1955. At the same time, work was conducted on the use of plant indicators in agricultural research in meadows and pastures. As an outcome of this work, L. G. Ramenskii (1956) produced ecological tables for diagnosing habitat conditions, which were published only after the death of this outstanding geobotanist.

The further development of research along these lines led to the origin in the Soviet Union of a special discipline, indicator geobotany, representing one branch of biogeography and having as its aim the study of the theory and practice of using vegetation as an indicator of different physical and geographical conditions. At the All-Union Research Institute of Hydrogeology and Engineering Geology (VSEGINGEO), a special laboratory concerned with indicator geobotany research has been established. There are groups of scientists working in the same field at the Institute of Geography of the USSR Academy of Sciences, the Laboratory of Aerial Methods at the All-Union Aerogeological Trust, Moscow University, and other institutions and organizations in the country. Meetings and Conferences on indicator geobotany have been held since 1956. The 1951 meeting, the proceedings of which are contained in the present collection, was the broadest in its program and the most representative.

The most actively developing trends in indicator geobotany in the USSR at present are the use of hydrologic indicators (in subsurface waters), lithologic indicators (in rocks), pedologic indicators (in soils), and halogenic indicators (for different types of saline substrates), as well as in prospecting for useful fossils. Aerial methods (the interpretation of aerial photographs and aerovisual observations) are of great significance in almost all the trends mentioned.

The considerable achievements attained in landscape studies in the USSR and the steady expansion of the use of aerial methods in geographical research have led to the gradual evolution of indicator geobotany into a new branch of science—indicator landscape studies. Basic to this branch of science is the concept that in each landscape we can differentiate readily observable (physiognomic) components and components accessible to observation only with difficulty (decipient components). There exist close and complex relationships between these groups of components which permit the use of physiognomic components as indicators of decipient components. This promising approach has already some successes to its credit. It is being most actively developed within the walls of the senior Society of our country, the Moscow Society of Naturalists.

The present collection, which comprises the Proceedings of the 1961 Conference on Indicator Geobotany, includes papers dealing with the main trends of indicator research. Unfortunately, there is a lack of papers devoted to the use of indicators of frozen ground, and research work on indicator research for useful fossils is inadequately represented, the latter problem having been frequently surveyed in conferences on biogeochemistry. Nevertheless, the articles included in the collection provide a good picture of the trends of indicator work in the USSR, since they cover not only problems in the theory of indicator geobotany, but also researches on the use of indicators in soils, parent soil-forming rocks, subsurface waters, and, to some extent, useful fossils, as well as certain problems of procedure.

A. G. Chikishev

# PREFACE

The process of development of geographical research is introducing new ways for the application of research work in solving various scientific and national economic problems. One of the new trends in this field is geobotanical indicator research, the basis of which is the utilization of the vegetative cover as an indicator of varied conditions of the physical and geographical environment. This type of research is particularly topical at the present time, in view of intensive development of the vast, sparsely inhabited regions of Kazakhstan, Siberia, and the Far East, associated with the opening up of the virgin lands and the massive growth of new industrial construction. Also relevant is the problem of the irrigation of enormous pasture areas in Kazakhstan and the republics of Central Asia. Work on land irrigation and amelioration is on a very large scale. All these projects require rapid, cheap, and reasonably accurate methods for the evaluation of climatic, soil and terrain, and hydrogeological conditions. The widespread use of vegetation as an indicator of these or other components of the landscape may be regarded as one of the methods in this category. The role of geobotanical indicator research has shown an especially significant increase with the introduction of aerial survey methods into geological and geographical science.

Geobotanical indicator research (indicator geobotany) is at present being applied in resolving a wide range of scientific and practical problems. It is being used in the agricultural evaluation of territories (in connection with the compilation of soil maps), in engineering and geological surveys (for highway construction in swamps, deserts, and districts where soil and terrain are saline, for surveys in districts where the ground is frozen, and also for small-scale engineering and geological surveys), for elucidating hydrogeological conditions in irrigated districts, for studies on swamps intended for industrial and agricultural uses, in prospecting for certain species of useful fossils, and so on. A substantial amount of experience has been gained by research workers in the field of indicator geobotany, and this requires analysis and assessment. Accordingly, in February 1961, the Geographical Section of the Moscow Society of Naturalists, the All-Union Research Institute of Hydrogeology and Engineering Geology (VSEGINGEO), and the All-Union Aerogeological Trust (VAGT) convened a conference on indicator geobotany, in which numerous individuals working in the field participated. The present collection deals with theoretical problems of indicator geobotany, as well as including papers on indicator work with soils, parental soil-forming rocks, subsurface waters, useful fossils, the results of studies on the biology and ecology of plant indicators, and also some special methodological problems, such as the procedure for the compilation of indicator reference books for geological and hydrogeological purposes and the procedure for interpreting geobotanical maps for the agricultural evaluation of a territory. It is hoped that the publication of the Proceedings of the Conference will promote the introduction of geobotanical indicator methods into research practice in soils, agronomy, hydrogeology, and engineering geology, and will also promote closer study of plant indicators and precise analysis of the practical results attained through the use of this method.

# CONTENTS

# SOME PROBLEMS IN THE THEORY OF GEOBOTANICAL INDICATOR RESEARCH

## S. V. Viktorov, E. A. Vostokova, and D. D. Vyshivkin

A large number of scientists are now devoting their attention to the study of plant indicators. The field has gradually developed and in some case more or less final procedural techniques for discovering indicators and for utilizing them for various purposes have been derived.

A massive literature has appeared on plant indicators, both in the USSR and abroad, and a number of review-type papers have been published (Ramenskii, 1938; Viktorov, 1955; Sykora, 1959; Vostokova, 1961; Viktorov, Vostokova, and Vyshivkin, 1962).

Within the science of the vegetational cover—geobotany—there has gradually developed a special branch, which is known as indicator geobotany. Like any other young branch of research, indicator geobotany requires exact definitions of some of the basic concepts and terminology involved in the subject and the precise statement of the main problems with which it is concerned. We should like briefly to deal with these problems. While giving below a number of definitions dealing with the basis of indicator geobotany, the authors would like to point out that they do not consider these definitions to be completely final and assume that, in the course of the further development of indicator geobotany, many of the definitions provided will undergo various modifications.

Turning now to the presentation of definitions of some of the basic concepts of indicator science, the authors consider that the indicator geobotany should be regarded as the division of geobotany which is concerned with the study of the theoretical principles and practical applications of the vegetation cover and its component species as indicators of the conditions of the environment.

In its capacity as the science of the vegetation cover within the field of geobotany, indicator geobotany is very clearly linked with plant ecology, physiology, and biochemistry, on the one hand, and with the related geological and geographical sciences (geology, soil science, terrain studies, hydrogeology, and geochemistry), on the other.

The main concepts which have a bearing on indicator geobotany are the concepts "indicator" and "indication," as well as the objects of indicator research; importance must also be imputed to views regarding indirect, direct, universal, and local indicators.

Indicators in indicator geobotany may be regarded as plant species (or smaller intraspecific entities) or their natural regular combinations—plant communities—having a preferential association with definite ecological conditions that is sufficiently constant to be recognized from the presence of the particular species or communities.

Besides species and communities having indicator significance, indicator characteristics should also be differentiated. Indicator characteristics may be regarded as those factors of species and communities which can serve as indicators of definite environmental conditions, although the actual organisms and communities themselves are not indicators of the particular conditions.

The study of the ecology of indicators and indicator characteristics comprises the branch of science known as indicator geobotany, while those components of the geographic environment, or their individual properties, for whose determination the indicators are utilized may be defined as the indicator objects. The indicator objects can be either distinct types of natural matter and their components (for example, types, subtypes, and forms of soil), or their different properties (mechanical and chemical composition of soils or earths, or mineralization of ground waters), or various elements and compounds distributed within the limits of the zone available to plants, specific features of the spacial structure (for example, the depth of the water table) and the processes going on in it, and, finally, the practical suitability of the area for particular forms of economic activity (e.g., plowing, the sowing of definite crops, and so on). The link between the indicator and the indicator object is based on the existence of an association between them, which may be termed indicational. The indicational association arises on the basis of definite physiological, biochemical, and biogeochemical processes, during the course of which the plant enters into a close interrelationship with the surrounding environment. The analysis of indicational associations provides the data on which is based the theoretical possibility of indicating specific components of the environment.

Direct indicators should clearly be regarded as those whose presence is directly determined by the presence of the given indicator object and depends on it. Indirect indicators are those which do not have a direct association with the indicator object but which are related to some condition associated with the indicator object; there is thus no direct relationship between an indirect indicator and the indicator object, but merely the concurrent presence of the two, controlled by some other indicational association.

Another essential difference between "direct" and "indirect" indicators consists of the fact that although the indicational significance of direct indicators can fluctuate somewhat under varied physical and geographical conditions, they do, as a rule, constitute indicators of the corresponding indicator objects, whereas the indicational significance of indirect indicators changes markedly from district to district.

The association between indicator and object may be maintained throughout the entire area of distribution of the indicator or may exist in some defined region, representing part of the area. In the first instance, the indicator is designated as universal,* and in the second, as local.

Universal indicators are encountered rarely. In the main, they are either species with an area which is very restricted, both geographically or ecologically, or communities in which such species dominate. Their significance in practical indicator research is still very small, although they are encountered in each group of indicators. An example of a universal hydrologic indicator, in the widest sense of the word, is the common reed (Phragmites communis Trin.), which throughout the entire range of its vast distribution area develops an overgrowth only under conditions where there is an association between its subterranean part and water. Many landscape sand types are among the universal geologic indicators. Certain halophytes also have a considerable degree of universality.

The brief definitions of these basic, most widely used concepts of indicator geobotany suggested by the authors should lead to the attainment of unified terminology and usage.

The following are the most important problems under study in indicator geobotanical research:

1. discovery of indicators;
2. evaluation of the effectiveness and reliability of different indicators;
3. the study of the biology and ecology of indicators and the nature of their relationship with the object of indication;
4. the study of the geographical dependence of indicator associations (indicator zonation);
5. the cartographic representation of indicator associations (indicator mapping);
6. the search for more effective ways of using indicators in the national economy.

The most accurate methods for discovering indicators found by the present authors in the course of their research work is the technique of standardized plots and the profile method. The standardized plot method is based on the description of a series of experimental plots, which as a rule include a prospecting shaft, borehole, or a mountain site, where there is known to be a particular indicator object, and the statistical treatment and

---

*It would be more correct to call the universal indicators "pan-areal," but we shall retain the previous term as it is more widely used.

comparative analysis of the data obtained with a view to revealing characteristic features of the vegetative cover associated with the desired object. Quite frequently, particularly in the case of searches for useful fossils, it is useful in addition to describe control plots where the particular indicator object is known to be absent: a comparison of the standardized and control descriptions permits indicators or indicator characteristics to be found with greater accuracy. In some cases (mainly when we are dealing with hydrologic, halidic, or lithologic indicator investigations) the descriptions of standardized plots may be effected in a somewhat different manner. It is as follows: a large number of plots are selected within the limits of a single community, and the purpose of the descriptions is to discover those soil, hydrogeological, and lithological conditions with which the community is associated. In both cases, the detailing of the standardized plots provides data the analysis of which can establish the relationship of the objects of interest with particular indicators.

Clearly related to the standardized plot procedure is the method of profiling, which in essence represents the fixing and analysis of natural ecological series. In the process of this analysis, the relationship of different indicator characteristics defined elements of the physical and geographical environment becomes evident.

The next problem of indicator geobotany is the assessment of the reliability and effectiveness of the indicators discovered. For this purpose, the authors suggest the employment of determinations of two values: validity and significance.

By validity, we mean the degree of concurrence of the indicator with the indicator object. We suggest as an appropriate index the ratio of the percentage of standardized plots where the indicator object and the indicator are encountered together, to the percentage of plots where the combination is not observed; 100% is taken to be the number of standardized plots described within the range of the particular indicator community. By significance, we understand the frequency with which the given indicator is found* together with the indicator object, expressed as a percentage of the total number of standardized descriptions made at the given object. Thus, in contrast to the definition of validity, 100% is here taken to be the sum of descriptions relating to a definite indicator object rather than to a specific indicator.

We thus see that validity determines the degree of stability of the association between the indicator and the indicator object, i.e., the reliability of the indicator; significance, however, indicates how frequently the indicator is encountered at the indicator object, i.e., it characterizes its practical efficiency. Scales of indicator validity and significance have been worked out (Viktorov, Vostokova, and Vyshivkin, 1962).

Investigation of the biological principles of the association between indicator and its object is still the least developed branch of indicator geobotany. Within the field of hydrologic indicator research, L. N. Beideman and his pupils have obtained data elucidating the ecological principles of this relationship. Contributions by biogeochemists of Academician A. P. Vinogradov's school have done much to provide a biological basis for the indication of useful fossils. But extremely little has been done on the biological principles of the indication of soils and soil-forming rocks and there are absolutely no defined concepts relating to the basis of the indication of the most recent tectonic processes. In all these research areas, there has up to now been merely the recording of concurrence of indicators and indicator objects without any fundamental biological study. The biological principles of indicator research, as a whole, still await clarification.

One of the most studied divisions of indicator geobotany at the present time is indicator mapping. Many years of experience in the compilation of detailed maps, during which different types of geobotanical indicator surveys were tried out, have convinced the present authors that the following two techniques are most accurate: specific mapping of indicators and indicator mapping on the basis of interpretation.

Specific indicator mapping is possible when the indicators of the objects of interest to the researcher have already been ascertained. In these cases, indicator mapping leads to a possibly more exact pinpointing of the distribution of these indicators on a topographical basis: the areas where the indicators are absent can be shown

---

*We do not here designate frequency as "frequency of occurrence" in the usual geobotanical sense, but rather as the frequency with which a given indicator (for example, a specific community) is found on the given object of indication (for example, on a specific rock formation or soil).

on the map in outline form only. This approach to mapping is most applicable in those cases where we are concerned with finding objects which do not have continuous distribution, but are spread in some sort of isolated, irregular plots (for example, in the case of searches for a particular useful fossil, when the researcher should really attempt to discover and map only those areas in which known indicators are present).

This compilation of indicator maps on the basis of interpretation can be applied in a much larger number of cases. This should be carried out by making an ordinary geobotanical survey and compiling a map, which is then transposed by interpretation into an indicator map on the basis of an "indicator scheme," i.e., tables in which the indicators are listed and their indicational significance is shown. This scheme may be provided either by preparing it simultaneously with the compilation of the map (using the methods for discovering indicators described above) or by preparation prior to the commencement of the survey. The compilation of indicator maps on the basis of the interpretation of the results of geobotanical surveys finds its widest application in those cases where we are dealing with the indication of objects having continuous distribution, for example, soils, soil-forming rocks, etc. It is inapplicable for the compilation of the type of indicator map where the indicators used are not plant communities, but individual species or some specific indicator characteristics (for example, the appearance of some anomalous forms and readings), i.e., those phenomena which are not covered by the usual geobotanical map.

The above reflections on the theory and procedure of geobotanical indicator research represent one of the first attempts to systematize definitions of the main concepts of this branch of geobotany, to describe the problems with which it is concerned, and to indicate some of the most accurate methods of research. It is fully evident that deep consideration of each of these questions was necessary.

Summary

Definitions are given of some of the basic concepts of indicator geobotany, which the authors regard as the branch of geobotany concerned with the study of the vegetative cover as an indicator of the physicogeographical environment and its controlling factors. The major problems in this field of research are: (1) discovery of indicators, (2) estimation of their effectiveness and reliability, (3) investigations on the biology and ecology of indicators and the nature of their relationship to the indicator object, (4) indicator zonation, (5) indicator mapping, and (6) the possible use of indicators in the national economy.

The concepts of indicator significance (the frequency of occurrence of the indicator on a given object of indication) and validity (the ratio of the number of test plots where the indicator occurs together with the indicator object to the total number of plots described within a given indicator community) are suggested. The concepts of "indicator" (direct, indirect, local, and universal) and "indicator object" are amongst several others defined.

Literature Cited

Viktorov, S. V. 1955. The Use of the Geobotanical Method in Geological and Hydrogeological Research. Moscow, Izd. Akad. Nauk SSSR.
Viktorov, S. V., and Vostokova, E. A. 1961. Principles of Indicator Geobotany. Moscow, Gosgeoltekhizdat.
Viktorov, S. V., Vostokova, E. A., and Vyshivkin, D. D. 1962. Introduction to Indicator Geobotany. Izd. MGU.
Vostokova, E. A. 1961. Geobotanical Methods of Prospecting for Subsurface Waters in the Dry Regions of the USSR. Moscow, Gosgeoltekhizdat.
Ramenskii, L. G. 1938. Introduction to Complex Soil and Geobotanical Land Research. Moscow, Sel'khozgiz.
Sykora, L. 1959. Rostliny v geologickem vyzkumu. Praha.

# THE PRESENT STATE OF HYDROLOGIC INDICATOR RESEARCH

## E. A. Vostokova

The study of the relationships of individual plants and the plant communities they form to subsurface waters comprises two interrelated parts: the first deals with research on the influence of the level of subsurface waters and their chemical status on the distribution and species composition of the vegetation, while the second is concerned with understanding the influence of vegetation on the subsurface water regime.

Studies conducted in the first of these branches comprise the essence of hydrologic indicator research on vegetation, where the latter serves as an index (indicator) of the depth of occurrence and the degree of mineralization of subsurface waters.

In the present article, the results of hydrologic indicator studies will be surveyed, as well as some problems associated with the closer study and wider introduction into practice of hydrogeological and ameliorative surveys. An examination of the influence of vegetation on the subsurface water regime represents an important but independent problem which we shall not deal with.

Hydrologic indicator investigations are by now quite widely used in practice. The results of hydrologic indicator observations have been most widely applied in the search for fresh water close to the soil surface in arid regions, as for example in seeking freshwater lenses, necessary for expanding the network of wells in districts of range livestock farming poorly supplied with water.

V. N. Kunin (1957, 1959), K. F. Orfanidi (1957), and others have referred to the great practical value of the lenses of local subsurface waters for supplying water in desert areas. The value of using hydrologic indicator observations for discovering lenses and shallow-lying ground waters has been demonstrated in practice. This type of work has been carried out on chernozems (Demidova et al., 1955; Levin, 1960; Rodman et al., 1960). in the North Caspian region (Ivanova, 1959, 1960; Tagunova, 1961; Khudyakov, 1961) in Central and Eastern Kazakhstan (Viktorov, 1959, 1960; Vinogradov, 1958; Ostrovskii, 1959; Shavyrina, 1959), and in Kashgaria (Panin, 1960). All these studies involved normal hydrologic indicator mapping, the methods of which have by now been more or less made clear (Vostokova, 1961). The widespread use of geobotanical methods in hydrogeological studies is associated with the introduction of aerial methods into practical hydrogeological research, since vegetation provides features graphically and clearly for hydrogeological deciphering (Viktorov, 1955; Viktorov and Vostokova, 1959; Vinogradov, 1958, 1959, 1961; Vostokova, 1960). The application has markedly increased the speed and precision of hydrologic indicator mapping.

In desert regions, hydrologic indicator research in combination with aerial methods has found wide application not only in seeking small subsand or subtakyr lenses, but also for identifying zones of discharge of large freshwater lenses. As an example, the entire course of the discharge zone of the very large Yaskhan lens was traced by fragments of tugaic vegetation, distributed along the southern edge of the Uzboi. Plots carrying this vegetation were described geobotanically by A. A. Nitsenko (1956) and I. G. Rustamov (1954) and were mapped by deciphering aerial photographs. It proved possible to determine the area of discharge of the lens from the map compiled. The zone of exudation of the Karabil'sk (southern Turkmenia) lens was characterized by a broad belt of black Haloxylone, which were studied in detail from the phytocenological point of view by a number of workers (Leont'ev, 1954; Petrov, 1958; Yagdyev, 1958). A map was prepared of this zone from aerial photographs and this illustrated the change in salinity in subsurface waters in the area where the freshwater lens petered out.

The ever-increasing employment of the geobotanical indicator method in hydrogeological survey or prospecting work has led to the use of new procedural approaches and to new forms of organization of work. For example, V. N. Ostrovskii (1959) made wide use of hydrologic indicator mapping for hydrogeological mapping. As has been shown by the experience of VSEGINGEO in Turkmenia, complex programs applying geobotanical and geophysical methods are highly effective in prospecting for subsurface waters. Areas with shallow-lying subsand or subtakyr lenses can be identified quite clearly by their vegetation, while geophysical methods can be used to define the limits of water with minimal mineralization.*

Hydrologic indicator observations are used in provinces with a temperate climate for finding localities with an accumulation of leakage water, particularly in those cases where there is a resemblance to ground waters in that the water is retained for a more or less prolonged period. In these instances, plots of accumulation are revealed by direct hydrologic indicators. A detailed comparative study of plots with and without leakage water shows that in these cases much attention should be paid not only to normal hydrologic indicator methods, but also to phenological observations, i.e., differences in the dates and progress of phenophases in plants in individual plots. Only the first steps have as yet been taken along these lines. Thus, the observations of N. G. Moskalenko (VSEGINGEO) confirm that plants flower for a longer period in plots with an accumulation of leakage water, while in plots without any such accumulation flowering terminates earlier.

Jakucs (1957) applied the method of coloring water for showing up indicators of the rate and direction of flow of water in a spring-fed marsh in Hungary. He identified communities which were indicators of strong, weak, and slow flow and the stagnant condition. Some new characteristics in hydrologic indicator communities were used by Ellenberg (1950, 1952), the frequency of finding certain phreatophytes and hydrophytes being applied in improving the accuracy of hydrologic indicator forecasts.

The broad development of geobotanical indicator research poses a series of problems, which can be divided into the following main groups.

1. Individual species of indicator plants were originally more widely used in hydrologic indication (Mager, 1912); this leads us to the type of problem in the first group, primarily concerned with the principles of using individual plant species or plant communities in the role of indicators.

Many researchers consider that hydrologic indicator deductions based on the presence of individual plants, without considering their abundance, vegetative vigor, and other ecobiological and phytocenological characteristics (Vostokova, 1961), are inappropriate.

Optimal results are given by using plant communities as indicators. It should be mentioned that sometimes such direct hydrologic indicator communities are composed of one or two species, i.e., they represent an almost pure single-species population. Their indicational significance is, however, equivalent to that of a community.

It is, of course, necessary to carry out a sufficiently comprehensive ecological analysis of the floristic composition of the community, the characteristics of the vegetative vigor and vitality of its component plants, and to sudy the structural features of the phytocenosis (the type of plant distribution in the cenosis, and the layer and sinusoid composition).

Indicator research on the basis of communities has great advantages, especially for the hydrologic-indicator deciphering of aerial photographs and careful aerovisual indicator observations, for which a knowledge is required of deciphering characters in aerial photographs, even if we are dealing with the most widely distributed hydrologic indicator communities. There is no exhaustive monograph, which describes deciphering and aerovisual features even for arid zones, although the problem has been partially dealt with in certain papers (Vinogradov, 1958, 1959, 1961; Viktorov, 1955; Vostokova, 1960; Viktorov and Vostokova, 1959).

---

*V. N. Kunin and G. T. Leshchinskii (1960) established that mineralization of subtakyr lenses changes very gradually over an area, as a result of which it is difficult to define precisely, through the vegetation, the dividing line between fresh (up to 1 g/liter) and weakly saline (up to 3 g/liter) water.

In this connection, more accurate knowledge on the size of the plant indicator communities is very important. The use, as indicators, of large phytocenological units, such as formations or groups of associations in which phreatophytes are dominant, can obviously give merely a general picture of the presence of water. Data on the degree of mineralization of water, on the depth intervals, and some other necessary details, can be obtained by using smaller vegetation units as indicators (Vostokova, 1952).

Among the problems in this group, it is most important to determine the validity and significance of hydrologic indicator communities by means of the analysis of comprehensive hydrologic indicator data. A particularly acute problem at present is finding the most valid hydrologic indicators of subsurface waters at different depths, since this is necessary for the large-scale introduction of geobotanical indicator data into the practical work of hydrogeologists and of specialists in amelioration and water supply.

2. The next problem of hydrologic indicator research is the ecological study of direct hydrologic indicators and the problem of their ecological replaceability.

Many aspects of the ecology of the phreatophytes and trichohydrophytes most frequently used in direct hydrologic indication have not been sufficiently studied up to the present. In particular, there is still little information on the range of depths of subsurface waters indicated by particular hydrologic indicators. The root systems of this group of plants have been studied partially (Iordanskaya, 1958, 1960) but information is still inadequate, and far from all phreatophytes have been studied. Data on the depth of subsurface waters indicated by many hydrologic indicators are very contradictory.

All these aspects of the ecological study of phreatophytes are closely associated with the determination of the degree of validity and significance of edificator plants in indicator communities.

A more detailed study of the ecology of phreatophytes and trichohydrophytes would offer the possibility of more accurate predictions of the depth of groundwater tables and their degree of mineralization on the basis of vegetation.

The ecological study of phreatophytes and trichohydrophytes is associated with the problem of the ecological replaceability of indicators.

The ecological replaceability of hydrologic indicators implies the replaceability of indicators at a particular indicator object in a geographically uniform zone. Its study is closely linked with the elucidation of the ecological range of hydrologic indicators.

Particular interest in this context is presented by research on the relationship of vegetation to a definite type of subsurface water and on the ecological range of hydrologic indicators in relation to this factor.

A rather substantial amount of factual material is now available on the ecological replaceability of direct hydrologic indicators in the arid regions of the Soviet Union. Unfortunately, a substantial proportion of this material has still not been published.

One of the aspects of this problem is the study of the effect of the chemical status of ground waters on the concentration of salts in the cell sap of phreatophytes. A. V. Shavyrina's studies (1962) showed the existence of biohydrochemical correlations, which are of great interest for hydrologic indicator research.

3. The third important branch of hydrologic indicator research is the study of the geographical replaceability of hydrologic indicators. This began to attract attention principally because of the transfer of hydrologic indicator research from the semidesert and desert zones (where it had previously been located) to the forest and forest-steppe zones.

Thus, for example, springs arising in both the desert and forest zones are indicated by characteristic phytogenic hillocks with groups of hydrophytes on the summit. In the desert, such hillocks are called bulak hillocks, or chukalaks, while the term spring hill-swamps is used in the forest zone (Bogdanovskaya-Gienéf, 1926; Kushnar', 1939; Vostokova, 1956).

Depressions (or "saucers") caused by settling following solution or mechanical undermining, with a concentrated microcomplex of hydrologic indicators in the forest zone are obviously homologous to the sagging depressions in the desert, which also have concentrated belts of hydrophilic vegetation.

The problem of the geographical replaceability of hydrologic indicators is still in the initial stages of analysis. There are thus as yet no clear concepts concerning either the nature of replaceability or homologous landscape elements.

Of no less importance is the study of changes in the hydrologic-indicator significance of particular hydrologic indicators at the borders of the distribution area of the dominant phreatophytes in the community. This phenomenon occurs widely. As an example, growths of Halostachys caspica (Pall.) C.A.M. at the center of its area are indicators of bitter saline waters at a depth of 10-15 m, while at the northwestern limits of the distribution area the corresponding depth is 5 m. This is the most simple case of changes in indicator significance; more complicated cases are possible, however, especially in a polydominant community. The solution of the problem of the geographical indicational variability of hydrologic indicators requires the collection of a large amount of factual data for statistical analysis.

In conclusion, it should be noted that hydrologic indicator investigations have become firmly incorporated into practical hydrogeology. Up to the present, they have mainly been carried out by special geobotanical units. Particular attention, however, must now be devoted to working out ways and methods for the wider introduction of the results of these investigations into commercial hydrogeological expeditions and hydrologic regime stations, and also to acquainting large numbers of hydrogeologists and other specialists with the hydrologic indicator technique.

One of the appropriate ways of doing this is the compilation of indicational reference books, presenting indicator data in a systematic fashion.

An indispensable condition for the further development of hydrologic indicator research is the deepening and widening of our knowledge on the interrelationships of the vegetation with subsurface waters, and this requires the organization of special hydrologic indicator research.

## Summary

The author considers the following main methods of practical application of hydrologic indicator research: mapping of subsurface water lenses and underground flow, and hydrogeological zonation based on geobotanical data. The principal tasks of hydrologic indicator research are pointed out, as follows: 1) elucidation of the validity and significance of indicator communities, based on statistical analysis of hydrologic indicator data; 2) investigation of the ecological and geographical replaceability of indicators; 3) study of the ecology of the species of greatest importance for hydrologic indicator research (especially their root systems), and of their developmental rhythms (indicator phenology); and 4) investigation of the correlation between the chemical composition of the tissues of phreatophyte plants and the mineralization of ground waters.

## Literature Cited

Bogdanovskaya-Gienéf, I.D. 1926. "Spring-fed swamps of the Kingisepp district of Leningrad Province." Zh. Russk. bot. obshch., Vol. 11, No. 3-4.

Viktorov, S. V. 1955. "Vegetation as an indicator in the hydrogeological deciphering of aerial photographs." Geogr. sborn., VII. Geogr. obshch. SSSR. Moscow—Leningrad, Izd. Akad. Nauk SSSR.

Viktorov, S. V. 1959. "Plant communities as indicators of groundwaters in meadows of the Turgai valley." Vestn. MGU, No. 2.

Viktorov, S. V. 1960. "Communities as indicators of groundwaters in the Turgai-Irgizsk semi-desert." Byull. MOIP. Nov. seriya, otd. biol., Vol. 65, No. 5.

Viktorov, S. V., and Vostokova, E. A. 1959. "Experience in the use of aerial methods in geobotanical observations carried out within the complex of geological and hydrogeological research." Tr. Labor. aéromet. Akad. Nauk SSSR, VIII Moscow, Izd. Akad. Nauk SSSR.

Vinogradov, B. V. 1958. "On the association of vegetation with groundwaters in the steppe landscapes of Northern Kazakhstan and the use of vegetation in the role of indicator in the hydrogeological deciphering of aerial photographs." Izv. Akad. Nauk SSSR, seriya geogr., No. 1.

Vinogradov, B. V. 1959. "Deciphering the vegetation of arid and subarid zones." Tr. Labor. aéromet. Akad. Nauk SSSR, VIII. Moscow, Glavgeoltekhizdat.

Vinogradov, B. V. 1961. "The use of vegetation as an indicator in deciphering aerial photographs of desert landscapes in Western Turkmenia." Izv. Vses. geogr. obshch., Vol. 93, No. 1.

Vostokova, E. A. 1952. "Chievniks of Western Kazakhstan." Byull. MOIP. Nov. seriya, otd. biol., Vol. 57, No. 1.

Vostokova, E. A. 1956. "Chukalak hillocks as indicators of conditions of humidity and salinity." Byull. MOIP. Nov. seriya, otd. biol., Vol. 61, No. 6.

Vostokova, E. A. 1960. "The use of geobotanical characters in the hydrogeological deciphering of aerial photographs in the arid provinces of the USSR." In book: Problems of Indicational Geobotany. Moscow, Izd. MOIP.

Vostokova, E. A. 1961. Geobotanical Methods of Searching for Subsurface Waters in Dry Regions of the Soviet Union. Moscow, Gosgeoltekhizdat.

Demidova, L. S., Shavyrina, A. V., et al. 1955. "Experience in using the geobotanical method in hydrogeological investigations in the Chernozems." Tr. Vses. aérogeol. tresta, No. 1. Moscow, Gosgeoltekhizdat.

Ivanova, L. S. 1959. "The use of aerial methods in finding leakage water lenses in the semidesert." Izv. Akad. Nauk SSSR, seriya geogr., No. 4.

Ivanova, L. S. 1960. "Leakage waters of the dry steppes and their association with the vegetation cover, relief, and nature of the ground." In book: Problems of Indicator Geobotany. Moscow, Izd. MOIP.

Iordanskaya, N. N. 1958. "Some data on the root systems of phreatophytes in the chernozems." Byull. MOIP. Nov. seriya, otd. biol., Vol. 63, No. 1.

Iordanskaya, N. N. 1960. "The root systems of some hydroindicator plants of the Kalmytskii steppes." In book: Problems of Indicator Geobotany. Moscow, Izd. MOIP.

Kunin, V. N. 1957. "Some results of studies of local (subsurface) waters in deserts." Izv. Akad. Nauk SSSR, seriya geogr., No. 4,

Kunin, V. N. 1959. Local Waters in Deserts and Problems in Their Utilization. Moscow, Izd. Akad. Nauk SSSR.

Kunin, V. N., and Leshchinskii, G. T. 1960. Temporary Surface Runoff and the Artificial Formation of Groundwaters in the Desert. Moscow, Izd. Akad. Nauk SSSR.

Kushnar', S. A. 1939. "Sandy chukalak hillocks in the Kyzyl-kum desert." Razvedka nedr, No. 1.

Levin, V. L. 1960. "Experience in applying geobotanical research in hydrogeological studies in the northwestern Caspian region." In book: Problems of Indicator Botany. Moscow, Izd. MOIP.

Leont'ev, V. L. 1954. Haloxylon Forests of the Kara-kum Desert. Moscow—Leningrad, Izd. Akad. Nauk SSSR.

Nitsenko, A. A. 1956. "The vegetation of natural oases of the western Kara-kum and its significance in planning the transformation of the desert." Vestn. LGU, No. 15.

Orfanidi, K. F. 1957. "Conditions for the formation of fresh groundwaters in dry districts and their significance." Byull. nauchno-tekhn. inform., No. 4, p. 9.

Ostrovskii, V. N. 1959. "Some experiences in applying geobotanical methods in hydrogeological investigations in the southwestern part of the Zaipansk depression." Vestn. Akad. Nauk KazSSR, No. 8, p. 173.

Panin, P. S. 1960. "Some features of the vegetation cover of Kashgaria." Bot. zh. SSSR, Vol. 65, No. 4.

Petrov, M. P. 1958. "Black Haloxylons along the course of the Kara-kum canal." Priroda, No. 11.

Rodman, L. S., Levin, V. A., and Polikarpova, L. D. 1960. "Experience in the quantitative characterization of the hydroindicational significance of the vegetation of the northwestern Caspian region." Nauchn. dokl. vyssh. shkoly, seriya biol., Vol. 4, No. 3.

Rustamov, I. G. 1954. The Vegetation of the Central and Lower Parts of the Western Uzboi. Author's Abstract of Candidate's Dissertation, Ashkhabad.

Tagunova, L. N. 1961. The Development of the Vegetative Cover of the Northeastern Shore of the Caspian Sea (in Connection with Conditions of Acidity in Soil-Forming Rocks). Author's Abstract of Candidate's Dissertation, Moscow.

Khudyakov, I. I. 1961. "The vegetative cover as an indicator of the chemical composition and the depth of groundwaters." In book: Problems of Indicator Geobotany.

Shavyrina, A. V. 1959. "The use of geobotanical characteristics in searching for water in the virgin lands of Kustanai province." Razvedka i okhrana nedr, No. 1.

Yagdiev, N. 1958. "Types of black Haloxylon in the central part of the southeastern Kara-kum." Izv. Akad. Nauk TurkmSSR, No. 3.

Ellenberg, H. 1950. Unkrautsgemeinschaften als Zeiger für Klima und Boden. Stuttgart-Indwigsburg.

Ellenberg, H. 1952. "Auswirkung der Grundwasserabsenkung auf die Wiesengesellschaften." Angew. Pflanzensoziol., No. 6. Stolzenau-Weser.

Jakucs, P. 1957. "Ökologische Untersuchungen der Mosaik-Komplexe von Quellmoor und Sumpfgesellschaften durch Wasserfärbung." Acta bot. Acad. sci. hung., Vol. 3, No. 1-2.

Mager, H. 1912. Les moyens de découvrir les eaux souterraines et de les utiliser. Paris.

10

# EXPERIENCE IN HYDROLOGIC INDICATOR ZONATION
# OF THE NORTH CASPIAN REGION IN THE SEARCH
# FOR FRESH AND BRACKISH WATERS

## V. I. Levin

The geobotanical hydrologic indicator method of prospecting and mapping shallow-lying fresh and saline groundwaters* has been very widely used in the last decade in hydrogeological work undertaken by geological organizations in the North Caspian region. As a result of the carrying out of geobotanical hydrologic indicator investigations, which are of commercial importance for the northern part of the Caspian lowlands (an area delimited from the south by the Kuma and Émba Rivers and the Caspian Sea) geobotanical traits of the shallow-lying fresh and saline waters have been discovered and studied, and for the greater part of this district hydrologic indicator maps showing the distribution of these traits have been compiled.

The features used in searches for shallow-lying fresh and saline waters in hydrologic indicator mapping in the North Caspian region are as follows: 1) the presence of plant communities which are direct or indirect evidence of the salt-free nature of subsurface waters; 2) the absence from plant communities of those plant species which are direct or indirect indicators of the close presence of subsurface waters under specific terrain conditions; 3) definite geomorphological conditions, expressed mainly by the development of various depressions.

It was found during hydrologic indicator studies of the northern part of the Caspian plain that the geobotanical indicators of shallow-lying fresh and saline groundwaters vary considerably in different parts of the territory. Substantial variations occur both in the conditions of formation and distribution of shallow-lying fresh and saline waters, and in the very nature of the vegetation, regarded by us as evidence of these waters in different parts of the territory under study.

These differences are obviously due to variability in climatic conditions, expressed by an increase in the continentality and dryness of the climate moving from the northwest to the southeast, by the history of the territory's development, controlled by the features of geomorphology and hydrogeology of its different regions and, finally, by the botanical and geographical variations in the territory.

There is thus obviously a great need for the hydrologic indicator zonation of the North Caspian region, i.e., the identification of hydrologic indicator districts, characterized by the specific properties of geobotanical indication of fresh and saline groundwaters.

We shall attempt in the present paper to present a hydrologic indicator zonation of the northern part of the Caspian lowland.

The zonation suggested is of course merely an outline, since in the main it covers only the most obvious differences in characters used in this type of mapping.

We shall leave aside the problem of the indication of the depth of groundwaters, since the fresh and saline groundwaters are in this area usually at a relatively shallow position (1-10 m, rarely deeper) and are of interest for purposes of water supply irrespective of their depth.

---

*Fresh waters—waters with mineralization up to 1 g/liter; slightly saline waters—from 1 to 3 g/liter; saline—from 3 to 6 g/liter.

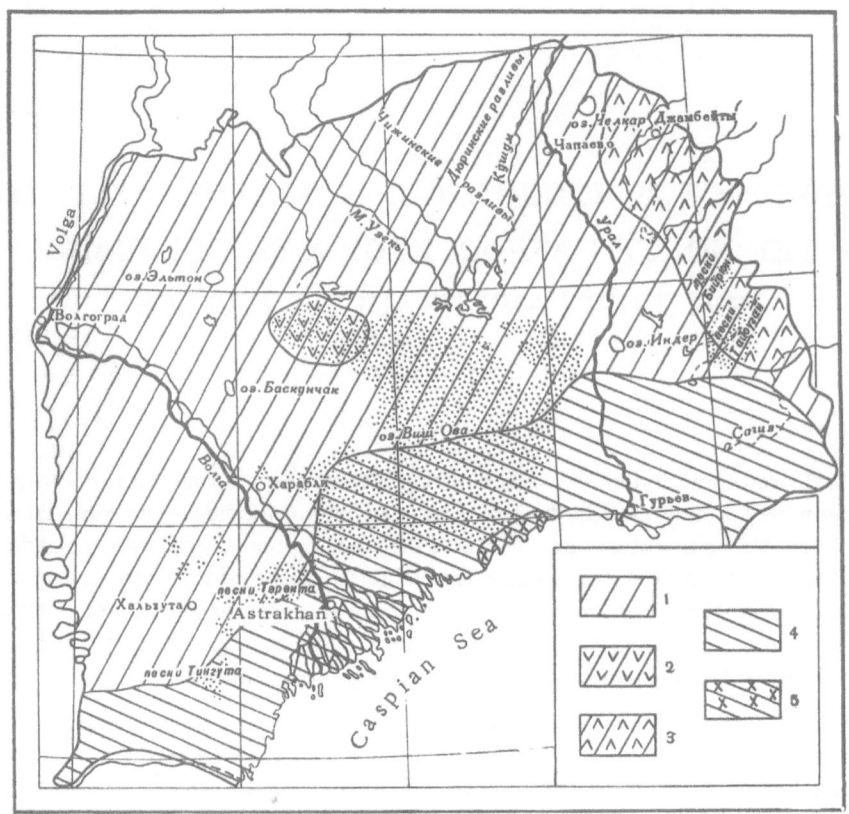

Hydrologic indicator mapping scheme of the North Caspian region: 1) northern hydrologic indicator district; 2) Urdinsk sands subdistrict; 3) northeastern subdistrict (Ural-Émba sands); 4) southern hydrologic indicator district; 5) sandy subdistrict.

Two main hydrologic indicator districts may be identified in this territory, and these will be designated, according to their position, as the northern and southern districts (see zonation scheme).

It should be noted that these districts differ not only in climatic conditions but are also controlled by a whole complex of physiogeographical conditions.

We shall deal briefly with a comparative description of hydrologic indicator conditions in the northern and southern hydrologic indicator districts. The distribution of shallow-lying fresh and saline waters in the North Caspian region is associated with specific landscape types and elements. These are as follows: sand steppes and sands of differing degrees of firmness; varying types of relief depressions: deflation basins, dips, large sink holes, gullies, old river-bed hollows, and fluvial plains. The types of direct or indirect indicators of fresh or saline groundwaters should therefore usually be plant communities with a predominance of glycophilic, sandy-steppe, psammophilic, meadow-steppe, and meadow species, and, more rarely, xerophilic shrubs and grasses. Glycophilic weeds can also be used as hydrologic indicators.

Shallow-lying fresh and saline groundwaters have a much wider distribution within the northern hydrologic indicator district than in the southern.

In hilly, loosely held sands and basins, the active deflation of shallow-lying fresh waters is indicated by thinnish pioneer groupings, made up of Elymus giganteus Vahl., Chondrilla ambigua Fisch., Artemisia arenaria D.C., Aristida pennata Trin., Agriophyllum arenarium M.B., and Corispermum aralo-caspicum Iljin.

A reliable indicator of fresh and saline waters in hilly sands of varying degrees of firmness is provided by associations of these species, frequently including Agropyron sibiricum (Willd.) P.B., Festuca beckeri Hack., and Calligonum aphyllum (Pall.) Guerke.

These associations indicate fresh water under hilly, semifirm sands in the Urdinsk sands, distributed in the northwestern part of the Volga-Ural sandy massif, and in the sands of the northeastern Caspian region (Kukuzyak-Kum, Bairyuk, and Taisugan).

Slightly saline and saline waters are encountered under these associations in the Volga-Ural sand massif. Fresh waters in this area are usually associated with loosely held sands and deflation basins, being indicated by associations of psammophytes with the inclusion of mesophytes, as follows: Calamagrostis epigeios (L.) Roht., Holoschoenus vulgaris Link., Glycyrrhiza glabra L., and the shrubby willows Salix rosmarinifolia L. and S. caspica Pall.

Fresh and slightly saline groundwaters, associated with sand and sandy-loam steppe landscapes, are widely distributed within the northern hydrologic indicator district. Thus, indicators of shallow-lying fresh and slightly saline waters in the sandy-loam and sandy plains of the northeastern Caspian region are associations in which sandy-steppe grasses are dominant: Stipa ioannis Čel., S. capillata L., Agropyron sibiricum, and fescues, with a small or large proportion of psammophytes: sand sagebrush, Euphorbia seguieriana Neck., and Achillea; and the mesophytes: Glycyrrhiza glabra L., Sophora alopecuroides L., Lasiagrostis splendens (Trin.) Kunth, Calamagrostis and, more rarely, associations of Artemisia lercheana Web. In the Volga-Ural interfluve, on the sandy plains delimiting the Urdinsk sands from the north, a relationship is observed between fresh and, more frequently, slightly saline groundwaters with associations of Artemisia lercheana, together with plants linked with light soils: Artemisia scoparia W. et K., Euphorbia seguieriana, Ephedra dystachia L., and Helichrysum arenarium (L.) Moench. These associations are apparently modifications of the above-mentioned associations of sandy-steppe grasses and have developed as a result of depression. Within the northern hydrologic indicator district (in its northwest part), fresh and, more frequently, slightly saline waters are also observed in basins of attenuated deflation, where they are indicated by associations with dominance either of a significant proportion of sagebrush, usually with a greater or smaller number of psammophytes (Artemisia arenaria, Lasiagrostis splendens, and Coriospermum aralo-caspicum), or a mesophytic, frequently weedy, mixed grass population: Alhagi pseudalhagi (M.B.) Desv., Potentilla bifurca L., Malva pusila Sm. et Som., and others. In addition, both Artemisia lercheana Web. and A. astrachanica P. Pol. normally have a less xerophilic appearance, more vigorous vegetative organs and reduced pubescence, as a result of which they acquire a bluish tone, differentiating them from typical specimens. It is probable that these species are represented by more mesophyllic ecological forms under conditions of shallow groundwater. In this connection it should be mentioned that M. Il'in and G. Rozhevits (1928) observed, in the northeastern Caspian region, a special form of Artemisia maritima f. amarissima L., associated with the slopes of drained depressions, and having a bluish tint.

The widely distributed lenses of fresh and slightly saline groundwaters in the northern hydrologic indicator district associated with dips, large sinkholes, estuaries, and low, dry water-courses and gullies. The indicators which serve as evidence for demineralization of groundwaters under these geomorphological conditions are associations with dominance of steppe and meadow grasses: quack-grass (Agropyron repens), A. pectiniforme Roem. et Schult., Bromus inermis Leyss, Festuca sulcata Hack., Stipa, together with Glycyrrhiza glabra, Potentilla and, sometimes, steppe shrubs such as Spiraea; associations of Artemisia austriaca Jacg. together with meadow-steppe mixed grass; associations of A. arenaria, Glycyrriza glabra, and Calamagrostis; and, in the northeastern Caspian region, associations of Elymus giganteus together with glycophilic mixed grasses. Indicators of slightly saline, and, more rarely, fresh waters are associations of Artemisia lercheana, Agropyron repens, Stipa, Festuca sulcata, and meadow-steppe mixed grass. Evidence for the very close proximity of fresh, and, more frequently, weakly saline and saline waters, is provided by associations of Agropyron repens + Bekmannia eruciformis L. Host.; A. repens + sedges; and associations of Phragmites communis Trin. together with meadow-swamp mixed grasses, such as Lythrum virgatum L. and Euphorbia seguieriana.

Shallow fresh and slightly saline groundwaters in the river floodplains of the northern hydrologic indicator district are indicated by mixed grass-meadow associations with dominance of A. repens, Bromus inermis, Agropyron pectiniforme, and Calamagrostis, together with meadow-type mesophytic mixed grasses: Glycyrrhiza glabra, Althaea officinalis, and Euphorbia seguieriana; where the groundwater table is very shallow, there are sedge and reed-sedge mixtures. On fresh and slightly saline waters in the floodplains of small rivers, there are associations with dominance of Artemisia dracunculus L., Agropyron pectiniforme, Bromus inermis, and A. repens, sometimes together with Elymus giganteus and Sophora alopecuroides L. (the last two species are found in the northeastern Caspian region). Woody and scrub vegetation has developed in the floodplains of the Volga and Ural Rivers, in the northern part of the district.

Coming now to a survey of the hydrologic indicator conditions of the southern hydrologic indicator district, it should be noted that shallow-lying fresh and slightly saline groundwaters are much less widely distributed here than in the northern district, while the range of landscape and vegetation communities with which fresh and slightly saline groundwaters are associated is substantially narrower.

Thus, there is an almost complete absence of fresh and slightly saline waters in relief depressions in the plains. Meadow-solonchak or solonchak vegetation is generally dominant in lowland liman and liman-like depressions in the area. Fresh and slightly saline waters are much less widely distributed in the floodplains and deltas of rivers, where instead meadow-solonchak associations, developed on saline waters, are dominant.

In the southern hydrologic indicator district, fresh and slightly saline groundwaters are most widely distributed in association with sandy massifs; here also, however, they occur much more rarely than in similar geomorphological conditions in the northern hydrologic indicator district. Thus, lenses of fresh and slightly saline waters in the sands of the southern district are found under sandy hillocks devoid of vegetation, and in loosely held hilly sands and basins of active deflation with sparse pioneer groupings of Elymus giganteus, Artemisia arenaria, Chondrilla pauciflora, Melilotus polonicus (L.) Desr., Aristida pennata Trin., Corispermum aralo-caspicum Iljin., and Agriophyllum arenarium M.B. In contrast to the northern hydrologic indicator district, psammophyte associations, which develop by overgrowing hilly sands and basins of attenuated depressions, and include associations of Artemisia arenaria + ephemerals like Bromus tectorum L. and Alyssum desertorum Stapf; associations of Artemisia arenaria with A. astrachanica P. Poll. and A. lercheana; associations of A. arenaria + Calligonum aphyllum; and associations of Elymus giganteus + Artemisia, are usually developed on saline and, more rarely, slightly saline groundwaters. The presence in psammophyte-dominating associations of such species as Artemisia astrachanica, A. lercheana, Agropyron sibiricum, and Calligonum aphyllum, even in small numbers and in geomorphological conditions normally suitable for the formation of fresh lenses is, in the southern district, an indication of salination of groundwaters.

In river floodplains and deltas, fresh waters are found more frequently under associations of Agropyron repens + glycophilic mixed grasses: A. repens + sedges; associations of aerial reed grass (Calamagrostis) and C. pseudophragmites Koel.; associations of reeds (Phragmites) and meadow mixed grasses; reeds and Carex acuta L.; Glycyrrhiza glabra associations; and Bromus inermis associations, while indicators of groundwater salinity are provided by pure reed and reed—reed mace (Typha) associations, and Juncus meadows, in addition to meadow-solonchak associations.

We thus see that many species and associations which are distributed on fresh and slightly saline waters in the northern hydrologic indicator district become indicators of groundwater salinity in the southern district.

It is possible to isolate, within the above-described districts, hydrologic indicator subdistricts, differing either in the floristic composition of the vegetational communities serving as indicators of fresh waters or by some other feature of fresh and saline water indication. For example, in the northern district one can isolate the Ural-Émba sands sector, which is characterized by the wide distribution of fresh waters under sandy-steppe vegetation and by the presence of such plant indicators as Elymus giganteus and Sophora alopecuroides, absent in the other districts. One can also differentiate the Urdinsk sands, where, in contrast to the remaining Volga-Ural sand districts (lying within the northern hydrologic indicator district), it is primarily fresh and not slightly saline and saline waters which are developed under hilly sands with dominance of Artemisia arenaria and a substantial quantity of sandy-steppe grasses and Calligonum aphyllum. In the southern hydrologic indicator district, one can isolate the coastal subdistrict in which fresh waters are found in deflation basins among spring-fed sands, under sparse psammophytic groupings, which in this subdistrict lose their significance as indicators of groundwater salinization.

As already noted, the reasons for the above-described phenomena can be found in either the history of the geological development of the territory or in differences in the hydrogeological and climatic conditions of its different districts. Thus, in the southern hydrologic indicator district, lenses of fresh or slightly saline waters are formed in conditions of geologically young, very strongly saline land and the minimal amount of precipitation in the district. The occurrence of these lenses can thus be linked only with exceptionally favorable combinations of conditions, in which certain sectors of the territory are completely desalinated and readily concentrate and infiltrate the atmospheric precipitation which feeds the freshwater lenses. Fresh waters do not form where there are some difficulties in conditions for the infiltration of groundwaters, related to heavier soil types, ground compaction because of overgrowth of the sand, or some other reasons. The range of indicator communities of fresh or slightly saline waters in the southern district is therefore restricted to the rather small number of psammophyte and meadow communities which develop where the formation of lenses is not difficult. Differences in the conditions of hydrologic indicator research, governing the differentiation of subdistricts, have more specialized causes. For example, the wide distribution of fresh waters under sandy-steppe associations in the northeastern Caspian region is governed by the fact that fresh waters do not have here a purely infiltrational origin, but are formed from the inflow of water from a chalky water-carrying horizon. The presence of fresh waters in deflation basins among the spring-fed sands of the coastal district, where the grassy population includes the halophytic Nitraria schoberi L. can evidently be explained by the fact that this species has a secondary, anthropogenic origin in the highly populated coastal strip.

## Summary

The significance of hydrologic indicator communities may vary in different parts of the North Caspian region. Experience in the search for fresh water with the aid of geobotanical methods has shown that it is convenient to divide the territory under investigation into two districts, with specific significance for hydrologic indicator research in each. The northern district is characterized by the widely distributed fresh and slightly saline waters, while in the southern districts such waters are infrequent and confined solely to sands,

## Literature Cited

Il'in, M., and Rozhevits, G. 1928. "An outline of the vegetation of the sources of the Émba, Temir, and Chegan Rivers." In book: Report of the Work of the Soil and Botany Division of the Kazakh Expedition of the USSR Academy of Sciences, 1926, No. 3. Materials of the Commision of Expeditionary Investigations, No. 5, Leningrad.

# THE VEGETATION COVER AS AN INDICATOR OF THE CHEMICAL COMPOSITION AND DEPTH OF GROUNDWATERS

## I. I. Khudyakov

In 1952, a hydrogeological survey was carried out in Volgograd province. In addition to several hydrogeological teams, the expeditionary group included a division of soils and botany, whose task was to carry out research on the association of vegetation with groundwaters.

We present below the results of the work, obtained in a territory of some 20,000 km² in 1952.

The reed group of associations was represented by the following main species: Phragmites communis Trin., Agropyron repens (L.) P.B., Aeluropus litoralis (Gouan.) Parl., Rhaponticum salinum (Spreng.) Less., Chenopodium rubrum L., Echinopsilon hyssopifolium (Pall.) Moq., and others. The species Phragmites communis has a low vigor index, 2-3 (on a five-point scale), this being evidently associated with the depth of the water table (2 m) and its marked salinization (dry residue, 4.6 g/liter). The waters under this group of associations are chloride-sulfatic.

Rather similar to the reed group of associations as regards chemical composition and depth of groundwaters are hydrohalophytic (Atriplex, Aeluropus and Puccinellia) and mesophytic (the Agropyron group of associations) communities, found on limans. The first three include in their composition the dominants Atriplex verrucifera M.B., Aeluropus litoralis (Gouan.) Parl., and Puccinellia sclerodes Krecz., and in addition, the following principal species: Phragmites communis, Artemisia monogyna, Echinopsilon sp., Limonium gmelinii (Willd.) Ktze., and others. The Agropyron associations include A. repens, A. pectiniforme Roem. et Schult., Ph. communis, Artemisia paniculata Lam., Acroptilon repens (L.) D.C., Echinopsilon sedoides (Pall.) Moq., Aeluropus litoralis, and others. The groundwaters under these groups of associations are chloridic, the dry precipitate being more, but the $MgSO_4$ content less, than in the case of the reed group of associations.

The decisive factors in evaluating the indicator significance of a community are the components of the particular association other than the dominant. Thus, the group of Agropyron—mixed grass associations, found on dark meadow sinkhole soils among fescue and fescue—Stipa associations have the following composition: A. repens, Alopecurus pratensis L., Galium vernum Scop., Allium decipiens Fisch., Phlomis tuberosa L., and others. This group of associations is linked to a different groundwater chemical composition and depth than the group of Agropyron associations. Besides NaCl, $CaHCO_3$, $MgCl_2$, and $MgSO_4$ are predominant in the groundwater at a depth of 4 m, while the dry residue is only 0.6 g/liter. The water is fresh and is primarily hydrocarbonate—sodium—magnesium in composition. A similar phenomenon was also noted with the Agropyron sibiricum—mixed grass and A. sibiricum—Stipa capillata groups of associations.

The steppe xerophytic group of associations, Festuca sulcata, F. sulcata—S. ioannis, and F. sulcata—S. capillata, found on light chestnut-colored, slightly saline soils in microdepressions, have the following species composition: F. sulcata Hack., S. lessingiana Trin. et Rupr., S. capillata L., and Agropyron sibiricum (Willd.) P.B.: Artemisia lercheana Web., A. austriaca Jacq., Salvia tesquicola Klof. et Pobed., Phlomis pungens Willd., and Carduus uncinatus M.B.

Groundwaters under these communities are chloridic, occur at a depth of 5 m, and have a dry residue of 4.5 g/liter and a characteristic high content of $MgCl_2$. In close proximity to this association group in respect to groundwater depth and chemical composition is the Glycyrrhiza glabra L. community. A completely different pattern is observed in another steppe association, comprising Festuca beckeri Hack., which is linked with light-chestnut sandy soils. The groundwaters in this case are fresh, chloride-sulfatic, with a dry residue of 0.4 g/liter; they occur at a depth of 16.5 m.

The desert group of associations, Artemisia incana and Kochia prostrata (L.) Schrad., are linked with light-chestnut, strongly saline, loamy soils in microdepressions. They contain Artemisia lercheana, Kochia prostrata, Agropyron ramosum (Trin.) Richt., Festuca sulcata, Agropyron sibiricum (Willd.) P.B., Pyrethrum achillei-folium M.B., Camphorosma monspeliacum L., Limonium gmelinii, and others. The chemical composition of groundwaters under the Artemisia incana and other associations of this group is rather similar to that of waters under steppe communities. The difference lies in the higher degree of mineralization (dry residue = 6.4 g/liter) and the greater depth of the water table (7 m).

When mentioning the white wormwood, it is important to remember that in the Caspian lowlands two clearly distinct forms are differentiated by Keller: Artemisia incana v. erecta Kell. and A. incana v. nutans Kell. The first is associated with soils of heavy mechanical structure—clays and loams—and the second with light soils—sands and sandy loams. The groundwaters under the latter are different in chemical structure from those under the former. Thus, under a depression with light-chestnut, sandy-loam, weakly saline soil, covered with A. incana v. nutans Kell., Festuca beckeri, Koeleria glauca DC., Syrenia siliculosa (M.B.) Andrz., Heli-chrysum arenarium (L.) Monch., and Astragalus virgatus Pall., groundwater occurred at a depth of 5.5 m and was fresh and hydrocarbonate—sodium—calcium in type, with a dry residue of 0.6 g/liter.

Also widely distributed in the district described are associations transitional from steppe to desert types, such as Festuca sulcata—Artemisia incana, F. sulcata—A. incana—Kochia prostrata, Stipa—F. sulcata—A. in-cana, Agropyron ramosum—Artemisia incana—F. sulcata, and others. The chemical composition and depth of the groundwaters in these instances are close to the situation obtaining with the steppe, or desert (A. incana), groups of associations.

Also widely distributed and highly varied in its species composition is the Artemisia pauciflora Web. group of associations, which include A. pauciflora, A. pauciflora—Kochia prostrata L. (Schrad.), A. pauci-flora—Atriplex cana C.A.M., A. pauciflora—Camphorosma monspeliacum L. groupings, and others. The floris-tic composition of this group of associations includes the following main species: A. pauciflora, A. lercheana Web., Kochia prostrata, Atriplex cana, Camphorosma monspeliacum, Linosyris tatarica C.A.M., Agropyron ramosum (Trin.) Richt., Limonium gmelinii (Willd.) Ktze., Salsola tamariscina Pall., S. foliosa (L.) Schrad., Petrosimonia monandra (Pall.) Bge., and P. glaucescens (Bge. Iljin.). This group of associations is linked to microelevations and the margins of floodplain terraces and saline lakes. The chemical composition of the groundwaters under these associations differs substantially from that obtaining in the case of the Artemisia incana group, but is very similar to the next group of associations (Anabasis salsa and Atriplex cana), to which these communities are genetically close. The groundwaters are chloridic-sulphatic, strongly mineralized (dry residue = 16 (10-18) g/liter), and occur at a depth of 8 (5-13) m. In contrast to other desert and steppe associa-tions, $MgCl_2$ is absent (or is present in very small amounts) while $Na_2SO_4$ is present to the extent of 2-21% meq.

The desert xerohalophytic group of associations (Anabasis salsa (C.A.M.) Beth, Atriplex cana, A. cana—Artemisia pauciflora, and Anabasis salsa—Salsola foliosa (L.) Schrad.) is characterized by the presence of the following species: A. salsa, Atriplex cana, Artemisia lercheana, Camphorosma monspeliacum, Kochia pros-trata, Linosyris tatarica, Pyrethrum, Agropyron ramosum, Limonium suffruticosum (L.) Ktze., L. gmelinii (Willd.) Ktze., Petrosimonia brachiata (Pall.) Bge., Salsola foliosa, and S. laricina Pall. This group of associa-tions is encountered on floodplain terraces, around saline lakes, and on microelevations around Artemisia pauci-flora associations. Groundwaters under communities of this group are chloridic, strong mineralized (dry residue = 34 g/liter), and occur at a depth of 9 (5-14) m.

The following brief conclusions may be enumerated on the basis of the above:

1. The vegetative cover can serve as an indicator of the chemical composition and depth of groundwaters in the first water-carrying horizon (probably only within a territory which is uniform in respect of physical and geographical conditions).

2. Geobotanical data facilitate the selection of places for making boreholes, and reduce the cost and increase the speed and accuracy of hydrogeological surveys and the compilation of hydrogeological maps; the precise appraisal of soils for amelioration purposes and the determination of the sequences of watering and irrigation programs are also facilitated.

## Summary

The most widely distributed indicator communities in the Lower Volga region are discussed and their importance in prospecting for subsurface waters is pointed out. Waters of best quality were discovered under associations with dominance of Alopecurus pratensis, Agropyron pectiniforme, and Stipa capillata. This is related to the considerable degree of salt leaking from the soils under these communities.

# ON THE POSSIBILITY OF USING THE GEOBOTANICAL METHOD
# IN THE SEARCH FOR FRESH WATER
# IN SOUTHERN DESERTS

## A. V. Shavyrina

The possibility of using vegetation as a character in prospecting for fresh water in the desert is governed by the constant water deficit during the greater part of the growth period, as a result of which the vegetation is clearly differentiated in relation to the water-supply conditions. Toward midsummer, plants depending on atmospheric precipitation wither up or suspend their development, and only phreatophytes, which use subsurface waters, are able to continue their development. For prospecting purposes, therefore, the greatest importance attaches to communities of phreatophytes, which are associated with weakly mineralized subsurface waters. A reduction in the level of groundwaters below the optimal for a particular species, a decrease in the area of sub-surface-water lenses for a long period, and an increase in water mineralization, all of which reduce the vitality of phreatophytes and cause changes in the species composition of structure of a community, lead to a change in the entire appearance of the community. All these aspects must be taken into account when using the vegetation as a prospecting characteristic.

The present article presents the results of a study of some geobotanical features of weakly mineralized waters, carried out within the southern-type deserts in 1958-1961, in the Lake Tashkan district, the Karabil' upland, the Chardzhou district, and in the Central Kara-Kum.

The most widely distributed landscapes in which the presence of shallow-lying groundwaters is possible within the territory under study are individual sectors of sandy massifs (under which "subsand lenses" are located), takyrs (where subtakyr lenses exist), valleys of extant rivers and dry watercourses, and hollows. We studied 30 standardized plots with communities of phreatophytes.

## Indicators of Lenses in Sands

Indicators of the presence of fresh waters in sands at a depth of up to 1-2 m are dense growths of profusely developed reed (Phragmites communis Trin.), reaching 2.5-3 m in height and fruiting abundantly. An increase in the depth of the water table to 3 m or above leads to a decrease in the height of the reeds to 1-1.5 m, and a reduction in the number of fruit-bearing specimens. Where there is an increase in mineralization of the waters so that they are strongly salinated and saline, the reeds acquire a spreading habit and saltworts (Salsola spp.) appear in the herbage. Reed growths are most frequently located in deep but smallish basins, or occupy the lowest section of wide intermontane depressions, indicating localities with water closest to the surface.

Waters lying at a depth of 2-5 m (more rarely, 8-10 m) are also indicated by communities with dominance of Alhagi persarum Boiss. and A. pseudalhagi (M.B.) Desv. However, because of the great ecological lability of Alhagi, the mineralization of the subsurface waters indicated by these plants may vary substantially. The most favorable conditions for their growth occur on plots under which fresh or weakly mineralized waters are sited. Under these conditions, Alhagi develops luxuriant, dense growths, averaging 60-80 cm in height and with a bush diameter of 80-100 cm. Where the depth of the water table is more than 8 m, Alhagi growths thin out, their height drops to 20-40 cm, the bush diameter is sharply reduced, and the plants acquire a scraggly shape. An increase in mineralization of the water to the point of strong salinization depresses Alhagi, causing a reduction in size, the formation of small leaves and the development of yellowish coloration, and leads to the appearance within the community of salt-tolerant species (Salsola spp. and some species of Artemisia).

Other communities indicating lenses of fresh and weakly mineralized waters in sands are those of Tamarix ramosissima Ldb., in a well-developed condition and reaching a height of 2-3 m (there being no halophytic plant species in the communities). Their floristic composition frequently includes specimens of well-developed Haloxylon aphyllum (Minkw.)Iljin., and groups of Halimodendron halodendron (Pall.) Voss. With an increase in mineralization until the water is strongly salinated and saline, Tamarix ramosissima is replaced by T. hispida Willd., while halophytes appear in the herbage, such as the following: Kalidium foliatum (Pall.) Moq., Halostachys caspica (Pall.) C.A.M., and sometimes Halosnemum strobilaceum M.B.

In loosely held sands, groups of T. ramosissima are distributed in the lower parts of deflation basin slopes, while in firmly held sands they occupy the bottoms of depressions.

Communities with dominance of Haloxylon aphyllum may serve as indicators of weakly mineralized waters in sands. The depth and mineralization of waters under H. aphyllum can vary substantially, this being reflected in the density of the growth and the degree of development of the plants. The most reliable evidence of drinking waters are vigorously developed H. aphyllum plants, varying from 2 to 4 m (more rarely up to 5-6 m) in height and from 2-3 to 5-6 plants per 100 m$^2$ in density. These H. aphyllum populations are most frequently associated with lenses of fresh or slightly saline waters lying at a depth of 10-15 m; sometimes the depth of the water table increases to 20-25 m, this having been observed by us in certain plots in the sandy massifs of the Zaungusk Kara-Kum. Where mineralization increases until the waters are strongly salinated or saline, the growths of H. aphyllum thin out, plant height falls to 1-1.5 cm, and Halostachys caspica appears frequently in the community.

H. aphyllum growths most frequently occupy intermontane depressions of firmly held montane-hilly sands or the margins of unstable and loosely held sandy massifs. Individual groups of H. aphyllum, sometimes including Tamarix, frequently border salinated depressions in sands, indicating a zone of seepage from sands to the salinated depression. It is probable that this occurs in the well-known Yaradzhinsk massif which is covered with H. aphyllum.

In hilly and porous hilly sands in the northwestern border of the southern deserts, indication of the presence of lenses of weakly mineralized waters is also provided by communities of Artemisia arenaria D.C., which has penetrated here from the north and occupies depressions between hills, and deflation basins. The depth of groundwaters under these communities averages 6-10 m, while the degree of mineralization ranges from fresh to salinated (dry residue up to 3-5 g/liter). In the southern deserts, the distribution of A. arenaria is restricted to the area of the Caspian plain. We found overgrowths of the plant in the Oktumkum sands.

The above-described communities, which are indicators of shallow, weakly mineralized waters, are composed of phreatophyte plants, directly associated with the groundwaters and are thus direct hydrologic indicators. Evidence of the presence of shallow waters can in some cases be gained from vegetational groupings, consisting of species whose actual development is not associated with groundwaters but which show, in an environment of increased moisture, deviations from normal development in the form of a sharp increase in the green mass of the plant or copious, prolonged fruiting. Thus, at sites of shallow freshwater lenses in the sands of the Lower Kara-Kum (in the Lake Chokrak district), we observed a substantial increase in the length of annual green secondary branches (shoots) and more abundant flowering in buckwheat plants, as compared with specimens growing away from the lens area.

A ring of very profusely developed Aristida pennata Trin. was observed around basins containing shallow freshwater lenses in the sands of the Caspian Kyzyl-Kum.

Indicators of Subtakyr Lenses

For weakly mineralized waters in takyrs, the most typical communities are those with dominance of Alhagi and Peganum harmala L., and consisting of well-developed specimens retaining their green color for almost the entire vegetative period. The most profuse development of the herbage is attained with a water-table depth not exceeding 8-10 m, and in a plot whose area corresponds to that of a lens wherein the fresh water is retained even in the driest years. The average height of Alhagi on such plots generally exceeds 50-70 cm, while P. harmala forms pillowlike beds more than 0.5 m in diameter. A gradual increase in water mineralization at the periphery of the lens leads to thinning out of the herbage and a decrease in plant vigor. Individual retarded plants of Alhagi and P. harmala can exist if there is slight moistening of soil layers, but in this case they do not constitute indicators of water in the proximity.

Groups of profusely developed plants of T. ramosissima are indicators of fresh waters at a depth not greater than 10-15 m in takyrs. A narrow, irregular belt made up of individual plants of Tamarix and Alhagi, which can occur on the edge of a small takyr, is usually evidence that the lens of weakly mineralized waters is small.

## Indicators of Fresh Waters in River Valleys, Dry Watercourses, and Ravines

Evidence of weakly mineralized waters close to the surface (at a depth of up to 0.5 m) is provided by communities of well-developed reed and reed mace (Typha latifolia L. and Th. angustifolia L.), the average height of the reeds being 2-2.5 m. Where the depth of the groundwater table increases to above 0.5-1 m, reed mace disappears, and the reed communities become thinner, the height dropping to 1-1.5 m. A further increase in the depth of weakly mineralized waters to 1-1.5 m leads to the replacement of reed overgrowths by groupings of Scirpus maritimus L. and species of rush (Juncus sp.), sedge (Carex sp.), sometimes together with Erianthus purpurascens Anderss.

Evidence of fresh and, more rarely, salinated waters at a depth of 1-3 m is provided by pure growths of E. purpurascens, reaching a height of 2-2.5 m, or by this species together with Glycyrrhiza glabra and by G. glabra—E. purpurascens communities, with G. glabra plants reaching an average height of 70-80 cm, sometimes incorporating individual trees of Populus diversifolia Schrenk.

As the depth of the water table increases to 3-5 m, Erianthus-dominant communities are replaced by G. glabra communities, but the vegetative vigor of the G. glabra plants is somewhat reduced.

At sites where weakly mineralized waters occur at a depth of up to 3-5 m, tugaic overgrowths are widely distributed, these being formed from woody and shrub species, frequently in combination with mesophytic mixed grasses.

The main components of tugaic communities are poplars (Populus diversifolius Schrenk and P. pruinosa Schrenk), willows (Salix spp.), and, more rarely, Elaeagnus angustifolia L., forming an upper layer at a height of up to 6-8 m; the shrub layer is formed from Halimodendron halodendron, Tamarix, and box-thorns (Lycium ruthenicum Murr. and L. turcomanicum Fisch. et Mey.). The grassy cover is usually dominated by Erianthus purpureus, Glycyrrhiza glabra, and Alhagi. Tamarix tugais are formed in those cases where the first layer in the tugais is composed of large specimens of Tamarix, reaching 3-4 m in height.

Evidence of weakly mineralized waters (fresh and salinated) at a depth of up to 5 m or, more rarely, 8 m, is provided by communities with dominance of Karelinia caspica Less. and Alhagi, frequently with the inclusion of Calamagrostis epigeios (L.) Roth. and reed, and sometimes incorporating individual trees of Populus diversifolia. Where the mineralization of the water increases to above 5 g/liter, poplar, Tamarix, and Calamagrostis epigeios drop out of the community and there appear such solonchak species as Limonium gmelinii (Willd.) Ktze., L. otolepis (Schrenk) Ktze., Nitraria schoberi L., and Kalidium caspicum (L.) Ung.-Sternb.

Weakly mineralized waters at a depth of 5-8 m (up to 10 m) are frequently indicated by communities with dominance of profusely developed Alhagi, reaching a height of 60-80 cm, and sometimes including Zygophyllum fabago L., Peganum harmala, Lycium, and Halimodendron halodendron.

In river valleys and ravines of loessial deserts, communities with dominance or substantial inclusion of Lagonichium farctum (Banks et Sol.) Bobr. may also serve as indicators of waters at a depth of not more than 15-18 m. The mineralization of waters under these communities can vary substantially.

Evidence of weakly mineralized waters at a depth of 5-10 m, sometimes 15 m, is provided by communities with dominance of well-developed, tall-stemmed Haloxylon aphyllum (Minkw.) Iljin., which are widely

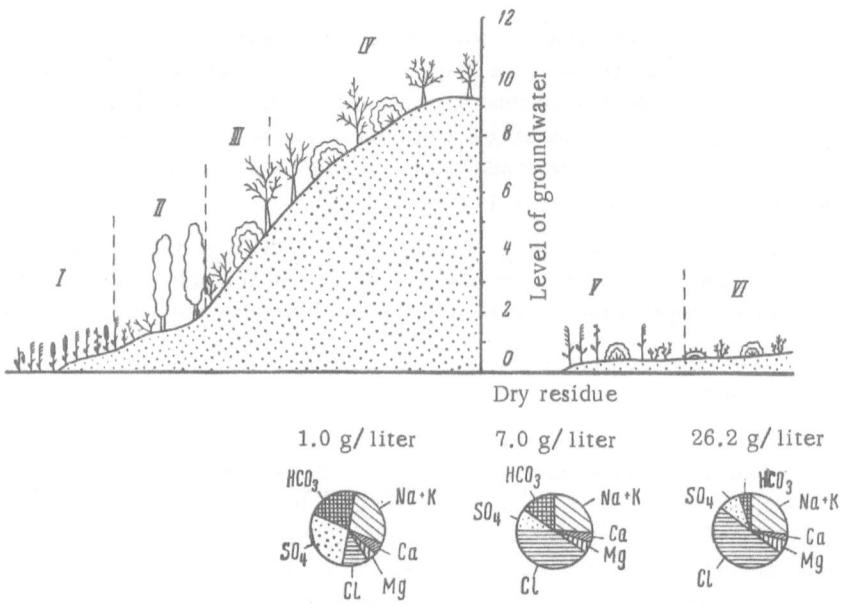

Generalized schematic profile of the vegetational pattern in relation to the depth and mineralization of groundwaters:  Left slope.  I, Typha-Phragmites associations;  II, growths of Populus diversifolia with Glycyrrhiza glabra–Alhagi associations in the grassy cover;  III, Tamarix–Alhagi associations;  IV, Tamarix–Haloxylon aphyllum associations.  Right slope.  V, Phragmites associations with Alhagi and Tamarix hispida;  VI, Halostachys caspica associations with Tamarix hispida and Halosnemum strobilaceum.

distributed in dry watercourses and ravines, and frequently include Tamarix, Lycium, Alhagi, and sometimes Anabasis aphylla L., the average height of the H. aphyllum plants being 2-3 m.

The accumulation of weakly mineralized waters at a depth of up to 10 and, more rarely, 15 m, associated with enclosed depressions, takyr-like surfaces, and the dry watercourses of old rivers, is indicated by Anabasis aphylla communities, and, more rarely, by communities with dominance of Peganum harmala.

An example of the changes in the composition of the vegetation in relation to changes in the depth and mineralization of groundwaters is provided by the schematic profile shown above (see figure).

In the coastal belt and the section contiguous to it, where the level of fresh waters is up to 0.5 m deep, reed and reed-mace communities develop (I).  As the depth of the water table increases to 2-2.5 m, they are replaced by groupings of Populus diversifolia, with the inclusion of G. glabra and Alhagi in the grassy cover (II). When the depth of the water table increases to 3-5 m, communities of Tamarix hispida and Alhagi acquire dominance (III):  with a further increase in the depth to the groundwater surface (to the range 5-9 m), Tamarix growths are replaced by Tamarix–Haloxylon aphyllum communities (IV).

Reed communities become thinner, while reed-mace disappears in a plot where shallow salinated waters are present (right part of the profile:  Alhagi is still present;  Tamarix ramosissima is replaced by T. hispida (V). With a further increase in water mineralization, these communities are replaced by communities of solonchak species:  Halostachys caspica, T. hispida, and Halosnemum strobilaceum (VI).

By using ecological series of hydrologic indicator communities, it is thus possible to discover sites with shallow-lying subsurface waters and make a forecast of their degree of mineralization.

## Summary

This paper deals with hydrologic indicator communities of freshwater lenses beneath sands and clay depressions (takyrs) and with communities associated with fresh waters in the river valleys and oases of Turkmenia and Uzbekistan.

The author used as hydrologic indicators communities of Alhagi persarum, A. pseudalhagi, Haloxylon aphyllum, Tamarix ramosissima, T. hispida, Halimodendron halodendron, Kalidium foliatum, Halostachys caspica, and several other species.

# THE USE OF THE GEOBOTANICAL METHOD IN HYDROGEOLOGICAL RESEARCH IN THE DZHEZKAZGAN-ULUTAU DISTRICT OF CENTRAL KAZAKHSTAN

## V. N. Ostrovskii

From 1957 to 1960, we carried out hydrologic indicator investigations in Central Kazakhstan which demonstrated the possibility of applying geobotanical methods for hydrogeological work in this complex and peculiar region.

The territory where the work was done is located in the southwestern part of the Kazakh fold area and, from the geomorphological aspect, has a hummocky topography. The geological structure of the region is characterized by the presence of highly dislocated, ancient, sedimentary, magmatic, and metamorphic rocks, broken down by the numerous tectonic disturbances, which accounts for its exceptional complexity.

Subsurface waters are found in nearly all lithological-stratigraphic rock complexes and, depending on circulation conditions, are divided into interstitial, interstitial-karst and pore types. Interstitial waters are the most widely distributed, although subsurface waters of the interstitial-karst type are of primary significance, since they are characterized by very high productive capacity.

Vegetation permits the mapping both of localities with shallow-lying subsurface waters in indigenous rocks, and of sites where interstitial and interstitial-karst waters discharge under a loose layer. The discovery of inundated breaks on the basis of vegetation is of particular significance.

Investigations of the interrelationships of subsurface waters and vegetation was carried out by the procedure suggested by E. A. Vostokova (1955).

By means of describing the vegetation at points with subsurface waters of known depth and mineralization, the principal hydrologic indicator communities were discovered, as shown in the hydrologic indicator scheme (see table). Our studies showed that the vegetation can serve as a very sensitive indicator of mineralization of subsurface waters. The following scheme designates the relationship of vegetation to subsurface waters of varying degrees of salinity.

Fresh subsurface waters (solid residue up to 3 g/liter) are indicated by communities of Agropyron repens (L.) P.B., Phragmites communis Trin. with glycophilic species of willow (Salix serrulatifolia E. Wolf., S. wilgelmsiana and others), Rosa canina L., Lasiagrostis splendens Kunth., Glycyrrhiza uralensis Fisch., and some other vegetational communities. Salinated waters (mineralization of 3-5 g/liter) are typified by creeping Agropyron communities including Limonium gmelinii (Willd.) Ktze., Camphorosma monspeliacum L. and other halophytes, Lasiagrostis splendens—Phragmites communis and Elymus angustus Trin. associations, and groupings of Atriplex cana C.A.M. The indicators of strongly salinated subsurface waters (solid residue 5-10 g/liter) are A. cana—L. splendens, E. angustus—Ph. communis, and Ph. communis—akmamyk communities, though the Ph. communis in these communities is depressed. Saline subsurface waters with mineralization in excess of 10 g/liter are typified by groupings of Halosnemum strobilaceum, Aeluropus litoralis (Gouan) Parl., and depressed reed and Elymus giganteus.

Vegetation also permits the determination of even comparatively small changes in mineralization of subsurface waters, this being clearly illustrated by a hydrogeological-geobotanical profile across the Baikonur River valley (Fig. 1).

Hydrologic Indicator Scheme for the Dzhezkazgan-Ulutau District of Central Kazakhstan

| Depth of subsurface waters | Vegetational indicator communities of subsurface waters | | | |
|---|---|---|---|---|
| | fresh (with mineralization up to 3 g/liter) | salinated (with mineralization of 3–5 g/liter) | strongly salinated (with mineralization of 5–10 g/liter) | saline (with mineralization above 10 g/liter) |
| 0–1 | — | — | — | 1. Salicornia with Halosnemum strobilaceum<br>2. Aeluropus litoralis with depressed reed |
| 1–3 | 1. Agropyron repens—mixed grasses<br>2. Agropyron repens—reed (Ph. communis)<br>3. Reed<br>4. Reed—Glycyrrhiza uralensis<br>5. Reed—Lasiagrostis splendens<br>6. L. agrostis together with G. uralensis and A. repens<br>7. Growths of Salix and Rosa canina | 1. Agropyron repens with the inclusion of halophytes<br>2. Reed with inclusion of Glycyrrhiza uralensis and halophytes<br>3. Elymus angustus<br>4. L. splendens—reed with inclusion of halophytes<br>5. L. splendens with inclusion of halophytes | 1. Atriplex cana—L. splendens with inclusion of halophytes<br>2. E. angustus with inclusion of halophytes<br>3. E. angustus—reed with inclusion of halophytes<br>4. Reed—akmamyk* (depressed reed) | 1. Reed with halophytes (reeds very depressed)<br>2. Atriplex cana—Artemisia pauciflora with halophytes<br>3. Halosnemum strobilaceum with Salicornia and large-leafed Zygophyllum |
| 3–5 | 1. Artemisia pauciflora—Atriplex cana with Lasiagrostis splendens<br>2. A. pauciflora—Anabasis salsa with L. splendens | 1. Atriplex cana—Artemisia pauciflora with Elymus angustus and L. splendens | 1. Atriplex cana with inclusion of Anabasis salsa and soranga | 4. Elymus giganteus with halophytes |

* Akmamyk—apparently the genus Atropis. Atropis distans and A. convoluta are found in the trans-Volga region.

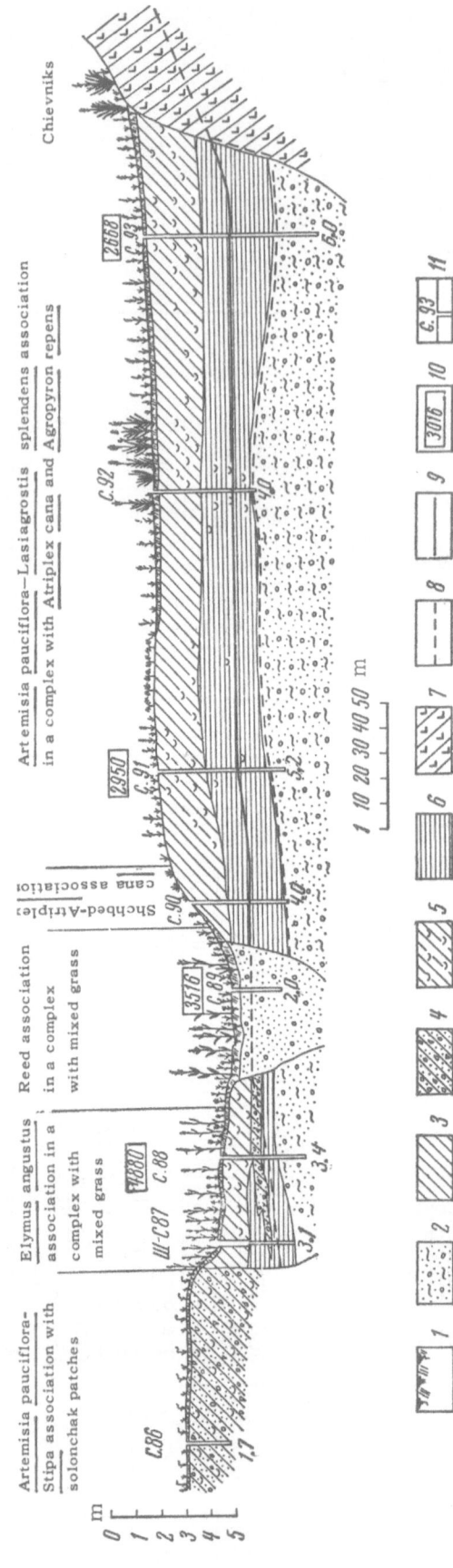

Fig. 1. Hydrogeological-geobotanical profile across the Baikonur valley 3 km west of the Sarysai estuary. 1) Soil-vegetation layer; 2) soil-gravel-pebble deposits with argillaceous material; 3) loam; 4) loam with inclusions of gravel, pebbles, and large-grained sand; 5) cemented loams; 6) clays; 7) porphyritoids of quartz-porphyries; 8) level of groundwaters; 9) established level of groundwaters, as revealed by wells; 10) mineralization of groundwaters, mg/liter; 11) number of well.

An analogous scheme could reflect the relationship of vegetation to differing depths of groundwater. In this case, the general pattern is expressed by the replacement of hydrophilic plants by more xerophilic species, as the depth of the water-carrying horizon increases.

The patterns of association between vegetation and subsurface waters shown in the hydrologic indicator scheme should not be used mechanically. Under conditions where the landscape consists of low, rounded, isolated hills, hydrophilic vegetation and phreatophytes are found quite frequently in localities with good surface moisture, comprising different hollows and sinks to which valley and rainfall waters flow. In this case, a thorough analysis of the moisture conditions is required.

We successfully applied geobotanical characteristics for discovering localities of subsurface water discharge into river valleys.

In the case of shallow-lying (0.5-1.5 m) water-carrying horizons in alluvial deposits, sectors supplied with subsurface river water are differentiated by the development of hydrophilic vegetation, which differs sharply from the communities ecologically characteristic for the particular conditions.

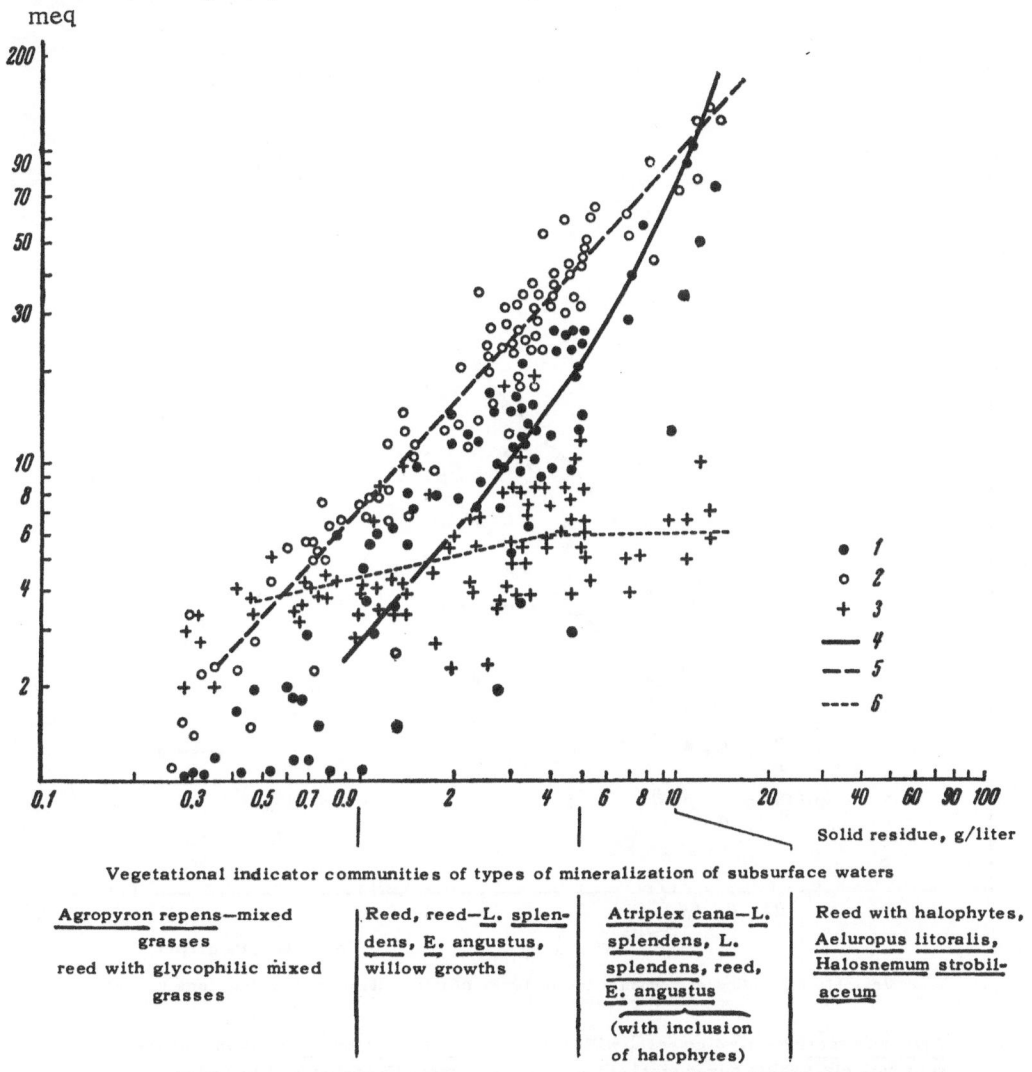

Fig. 2. Graph showing the relationship of the chemical composition of subsurface waters to their total mineralization in the northern part of Dzhezkazgan-Ulutau district, combined with the hydrologic indicator scheme. 1) Chloride ion; 2) sulfate ion; 3) hydrocarbonate ion; 4) line of change in content of chloride ions; 5) line of change in content of sulfate ions; 6) line of change in content of hydrocarbonate ions.

A geobotanical sign of river drainage of subsurface waters is also provided by the presence on the direct valley slope of _Lasiagrostis_ _splendens_, particularly where the level of groundwaters is above 2-2.5 m (Fig. 1).

Geobotanical characteristics are inapplicable when the discharge of subsurface waters occurs into the valley platform. Such sectors should be located by analysis of tectonic, hydrochemical, and other data.

We also attempted to discover the interrelationship of vegetation with the types of chemical composition of subsurface waters in the Dzhezkazgan-Ulutau district. It should be mentioned that geobotanical indication of chemical types of subsurface waters in Central Kazakhstan is rather difficult in view of the hydrochemical characteristics of the zone under survey.

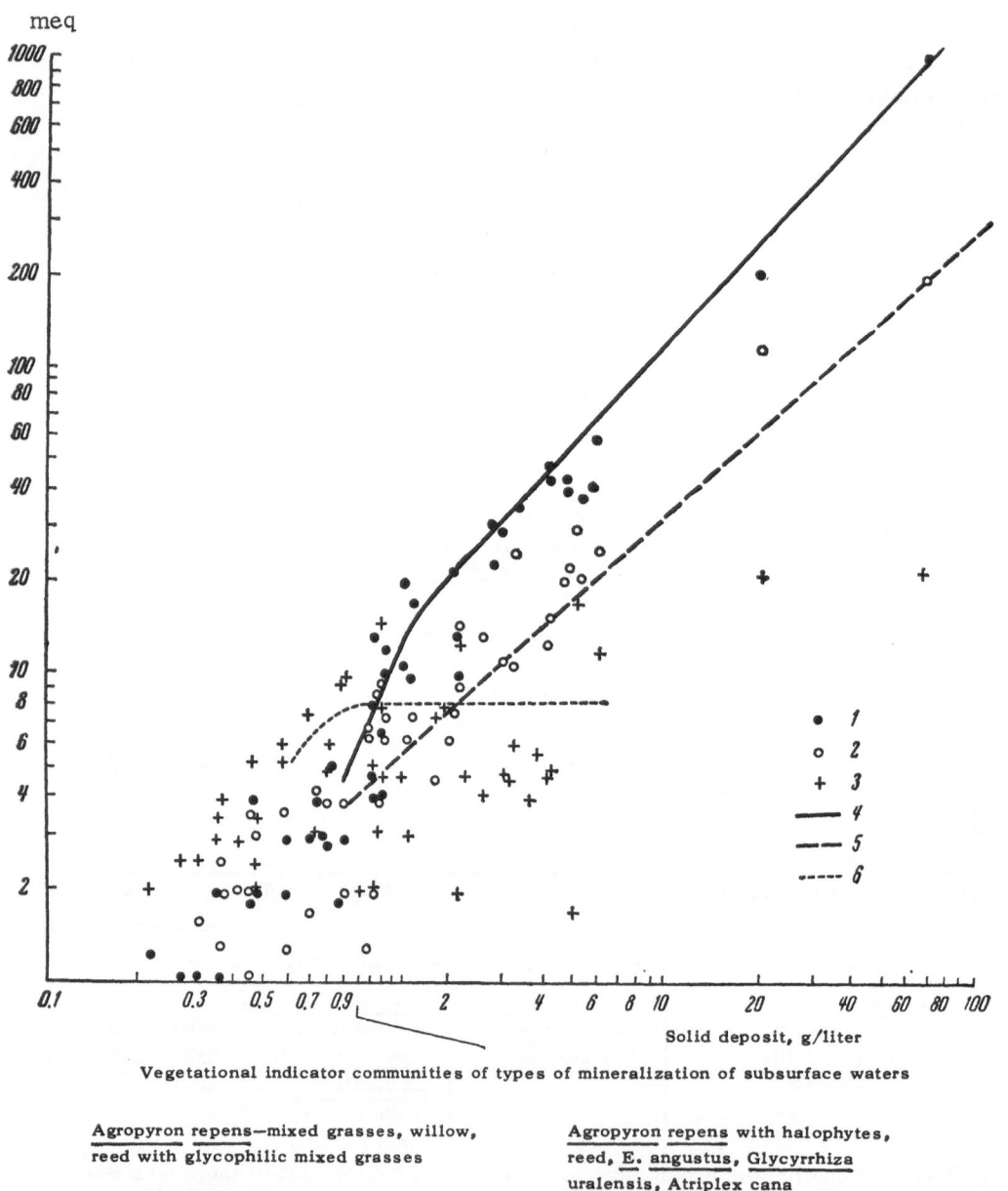

Vegetational indicator communities of types of mineralization of subsurface waters

Agropyron repens—mixed grasses, willow, reed with glycophilic mixed grasses

Agropyron repens with halophytes, reed, E. angustus, Glycyrrhiza uralensis, Atriplex cana

Fig. 3. Graph showing the relationship of the chemical composition of subsurface waters to their total mineralization in the northern part of Karsakpaisk district, combined with the hydrologic indicator scheme. Notations same as in Fig. 2.

It is enough to mention that the presence of sulfatic and chloridic waters with mineralization of 3-5 g/liter or less is typical for this territory, and thus geobotanical indication of chemical types by means of identifying plants which concentrate specific salts is not possible, since salt concentration does not occur in most phreatophytes.

A careful analysis of material available from geobotanical descriptions and hydrochemical analyses established that communities of Agropyron repens, willow, well-developed Ph. communis, and Lasiagrostis splendens are indicators of hydrocarbonate and sulfatic waters in the northern part of the Dzhezkazgan-Ulutau district. Groupings of reed, Elymus giganteus, and Glycyrrhiza uralensis are indicators of chloridic waters.

Somewhat different patterns were noted in the Karsakpaisk district. In this area, communities of A. repens and reeds with glycophilic mixed grasses are indicators of hydrocarbonate and sulfatic waters with mineralization up to 1 g/liter, while groupings of L. splendens, reed, and A. repens with halophytes, willow, and E. angustus are indicators of slightly mineralized sulfatic waters (solid residue 1-5 g/liter). Sulfatic waters with a moderate degree of salinity (5-10 g/liter) are indicated by communities of E. angustus, reed, and L. splendens with halophytes and Atriplex cana. Communities of depressed reed with halophytes and groupings of Halosnemum strobilaceum and Aeluropus litoralis are typical for chloridic waters, having mineralization in excess of 10 g/liter. It is obvious from these data that some phreatophytes are attracted to specific chemical types of subsurface waters, while other hydrologic indicators do not show this association.

Thus, L. splendens is a predominantly sulfatic species, while A. repens is a hydrocarbonate-sulfatic species. On the other hand, E. angustus is primarily a chloridic species in the northern part of the Dzhezkazgan-Ulutau district and a sulfatic species in the Karsakpaisk district. Schemes of geobotanical indicators of chemical types of subsurface waters should therefore be specially compiled for definite hydrochemical regions.

Since the chemical forms of subsurface waters constitute stages in the process of formation of their chemical composition, a study of these stages based on vegetational mapping should provide a picture of the process as a whole. In other words, the vegetation in a uniform hydrochemical situation can serve as an indicator of the process of formation of subsurface waters, and as an indicator of the orientation and dynamics of salt accumulation in shallow-lying water-carrying horizons. We have attempted to show this in graphs, with superimposed hydrologic indicator schemes, of the relationship between the content of individual ions and the increase in total mineralization, which thus represents the generalized pattern of the process of development of the chemical status of subsurface waters. The graphs indicate that where there is a change in the chemical composition of subsurface waters there is also a change in the composition of vegetational hydrologic indicator communities (Figs. 2 and 3).

The discovery of the regular patterns of geobotanical indication of chemical types of subsurface waters can be of great significance for compiling hydrologic indicator maps, since it allows them to be constructed with greater accuracy and with less expenditure of resources.

## Summary

Descriptions are given of a number of hydrologic indicator communities, the most frequent being associations with predominance of Agropyron repens, Phragmites communis, and Lasiagrostis splendens. The possibility of a rough evaluation of groundwater resources on the basis of hydrobotanical data is discussed.

It was found that fresh waters with a predominance of hydrocarbonates and sulfates are indicated by communities with prevalence of A. repens, whereas communities with dominant L. splendens tend to be associated with waters with sulfate predominance only.

## Literature Cited

Vostokova, E. A. 1955. "The application of the geobotanical method in hydrogeological investigations in deserts and semideserts." Tr. Vses. aerogeol. tresta, No. 1.

# GEOBOTANICAL FEATURES OF THE EVOLUTION OF THE MINERALIZATION OF FRESH-WATER LENSES IN THE SAM SANDS

## L. F. Voronkova

The sandy massif of Sam, which is located in the northwestern part of the Ustyurt plateau, is of great significance in the interprovince herding of cattle for the whole of Western Kazakhstan, since it is an area used for wintering.

The waters of the main water-carrying horizons in the Sam sands are strongly mineralized and therefore lenses of fresh or slightly salinated waters, flowing on the surface of waters of higher mineralization or lying in local water pores, are of particular significance for water-supply purposes. The origin of such lenses is associated with the processes of moisture condensation and infiltration into massifs of porous-ridge loosely held sands. Sites where the lenses formed in sands are located are indicated by specific vegetation communities; this phenomenon has been reported previously for various sandy massifs of the Lower Povolzh'e and Western Kazakhstan (Viktorov, 1955; Demidova et al., 1955). One of the most effective methods for finding lenses of fresh and slightly salinated waters under the conditions of the Sam sands is therefore the compilation of hydrologic indicator geobotanical maps, showing the distribution of vegetational communities which indicate water. In the course of a survey, the most important vegetational hydrologic indicator communities were discovered and their significance in searching for lenses of subsurface waters was elucidated (see table).

All the above-mentioned communities of phreatophytes, apart from Alhagi—Agropyron sibiricum, as already noted, are associated with small enclosed basins in sands; they thus form, in various combinations, a series of zonal complexes which do not emerge from within the confines of the basin. The communities constituting these complexes form ecological series, the analysis of which has led the author to the view that these series are a spacial reflection of a definite pattern of vegetational change with time, the change being caused by the evolution of water mineralization in the lens and the salinization of the ground containing the lens. The character of these changes is such that the process as a whole proceeds in the direction of a conversion of a deflation basin into solonchak; there is a corresponding salinization in the fresh lens in the basin. Some aspects of this process have been traced in other parts of the arid zone by various investigators.

The following four stages may be distinguished in this process.

1. At the initial stage, sites where lenses are located represent deflation basins, to which pioneer psammophytes gradually penetrate and form sparse groupings (mainly Artemisia arenaria D.C.). As the basin deepens on account of deflation, phreatophytes appear in addition to the pioneer psammophytes. The phreatophytes are initially those species which can endure deflation satisfactorily—Calamagrostis epigeios, Alhagi, and reed. Psammophyte-phreatophyte groupings can serve as indicators of fresh or slightly salinated waters at a depth of 3 m (rarely up to 6 m).

2. Subsequently, as the vegetation cover develops, some stability of the substrate is created and consequently conditions are more favorable for the development of rhizomatous grasses, and these spread to the bottom of the basin and gradually force out the last psammophytes at the periphery and then discharge Alhagi, also.

The species composition of the vegetation at this stage is very varied and mainly comprises elements of meadow and tugai flora. The main components are rhizomatous grasses such as Phragmites communis Trin., Elymus junceus Fisch., Bulboschoenus sp., Eragrostis arundinacea Roshev., Apocynum lancifolium Russan., and

| Communities | Habitat | Hydrogeological conditions |
|---|---|---|
| Dense, close, pure growths of reed (Phragmites communis Trin.) | Smallish enclosed deflation basins at the periphery of sand dunes or loosely held sands | Waters with mineralization not higher than 3 g/liter at a depth of 0.5–3.0 m |
| With dominance of Eragrostis arundinacea Roshev., Calamagrostis epigeios (L.) Roth., Equisetum sp., and with inclusion of Artemisia arenaria | — | — |
| With dominance of Alhagi pseudalhagi (M.B.) Desv., species of the genus Bulboschoenus, and Eragrostis arundinacea Roshev. | — | — |
| With dominance of Equisetum ramosissium Desf. | — | — |
| With dominance of Alhagi pseudalhagi (M.B.) Desv. and Agropyrum sibiricum (Willd.) P.B., sometimes incorporating Lasiagrostis splendens (Trin.) Kunth. | Sites with loosely held sands, adjacent to deflation basins | Waters with mineralization of 0.5–3.0 g/liter (rarely up to 5 g/liter) at a depth of 2.0 to 7.0 m |
| Reed and annual saltworts (reeds not higher than 0.8 m, depressed, thin) | Peripheral parts of solonchak depressions | Waters with mineralization of 12–48 g/liter at a depth of 1.0 m (rarely deeper) |
| Nitraria schoberi L. with inclusion of Halostachys caspica C.A.M., Halosnemum strobilaceum M.B., and several other succulent halophytes | Bottom of salinated basins among sands | Waters with mineralization of 12–48 g/liter at a depth of 1.0–3.0 m |
| Halosnemum spp. | The same | Waters with mineralization of 12–50 g/liter (and above) at a depth of 1.0–1.5 m |

Cynanchum acutum L. The basin may be termed a meadow type at this stage. Amongst the local inhabitants, such depressions are called "churots." Two main types of basin occur at this stage in the Sam sands. In one of these, the vegetation cover consists mainly of Equisetum, grass—Apocynum, and Apocynum—grass communities; an Alhagi—grass zone is virtually absent. The other vegetation cover comprises a bipartite zonal complex: a grass—Alhagi community at the periphery (zone width 25-30 m) and growths of phreatophytic grasses, sometimes including sedges, at the center of the basin (zone width 30-60 m). Communities which make up the vegetation cover at this stage indicate fresh or highly salinated waters at a depth of 3 m in both the center and the periphery of the basin. Similar basins are found with psammophyte—Artemisia—shrub vegetation in loosely held sands. The relative basin depths are 4-6 m; their sizes are highly variable.

The rhizomatous grasses in the central part of the basin form a thick turf layer. Sometimes marsh species, such as Poa palustris L. and sedges, begin to appear. The process of silting has considerable significance at this stage. It causes the mechanical structure of the soil to become firmer, increases the capillarity of the ground in the upper horizons, and consequently rising currents begin to play a significant role. It also leads to the vigorous development of the vegetation, particularly phreatophytes, species which evaporate a large amount of moisture, and the water consumption increases from year to year. At the same time, replenishment of the lens with fresh water lessens each year as the surrounding sands become firmer, the possibility of moisture condensation being reduced.

3. At the next stage of evolution, the increased pull of water to the surface and its evaporation lead to the gradual salinization of the ground waters. The lower the reserves of fresh water in the lens and the closer the mineralized water layer, the more rapidly is the process of increasing mineralization completed. Since the

groundwaters come closest to the surface in the central, most depressed part of the bottom basin (where the vegetation cover is most vigorous), it is at these sites that water salinization primarily occurs. As this proceeds, grasses, sedges, and Apocynum begin to disappear from the vegetation. Phragmites communis remains for longer than the other species. A solonchak is formed at the center of the basin, at the periphery of which thin, low-growing populations of reed together with annual species of the genus Suaeda and Salicornia herbacea L. are found. A 30-50 m strip of Alhagi and grasses still remains around the margin of the basin. The presence of sites with salt efflorescence and the existence of depressed reed with saltworts (Salsola) in the center of the trough are evidence of substantial mineralization of groundwaters and their closeness to the surface at the center of the basin. At the same time, the Alhagi—grass communities around the basin indicate a narrow strip of fresh waters, at a depth of 2-3 m, encircling the basin. At this stage of development, depressions are very suitable for the construction of wells, since they are most frequently located at sites which are readily accessible, with firmly held sands.

4. As the further development of the salinization process proceeds, the strip of grasses and Alhagi gradually disappears, and the entire basin is converted into a sor (salina), i.e., a solonchak, with its center void of vegetation and its margin encircled by a zone of perennial halophytes: Nitraria schoberi L., Halosnemum strobilaceum M.B., Kalidium caspicum (L.) Ung.-Sternb., and sometimes also Tamarix hispida Willd.

During this stage, the halophytes often form two independent strips: an inner one of Halosnemum strobilaceum and an outer one of Nitraria schoberi. Fragments of the Alhagi—grass community may still remain in some basins, but they have practically no indicational significance, fresh waters having virtually disappeared. Water under the solonchak lies at a depth of 1 m, but contains more than 12 g salts per liter.

It should be noted that the evolution of the vegetation of churots, or meadow-type basins, proceeds in a parallel manner with the overgrowth of sands as a whole. Solonchak basins are thus most frequently observed in firm sands.

Meadow-type basins or churots and the second and third stages of development are easily distinguished, even from distant sandy ridges, thanks to their clear green phreatophytes. They can also be well differentiated in aerial photographs by their darker image. The borders of basins are usually quite clearly demarcated in photographs.

Small basins with phreatophytic vegetation, which does not form a zonal complex, can be seen in the form of small, circular dark spots. The majority of large basins in the second stage of development, with a well-developed zonal complex, have an oval or elongated shape, but the zonal nature is not very evident, since the great density of the grassy growths in the center of the basin and the grass—Alhagi community at the periphery mean that the complex gives a monotonal image.

Zonality remains clear in aerial photographs even at the third stage of development, and in the largest basins. A narrow dark ring on the periphery of the trough usually corresponds to an Alhagi—grass community, indicating fresh water, while a whitish spot in the center represents a solonchak with very sparse, low-growing reeds, Salicornia herbacea, and annual saltworts. A whitish solonchak spot can be differentiated in the basin center at the fourth stage of development in aerial photographs; a somewhat darker strip round its edge is formed by bands of Halosnemum strobilaceum and Nitraria schoberi.

The above shows that the evolution of basins to some extent reflects the modes of development of fresh-water lenses as a whole and can provide information regarding the necessity of ameliorative measures for preserving fresh waters.


Summary

Lenses of subsurface fresh waters in the Sam sands (Ustyurt desert) are subjected to gradual salinization and intensive evaporation of water by plants. Different stages of increasing mineralization are associated with different plant communities in depressions among sands, where the lenses are located. Highly mineralized waters are indicated by thickets of Halosnemum strobilaceum while waters of good quality are indicated by communities with predominance of species pertaining to the genera Equisetum, Eragrostis, Phragmites, Bulboschoemus, and Apocynum.

## Literature Cited

Viktorov, S. V. 1955. Geobotanical methods of prospecting for subsurface waters. Exploration for and conservation of mineral resources, No. 4.

Demidova, L. S., et al. 1955. "Experience in using the geobotanical method in hydrogeological investigations in the Black Earths." Tr. Vses. aerogeol. tresta, No. 1.

# THE ZONAL DISTRIBUTION OF SUBSURFACE WATERS
# IN THE SOUTHWESTERN PART OF THE ZAISANSK DEPRESSION
# AND ITS REFLECTION IN THE PLANT AND SOIL COVER

## V. N. Ostrovskii

In a 1959 paper, a brief survey was given of experience in applying geobotanical methods in the compilation of distribution maps of subsurface waters in the southwestern part of the Zaisansk depression. We should like in the present paper to deal with some previously uninvestigated geobotanical and soil characteristics of this interesting territory, with reference to the hydrogeological conditions.

From the geological aspect, the Zaisansk depression represents a typical submontane depression. The submontane part of the trough is filled with large fragmental material. The latter includes ancient, so-called upper Gobi conglomerates and also more recent pebbles from debris cones and ancient terraces of the Bugaz, Bazarka, Katbaga, and Tebezge Rivers. The thickness of the pebbles apparently reaches several hundred meters.

Toward the central part of the depression, pebbles are gradually replaced by sandy-clay formations, whose thickness is usually not very great, rarely exceeding 10 m.

In the eastern part of the territory, we find outcrops of tertiary clays, and also hillocky sites, composed of granites of the Kezencu and Zhaksa-Arganata mountains. There has clearly occurred in this area a marked upheaval of the basic folding of the depression toward the surface. The southwestern part of the Zaisansk depression is characterized by a dryish climate, with a mean annual precipitation of 250-300 mm. Winter rainfall, which plays a role in supplying the subsurface waters, comprises not more than 20% of this total, and this is insufficient for the formation of the high-discharge water-carrying horizons which have developed in the Zaisansk depression. The main source of supply for the subsurface waters in the territory is consequently subsurface and surface drainage from the Western Tarbagatai ridge, where there is much more precipitation than in the plain.

Western Tarbagatai thus serves as the main region for the recharge of groundwaters in the Zaisansk depression.

As noted previously (Ostrovskii, 1959), we have differentiated the following zones in respect of groundwater distribution in the territory.

1. A zone of groundwater formation or transit, within which groundwater currents flow, at a depth of some tens of meters, toward the central part of the depression. This zone coincides in the main with the distribution of debris-cone pebbles and river terraces, and also upper Gobi conglomerates.

2. A zone of thinning-out of groundwater, which is clearly delimited only in the southeastern part of the district. Very marked thinning-out of the groundwaters occurs here, since the water consumption by the various groups of wells reaches hundreds of liters per second, evidence of the very substantial subsurface water resources.

3. A zone of shallow-lying (not deeper than 10 m and 3-5 m on average) groundwaters.

4. A zone of submerged groundwaters (depth of groundwaters in excess of 10 m).

The zonal distribution of groundwaters is disrupted in sectors where the basic folding of the depression has been uplifted. Thus, there are sites with shallow-lying groundwaters in the formation zone in the Zhaksa-Arganata mountain district.

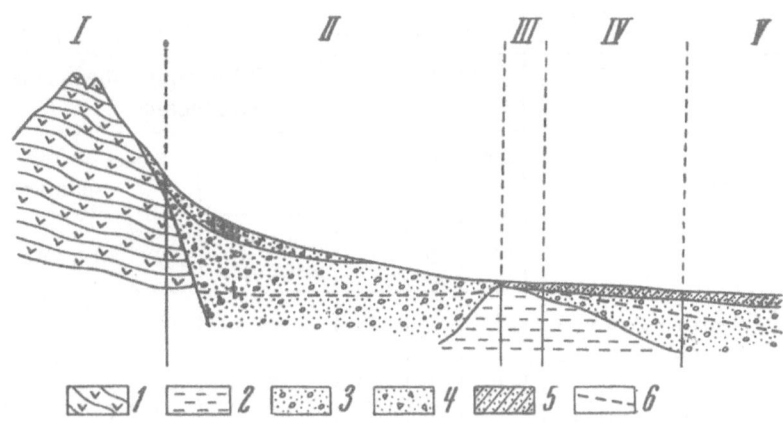

Schematic hydrogeological-geobotanical profile of the southeastern part of the Zaisansk depression. I-V) Zonal distribution of groundwaters; I) supply zone (Western Tarbagatai ridge); II) formation zone (vegetation cover comprising communities of Stipa capillata, Artemisia sublessingiana, and Festuca sulcata with inclusion of Spiraea hypericifolia; III) thinning-out zone (communities of Phragmites communis, Lasiagrostis splendens, and Elymus in the vegetation cover); IV) shallow-lying zone (communities of L. splendens, Atriplex cana, Elymus, Artemisia pauciflora, Stipa capillata, and Artemisia schrenkiana in the vegetation cover); V) submerged zone (communities of Stipa capillata, Artemisia sublessingiana, and Festuca sulcata, incorporating Spiraea hypericifolia in the vegetation cover). 1) Paleozoic rocks; 2) tertiary clays; 3) debris-cone pebbles and ancient terraces; 4) sills of temporary water-flow deposits; 5) sandy-clay deposits in peripheral parts of debris cones and ancient terraces; 6) level of groundwaters.

In the northern part of the district, close to the border of the depression with the hillocky area, the level of groundwater reaches 2-3 m in the submerged zone, as a result of being raised by the rock base of the small hillocks.

The mineralization of groundwaters in the southwestern part of the Zaisansk depression is not great and, according to data of our analysis, fluctuates from 600 to 1600 mg/liter. No marked increase in the salt content of groundwaters from the peripheral parts of the depression to its center is observed, which can be explained primarily by the good conditions of water exchange: even in the shallow water-table zone, water-holding rocks are represented by well-washed, large granular types (see figure).

The soil layer of the region is characterized by marked variability, but its distribution generally reflects the hydrogeological zonation.

Pale-chestnut soils are distributed in the zone of groundwater formation, according to the data of the soil-botanical expedition of the Ministry of Agriculture of the Kazakh SSR.

TABLE 1. Degree of Soil Salinization at Depth at Which Samples Were Taken

| Position of profile | Soil type | Depth of taking sample, cm | Solid residue, % | Water extract, mg | | | | |
|---|---|---|---|---|---|---|---|---|
| | | | | $Cl^-$ | $SO_4^{2-}$ | $Ca^{2+}$ | $Mg^{2+}$ | Total alkalinity |
| 2 km to the west of the Bazarok marker peg, on the plain | Solonetz steppe | 0-14 | 1.237 | 1.46 | 12.8 | 0.61 | 0.31 | 0.44 |
| | | 30-40 | 2.270 | 4.50 | 20.0 | 9.52 | 5.39 | 0.56 |
| | | 80-90 | 2.198 | 0.62 | 1.6 | 0.10 | 0.26 | 0.104 |
| | | 120-130 | 1.718 | 1.65 | 13.2 | 3.64 | 0.14 | 0.64 |

TABLE 2.  Differences in the Vegetation Cover of the Two Hydrogeological Zones

| Species | Number of plots (as %) in which the species shown were noted | | | | | | | |
| | zones of thinning-out and shallow bedding | | | | zones of formation and submersion | | | |
| | cop. | sp. | sol. | total frequency | cop. | sp. | sol. | total frequency |
|---|---|---|---|---|---|---|---|---|
| Stipa capillata L. | 61 | — | 39 | 26 | 66 | 11 | 23 | 84 |
| Artemisia sublessingiana (Kell.) Krasch. | — | — | 100 | 3 | 61 | 4 | 35 | 64 |
| Festuca sulcata Hack. | — | — | 100 | 2.5 | 60 | 15 | 25 | 34 |
| Spiraea hypericifolia L. | — | — | 100 | 0.7 | 42 | 6 | 52 | 14 |
| Kochia prostrata (L.) Schrad. | — | 10 | 90 | 7 | — | 19 | 81 | 19 |
| Eurotia ceratoides (L.) C.A.M. | — | 22 | 78 | 6 | 6 | 20 | 74 | 19 |
| Atriplex cana C.A.M. | 68 | — | 32 | 16 | — | — | 100 | 2.5 |
| Artemisia schrenkiana Ldb. | 17 | 3 | 80 | 29 | — | — | — | — |
| Artemisia gracilescens Krasch. et Iljin | 60 | — | 40 | 18 | 50 | — | 50 | 3 |
| Limonium gmelinii (Willd.) Ktze. | — | 78 | —22 | 8 | — | — | — | — |
| Suaeda physophora Pall. | 9 | 9 | 82 | 8 | — | 50 | 50 | 1.5 |
| Camphorosma monspeliacum L. | 10 | — | 90 | 7 | — | — | — | — |

TABLE 3.  Species Composition of the Vegetation of Zones of Groundwater Formation and Submersion

| Species | Abundance of species in associations | | | | | | |
| | Stipa capillata | Artemisia sublessingiana | Stipa capillata | Festuca sulcata-Spiraea hypericifolia | Artemisia sublessingiana-Spiraea hypericifolia | Artemisia sublessingiana-Stipa capillata | Artemisia sublessingiana-Festuca sulcata |
|---|---|---|---|---|---|---|---|
| Artemisia austriaca Jacq. | sol. | — | sol. | sol. | — | sol. | — |
| Artemisia scoparia W. et K. | sol. | — | — | — | — | — | — |
| Artemisia sublessingiana (Kell.) Krasch. | sp. | $cop._3$ | sol. | sol. | $cop._1$ | $cop._1$ | $cop._2$ |
| Caragana frutex (L.) Koch. | — | — | sol. | — | sol. | — | — |
| Ceratocarpus turkestanicus Sav.-Rycz. | sol. | sol. | — | — | — | — | — |
| Festuca sulcata Hack. | sp. | sol. | sp. | $cop._2$ | — | — | $cop._1$ |
| Galium verum L. | — | — | sol. | — | sol. | — | — |
| Kochia prostrata (L.) Schrad. | — | sol. | — | — | — | sol. | — |
| Nanophyton erinaceum (Pall.) Less. | sol. | — | — | — | — | — | — |
| Spiraea hypericifolia L. | sol. | — | $sp.-cop._1$ | — | $sp.-cop._1$ | sol. | sol. |
| Stipa capillata L. | $cop._2$ | sol. | $cop._2$ | sp. | sp. | $cop._1$ | sol. |

The zones of thinning-out and shallow-lying groundwaters are characterized by a more or less salinated soil cover.

The salinization of these soils can be explained by the shallowness of the water table, and is well illustrated in data from analyses of water extracts, the results of one of which are as given in Table 1.

The most marked salinization of the soil occurs in the 30-90 range, which may be explained by the active evaporation of groundwaters from the surface of the capillary zone. The lowest content of soluble salts was noted in the surface layer of the soil. A different distribution of soil salinization in vertical section is observed in the zones of groundwater formation and submersion. Here the greatest mineralization of water extracts, up to 0.365%, occurs directly in the soil surface in the 0-20 cm layer; this figure is 2-3 times less at a depth of 30-40 cm.

It is thus possible to differentiate in the soils of the southwestern part of the Zaisansk depression, as in other submontane depressions (Priklonskii, 1948), two directions in the migration of salts: one, descending, is in the zones of groundwater formation and submersion; the other, ascending, is in the thinning-out and shallow water-table zones. In comparison with the soils, the vegetation cover of the Zaisansk depression is still more closely associated with hydrogeological conditions. As is known, phreatophyte plants, whose roots reach the level of groundwaters on their capillary zone, have settled in many districts. This group of plants is represented in the Zaisansk depression by the following: Lasiagrostis splendens (Trin.) Kunth., Phragmites communis Trin., Glycyrrhiza glabra L., Halimodendron halodendron (Pall.) Voss., Calamagrostis epigeios (L.) Roth., and various species of Elymus L.

The depth of root-system penetration of these phreatophytes does not exceed 5 m, and therefore phreatophytes grow only in zones of shallow-lying and thinning-out groundwaters. In this context, it is useful to differentiate two groups of hydrogeological zones: the formation and submersion zones, in which the groundwaters do not influence the vegetation; and zones of thinning-out and shallow bedding, where the vegetation cover is formed under the influence of groundwaters.

The differences in the vegetation cover in these two hydrogeological zones are very obvious. The vegetation cover in the groundwater formation and submersion zones is represented by communities of Stipa capillata L., Artemisia sublessingiana (Kell.) Krasch., Festuca sulcata Hack., with the inclusion of Spiraea hypericifolia, Kochia prostrata (L.) Schrad., Eurotia ceratoides (L.) C.A.M., and other species (Table 2). The projective cover does not exceed 50-60% on the average. Table 3 shows that the species composition of the vegetation of this group of hydrogeological zones is characterized by marked barrenness.

The vegetation of the thinning-out and shallow-lying groundwater zones is much more varied (Table 4). There develop associations of phreatophytes—L. splendens, Ph. communis, Elymus giganteus Vahl., and also Atriplex cana C.A.M. and wormwoods (Artemisia gracilescens Krasch. et Iljin.and A. schrenkiana Ldb.). Stipa communities are also encountered, but their species composition is different from that in the groundwater formation and submersion zones (Table 4).

The second group of hydrogeological zones is characterized by complexity of the vegetation cover, determined primarily by the depth of the groundwater table and by soil conditions, and more rarely by the microrelief.

The projective cover of the vegetation is equal to 50-70% on average, reaching 100% in individual plots.

Still more graphic data indicating the effect of groundwaters on the formation of the vegetation cover are provided by an analysis of the distribution of dominant and subdominant species within the two groups of hydrogeological zones. As mentioned above, phreatophytes grow only in the zones of thinning-out and shallow-lying groundwaters. Of the 78 species encountered within the two groups of hydrogeological zones, only 22 have general distribution, giving a community coefficient equal to 28.2%.

The utilization of regular patterns in the association between vegetation and groundwaters has considerably facilitated the carrying out of hydrogeological investigations, and, in particular, the compilation of a hydrogeological zonation map of the southeastern part of the Zaisansk depression.

TABLE 4.  Species Composition of the Vegetation of Thinning-Out and Shallow-Lying Groundwater Zones

| Species | Abundance of species in associations | | | | | | |
|---|---|---|---|---|---|---|---|
| | Lasiagrostis splendens | Phragmites communis | Atriplex cana – Suaeda physopora | Elymus multicaulis | Artemisia gracilescens | Stipa capillata | Artemisia schrenkiana |
| Acroptilon picris C.A.M. | — | — | sol. | — | sol. | — | — |
| Aeluropus litoralis (Gouan.) Parl. | sol. | — | — | sol. | — | — | — |
| Agropyron repens (L.) P.B. | sol. | sol. | — | sol.-sp. | sol. | — | — |
| Anabasis salsa (C.A.M.) Benth. | — | — | sol. | — | — | — | — |
| Artemisia austriaca Jacq. | — | — | — | — | sol. | sol. | — |
| Artemisia gracilescens Krasch. et Iljin. | — | — | sol. | — | cop.3 | sol. | — |
| Artemisia schrenkiana Ldb. | sol. | — | sol. | sol. | — | — | cop.3 |
| Atriplex cana C.A.M. | sol. | — | cop.3 | — | — | sol. | — |
| Calamagrostis epigeios (L.) Roth. | sol. | sol. | — | sol. | sol. | — | — |
| Camphorosma monspeliacum L. | sol. | sol. | sol. | sol. | — | — | — |
| Ceratocarpus turkestanicus Sav.-Rycz. | — | — | — | — | sol. | sol. | — |
| Dodartia orientalis L. | sol. | — | — | — | — | — | — |
| Elymus giganteus Vahl. | sp. | — | — | cop.2 | — | — | — |
| Elymus angustus Trin. | sol. | — | — | sol. | — | — | — |
| Euphorbia gerardiana Jacq. | sol. | — | — | — | — | — | — |
| Eurotia ceratoides (L.) C.A.M. | sol. | — | — | sol. | — | — | — |
| Glycyrrhiza glabra L. | sol. | sol. | — | sol. | — | sol. | sol. |
| Halimodendron halodendron (Pall.) Voss. | sol. | — | — | — | — | — | — |
| Lasiagrostis splendens (Trin.) Kunth. | cop.3 | sol. | — | sol. | — | — | — |
| Limonium gmelinii (Willd.) Ktze. | sol. | — | sol. | sol. | — | — | — |
| Kochia prostrata (L.) Schrad. | — | — | — | — | sol. | — | sol. |
| Phragmites communis Trin. | sol.-sp. | cop.3 | — | — | — | — | sol. |
| Plantago salsa Pall. | sol. | sol. | — | sol. | — | — | — |
| Salsola brachiata Pall. | — | — | sol. | — | sol. | sol. | — |
| Sophora alopecuroides L. | sol. | — | — | — | — | — | — |
| Stipa capillata L. | — | — | sol. | sol. | sol. | cop.2 | sol. |
| Suaeda physophora Pall. | — | — | sol.-sp. | — | sol. | — | — |

Experience gained in hydrogeological work in the Zaisansk depression indicates the need for the wider application of geobotanical methods in studies of the hydrogeology of submontane and intermontane depressions in the arid zone.

## Summary

On the basis of the nature of the vegetation cover, it was found to be possible to define the outlines of the formation and submersion groundwater zones, as well as the thinning-out and shallow-lying zones.  In the first two zones, the vegetation is only slightly related to the groundwaters, whereas in the second zone the dominant species are phreatophytes, whose roots penetrate down to the level of the subsurface waters or to their capillary fringe.

## Literature Cited

Ostrovskii, V. N. 1959. "Experience in the application of geobotanical methods in hydrogeological research in the southwestern part of the Zaisansk depression." Vestn. Akad. Nauk Kaz.SSR, No. 8.

Priklonskii, V. A. 1948. "Some features of the formation of groundwaters in dry regions." Tr. Labor. gidrogeol. probl., Vol. 1, Moscow, Izd. Akad. Nauk SSSR.

# THE SIGNIFICANCE OF COEDIFICATORS
# IN THE INDICATOR PROPERTIES
# OF PLANT COMMUNITIES

## I. N. Beideman

Those vegetational communities whose combinations of plants reveal a particular depth or degree of mineralization of groundwaters have considerable significance as indicators of hydrogeological conditions. In each individual case, a different combination of plants is thus evidence of differences in the mineralization of the groundwaters. Thus, for example, a combination of plants, associated with fresh and saline groundwaters, can, within a single community, indicate that changes are in progress within a single plot, which may eventually lead to a change in the hydrogeological regime in the direction of demineralization or salinization of groundwaters. We obtained data, for different communities in the Kura-Araksinsk lowlands in the Zakovkaz'e, on the degree of mineralization and depth of groundwaters under different communities with dominance of a single edificator, but with varied combinations of coedificators and accompanying species.

We constructed a series of graphs on the basis of the data obtained, and these show how gradations in groundwater depth and mineralization change under similar communities. The graphs were constructed in the following way: the depth of groundwaters, in meters, is set along the ordinate, while the mineralization in grams of solid deposit per liter is placed on the abscissa. The marks shown represent the position of the particular community in the system of coordinates for each plot studied. The graphs obtained far from exhaust all the variability of naturally existing variations in a single community with different combination of species under varied habitat hydrogeological conditions, but do provide a basis for some preliminary assessments.

For illustrative purposes, we shall survey some of these graphs for different communities. Figure 1 shows the depth and mineralization of groundwaters under four communities, formed by the edificator Phragmites communis Trin.

1. Phragmites communis + Puccinellia gigantea (Fig. 1; 2). This community is widely distributed along the shores of the Caspian Sea in the Kura-Araksinsk lowlands, and fringes low parts of downstream valleys with growths of reed and water plants. The first layer is usually occupied by reed, and the second by Puccinellia gigantea Grossh. As shown in the figure, this community occupies areas where the ground waters are close to the surface, from zero to 1.5 m from the soil surface. The mineralization of the water lies within the range of 10-20 g solid deposit per liter.

2. Phragmites communis + Suaeda altissima (Fig. 1; 3). This community usually occurs in sites which have undergone a flood regime, and have rather deeper ground-waters and greater mineralization thanks to "biological salt accumulation." Such communities are found in muddy and slimy river flood areas, and they also arise secondarily after excessive watering and on formerly irrigated plots. The figure shows that the depth of groundwaters under these communities is 1.5-2.5 m, while the mineralization is 15-40 g of solid deposit per liter.

3. Phragmites communis + Atriplex tatarica (Fig. 1; 4). This community is close to the preceding one in habitat conditions. It covers the same type of sites, frequently in a complex with it. The depth of groundwaters under A. tatarica—reed communities fluctuates from 2 to 2.5 m, while the mineralization is 25-60 g solid deposit per liter.

4. Phragmites communis + Alhagi pseudalhagi (Fig. 1; 1). This variant of the reed communities normally occurs along the banks of streams and stagnant sites, where the water is somewhat deeper but is not yet highly

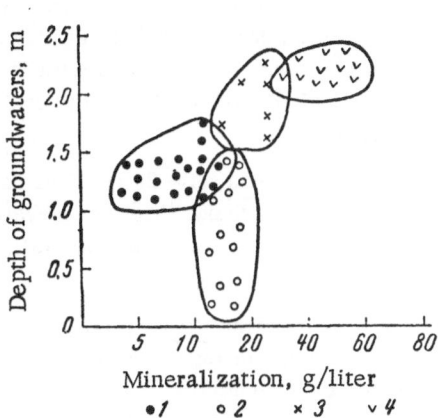

Fig. 1. Group of communities with the edificator Phragmites communis and their relationship to the depth and mineralization of groundwaters: 1) Phragmites communis + Alhagi pseudalhagi; 2) Phragmites communis + Puccinellia gigantea; 3) Phragmites communis + Suaeda altissima; 4) Phragmites communis + Atriplex tatarica.

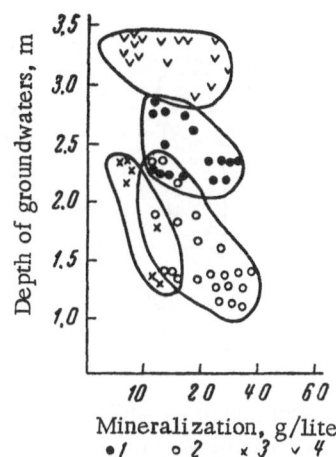

Fig. 2. Group of communities with the edificator Suaeda microphylla and their relationship to the depth and mineralization of groundwaters: 1) Suaeda microphylla + Salsola crassa; 2) Suaeda microphylla + Suaeda altissima; 3) Suaeda microphylla + Petrosimonia brachiata; 4) Suaeda microphylla + Tamarix ramosissima.

mineralized. The figure shows that the depth of groundwaters under Alhagi—reed communities fluctuates from 1 to 1.75 m, while the mineralization is within the range 2.5 to 20 g of solid deposit per liter.

An examination of Fig. 1, which demonstrates different community variations where the edificator is reed, shows how the hydrogeological conditions change under different combinations of species. Each species involved in the building-up of the community carries its individual features with respect to its indicator significance. In the particular hydrophyte communities covered by us, reed is combined with trichohydrophyte-halophytes: Puccinellia gigantea Grossh., Suaeda altissima (L.) Pall., Atriplex tatarica L., and the phreatophyte Alhagi pseudalhagi (M.B.) Desv. The combination of reed with P. gigantea (which usually occurs where groundwaters are in the close proximity) leads to the formation of a community which occupies localities with very close groundwaters of moderate mineralization. Ph. communis exists normally in this case. The combination of reed with halophytic annual saltworts produces a community which occupies sites with deeper groundwaters and substantially greater mineralization. In this community, the reed population is already somewhat depressed. The combination within one community of reed and the phreatophyte Alhagi pseudalhagi, both glycophilic species, results in this community being associated with areas with waters having the lowest mineralization. Each of these communities is a link in a chain of successive changes, which depend very much on the hydrogeological conditions, as well as other factors. As indicators of groundwaters, these communities complement one another. It is clear that reed in combination with A. pseudalhagi indicates fresher groundwater than does reed with annual saltworts. The presence of the latter means greater mineralization and a deeper water table, as compared with sites where reed grows together with Puccinellia gigantea.

Let us now examine some communities formed by the edificator Suaeda microphylla Pall. (Fig. 2).

1. Suaeda microphylla + Suaeda altissima (Fig. 2; 2). This community usually covers solonchak areas, located in some degree of deepening of the groundwaters, S. microphylla Pall. being the species which enters the community, previously formed only by S. altissima (L.) Pall. Groundwaters lie at a depth of 1-2.5 m, and the mineralization varies from 20 to 60 g solid deposit per liter.

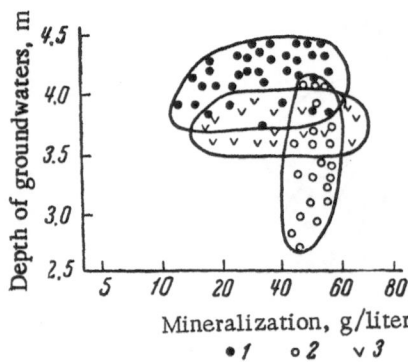

Fig. 3. Group of communities with the edifactor <u>Artemisia</u> <u>meyeriana</u> and their relationship to the depth and mineralization of groundwaters: 1) <u>Artemisia</u> <u>meyeriana</u> + <u>Salsola</u> <u>dendroides</u>; 2) <u>Artemisia</u> <u>meyeriana</u> + <u>Salsola</u> <u>crassa</u>; 3) <u>Artemisia</u> <u>meyeriana</u> + <u>Suaeda</u> <u>microphylla</u>.

2. <u>S.</u> <u>microphylla</u> + <u>Petrosimonia</u> <u>brachiata</u> (Fig. 2; 3). This combination of species in the community is the next link in the successional changes. It is located on drier solonchaks in plakors.* As compared with the previous community, the groundwater is somewhat deeper and lies at a depth of 1.25-2.5 m, while the mineralization is from 15 to 30 g solid deposit per liter.

3. <u>S.</u> <u>microphylla</u> + <u>Salsola</u> <u>crassa</u> (Fig. 2; 1). This community covers territories with still drier solonchaks. The groundwaters are even deeper, lying at a depth of 2-3 m, while the mineralization rises from 20 to 65 g solid deposit per liter.

4. <u>S.</u> <u>microphylla</u> + <u>Tamarix</u> <u>ramosissima</u> (Fig. 2; 4). This species combination produces communities occurring on areas with still deeper groundwaters, the depth reaching 2.75-4 m; mineralization fluctuates from 15 to 50 g solid deposit per liter.

All these combinations of <u>S.</u> <u>microphylla</u> Pall. with various other species clearly form a chain of interrelated communities, which replace one another as the groundwater regime changes.

Finally, we shall analyze some communities formed by the edifactor <u>Artemisia</u> <u>meyeriana</u> Bess.

Wormwood populations in the Kura-Araksinsk lowlands are a zonal formation, and the most typical vegetation cover under particular climatic conditions. Their development follows a long pattern from hydromorphic communities to xeromorphic communities. We have three different communities formed by <u>A.</u> <u>meyeriana</u> (Fig. 3).

1. <u>A.</u> <u>meyeriana</u> + <u>Salsola</u> <u>crassa</u> (Fig. 3; 2). The combination of <u>A.</u> <u>meyeriana</u> + <u>S.</u> <u>crassa</u> arises at still earlier stages of the formation of wormwood populations. The upper layers of the soil have already undergone solonetization but some of the solonchakization still remains. This community is distributed on plakor areas and the debris cones of rivers. The depth of groundwaters is already substantial (from 2.5 to 4.25 m), while the mineralization is from 40 to 60 g solid deposit per liter.

2. <u>A.</u> <u>meyeriana</u> + <u>Suaeda</u> <u>microphylla</u> (Fig. 3; 3). This community arises where there is deepening of groundwaters under <u>S.</u> <u>microphylla</u> and the introduction of <u>A.</u> <u>meyeriana</u> into the lower layer of this community. The subsequent pattern of development is a reduction of solonchakization in the soil, the depression of <u>S.</u> <u>microphylla</u> Pall. and its replacement by <u>Salsola</u> <u>dendroides</u> Pall., and finally evolution toward pure wormwood scrub. The depth of groundwaters under this association is equal to 3.5-4 m, while the mineralization is 15-65 g solid deposit per liter.

3. <u>A.</u> <u>meyeriana</u> + <u>Salsola</u> <u>dendroides</u>) (Fig. 3; 1). This community is very widely distributed in the Kura-Araksinsk lowlands, occupying plakor and slightly elevated positions. It is the last link in the chain of successive changes prior to the final formation of wormwood scrub. The upper horizons of the soil are solonetzed, and the quantity of readily soluble salts in them is reduced. Groundwaters are substantially deeper (to 3.75-4.5 m), while mineralization fluctuates from 10 to 60 g solid deposit per liter.

Our survey of the relationships of communities to the depth and mineralization of groundwaters shows that each combination of species is not fortuitous and is based on the dynamics of hydrogeological conditions. Each combination of species changes the hydrogeological conditions and is itself modified because of the new conditions. Each community thus indicates only some particular evolutionary stage of the landscape as a whole. If we know the ecology of the species and their relationship to moisture and salinity conditions, we can discover

---

*Plakor—flat and slightly undulating, well-drained lowland interfluves, the soils and vegetation of which reflect very clearly the zonal features of the landscape in the particular zone (for example, grassy steppes on chernozem soils in steppes; coniferous forests on podzolized soils in the targa; and so on).

the past history of the landscape and determine its future. Consequently, communities characterize not merely the simple position of the habitat conditions but also indicate the dynamics of these conditions with time.

Summary

A particular edificator of the phreatophyte group, in combination with different concomitant species, may be indicative of different depths of bedding and different degrees of mineralization of groundwaters. This confirms the great importance of a thorough analysis of the floristic composition of communities in hydrologic indicator work.

To illustrate the above, the author describes cases of changes in mineralization under communities with Phragmites communis, Suaeda microphylla, and some other species, in combination with various coedificators (see Figs. 1-3).

# ECOLOGICAL CHARACTERISTICS AND THE ROOT SYSTEM STRUCTURE OF SOME HYDROLOGIC INDICATOR SPECIES IN THE ALLUVIAL-DELTA VALLEY OF THE SYR-DAR'YA

## F. N. Chalidze

One of the current problems of hydrologic indicator research is the study of the biology of hydrologic indicator plants, since without a deep understanding of this it is impossible to make more accurate predictions of the depth of bedding and also the degree of mineralization of groundwaters. Data on the biology and ecology of hydrologic indicators are still far from complete and therefore observations on these aspects could be of some interest. The present author has studied the ecological characteristics and structure of the root systems of a number of plant species which are used for hydrologic indicator purposes in the alluvial-delta valley of the Syr-Dar'ya River, within Kzyl-Ordinsk province. The studies were carried out on communities of Alhagi pseudalhagi (M.B.) Desv., Halimodendron halodendron (Pall.) Voss., Lycium turcomanicum Fisch. et Mey., Pluchea caspica (Pall.) Hoffm., and tamarisks (mainly Tamarix ramosissima Ldb., distributed in the region of the Dingal well. For the root-system study, three trenches about 5 m in depth were dug, sited at a distance of 100 m from one another.

The root systems were pressed up against the wall of the trenches, stretched slowly, and measured. The groundwaters in the plot under study lay at a depth of 8 m.

The results of observations on the hydrologic indicator plants are described briefly below.

One of the widely distributed plants on the plot studied was H. halodendron (Pall.) Voss. Its appearance here in large numbers showed that N. P. Grave's data (1936), indicating that this species is associated with waters lying at a depth of not more than 5 m, are not fully accurate. It should be mentioned that this same species was also found by us in the Saikudak well district, where groundwaters lie at a depth of 12 m; the ecological range of the species, with respect to depth of groundwaters, is thus clearly rather wide.

The root system of H. halodendron (Fig. 1) was traced by us (in the Dingal well district) down to a depth of 4.35 m; at this depth the roots are 30 mm thick, the vertical root showing almost no branching and tapering being insignificant. All this indicates that the roots penetrate much deeper than the limit to which we uncovered them. We frequently noted root penetration along former root passages, filled with semidecaying residues of old roots. These residues form characteristic organic covers around individual parts of the living root system, as was noted for roots of this species in the Bol'shie Barsuki sands by N. T. Ageeva and N. B. Bulgakova (1955). This phenomenon apparently has definite significance for root development in lithologically varied, stratified alluvial-delta deposits, which rather frequently contain solid argillaceous layers.

The roots of the other landscape species in the plot studied, namely the Turkmenian box-thorn (Lycium turcomanicum Fisch. et Mey.), were traced to a depth of 2.5 m, where they were 20 mm thick (Fig. 2). They acquire this thickness at a depth of 1 m (above this level, the roots are markedly tapered) and retain it without any change for 1.5 m; the vertical direction of the roots and the absence of changes in its thickness suggest that the roots penetrate to a great depth. However, the branching of the root system is much greater on the whole than in H. halodendron, especially at depths of 2-2.5 m.

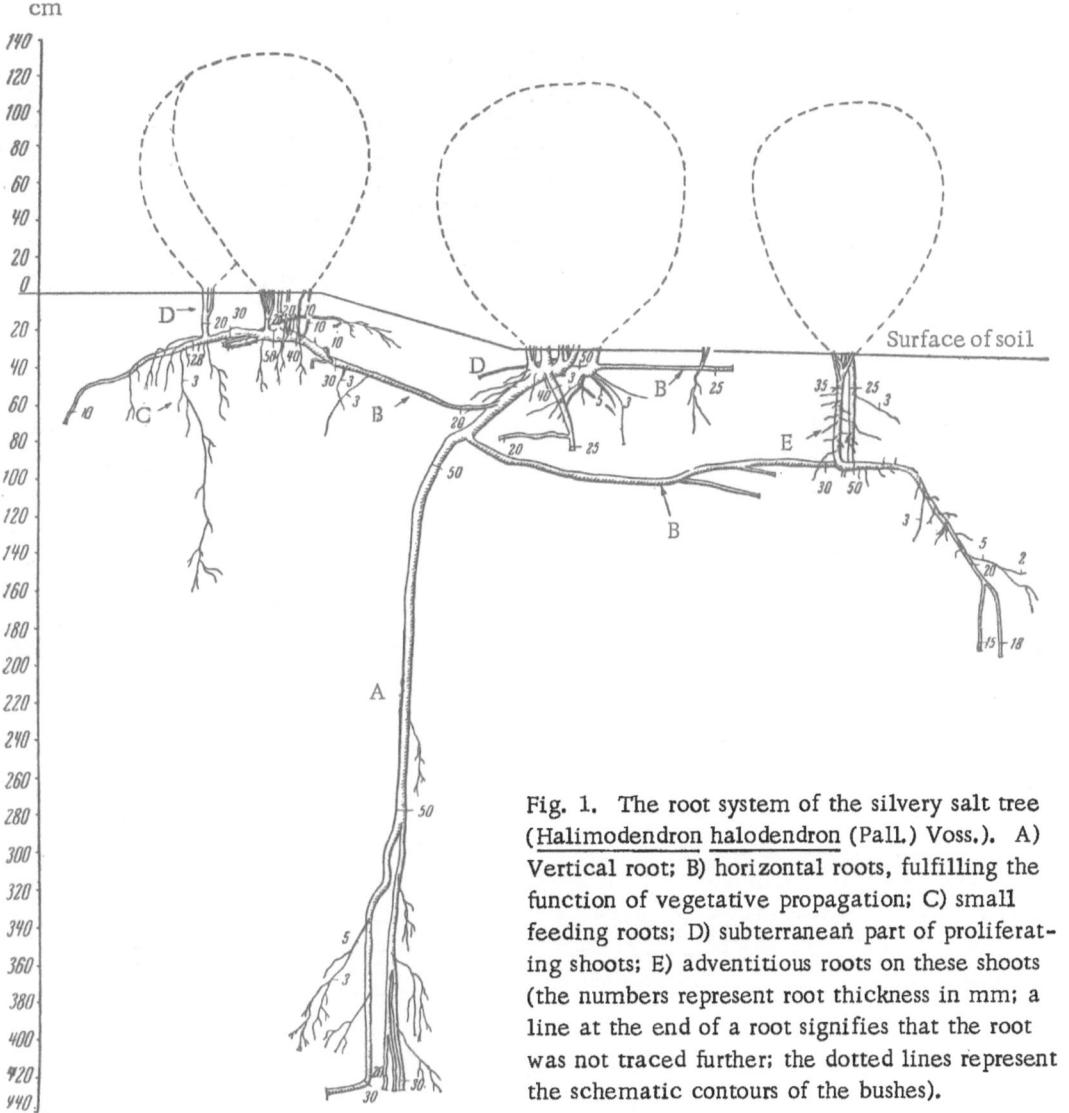

Fig. 1. The root system of the silvery salt tree (Halimodendron halodendron (Pall.) Voss.). A) Vertical root; B) horizontal roots, fulfilling the function of vegetative propagation; C) small feeding roots; D) subterranean part of proliferating shoots; E) adventitious roots on these shoots (the numbers represent root thickness in mm; a line at the end of a root signifies that the root was not traced further; the dotted lines represent the schematic contours of the bushes).

Finally, the roots of sea-lavender (Limonium otolepis (Schrenk) Ktze.) were uncovered down to a depth of 2.7 m; they attained a thickness of 6 mm, were very slightly tapered and did not have major branches indicating that, as in the preceding case, the roots penetrated to a level substantially lower than the range uncovered (Fig. 3).

The study of the uncovered root systems has allowed us to clarify some features of the structure of the subterranean parts of the phreatophytes studied. Thus, the following can be differentiated in the subsurface part of H. halodendron (see Fig. 1): a vertical root, which penetrates to the deepest level; horizontal roots, which are mainly responsible for vegetative propagation; small feed roots; the subterranean part of sprouting shoots; and adventitious roots on these shoots, which are partly of nutritional significance and partly serve to collect atmospheric precipitation. The following can be clearly distinguished in the root system of L. turcomanicum (Fig. 2): the main root of the maternal plant; horizontal roots responsible for vegetative propagation; their branches; and the subterranean parts of root suckers. The following can be observed in Limonium otolepis (Fig. 3): the main root, which penetrates to the deepest level; horizontal roots responsible for vegetative propagation; rhizomes; the subterranean parts of root suckers; feed roots of the IInd and IIIrd order; and bundles of ephemeral rootlets in the surface layers of the soil, which die off toward midsummer.

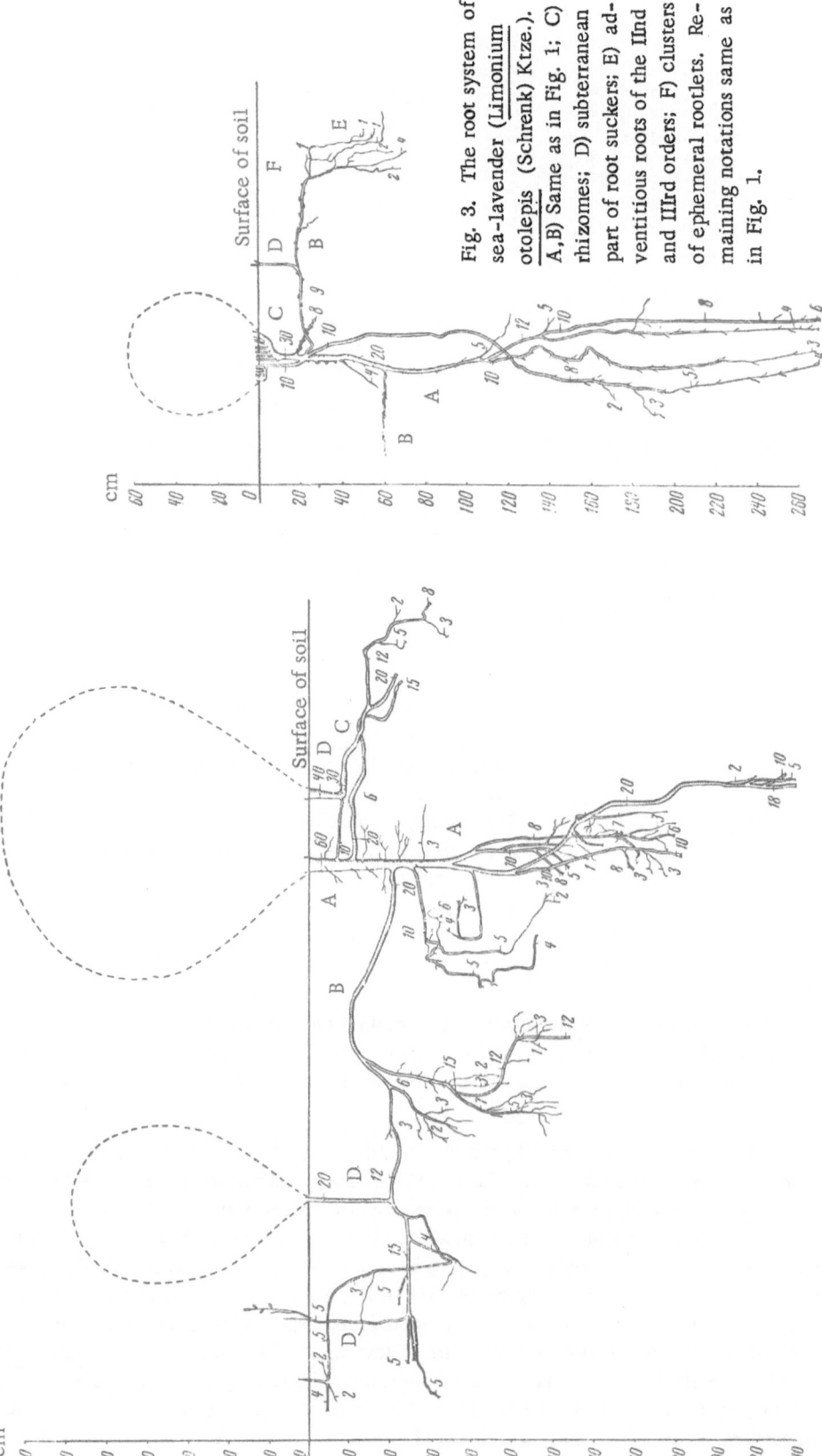

Fig. 3. The root system of sea-lavender (Limonium otolepis (Schrenk) Ktze.). A,B) Same as in Fig. 1; C) rhizomes; D) subterranean part of root suckers; E) adventitious roots of the IInd and IIIrd orders; F) clusters of ephemeral rootlets. Remaining notations same as in Fig. 1.

Fig. 2. The root system of the Turkmenian box-thorn (Lycium turcomanicum Fisch. et Mey.). Notations same as in Fig. 1.

The phreatophytes examined thus have some rather marked differences in the structure of their subterranean parts, mainly affecting their capacity for collecting water from atmospheric precipitation. Thus, H. halodendron is practically deprived of the possibility of collecting the latter, L. turcomanicum possesses this capacity to a greater degree, mainly on account of the greater branching of its root system, while in L. otolepis we find a special group of roots having the function of entrapping atmospheric waters. H. halodendron can therefore be classified as a permanent phreatophyte, and L. otolepis as a faculative phreatophyte; L. turcomanicum occupies an intermediate position in this series. When using the species as hydrologic indicators, these features should be taken into account.

## Summary

The author describes the root system of three phreatophytes not previously studied: Halimodendron halodendron, Lycium turcomanicum, and Limonium otolepis. The first two species, which have a very vigorous root system, may be encountered in both fresh and brackish water; the third species is associated primarily with water exhibiting high mineralization. L. otolepis was found to possess clusters of small rootlets lying near the surface of the soil and dying off when the summer drought commences.

## Literature Cited

Ageeva, N. T., and Bulgakova, N. B. 1955. "The root systems of woody and bush species in the Bol'shie Barsuki sands." Uchenye zap. Kazakhsk. Univ. XVII. Biologiya i pochvovedenie.

Grave, N. P. 1936. "Tugai jungles in the lower reaches of the Amu-Dar'ya." Izv. Kara-Kalpaksk. NIL

# THE INFLUENCE OF PLANTS AND PLANT COMMUNITIES
# ON THE SEASONAL DYNAMICS OF GROUNDWATERS

## I. N. Beideman

The seasonal changes in the level of groundwaters under different plant communities are closely associated with the transpirational activity of the plants. Several investigators have already studied this relationship (Ototskii, 1899, 1905, 1914; Meinzer, 1927; White, 1931, 1932; Znamenskii, 1938; Beideman, 1949; Kitredge, 1951; Molchanov, 1952; Robinson, 1957; Beideman and Filenko, 1959; Vostokova, 1961; Viktorov, Vostokova, Vyshivkin, 1962; and others). All these workers concluded that daily and seasonal fluctuations in groundwaters are clearly evident under phreatophytic and trichohydrophytic vegetation. Changes in the level of groundwaters in summer under ombrophytes are less marked, since their influence on groundwater is limited. They only slightly reduce the replenishment of groundwaters from rainfall, which is intercepted by their roots.

We made observations on the changes in groundwaters under different communities during the growing season and winter periods in the Kura-Araksinsk lowland and also in the Nogaisk steppe. Observations on the fluctuations in groundwater level in specially prepared holes under different communities were carried out by us during 1931-1933 and 1945-1946 in the Kura-Araksinsk lowland and the Mugan' steppe (Beideman, 1949). Almost all the edificators of these communities were phreatophytes and trichohydrophytes.

The 1931 observations were made on the following communities: Salsola dendroides + Petrosimonia brachiata + Suaeda confusa; Salsola crassa + Petrosimonia brachiata + ephemerals; in two communities with Artemisia meyeriana + Salsola dendroides; and in communities composed of Alhagi pseudalhagi + ephemerals, and Lagonychium farctum + ephemerals and Artemisia meyeriana + ephemerals.

The level of groundwaters in spring (April, May) is very close to the surface (from 30 to 180 cm below the soil surface) (Fig. 1). This is due to two reasons: the large amount of spring rainfall and the spring watering of cotton and grain-crop plantations; the irrigation waters replenish the groundwaters and the latter move up close to the surface. The highest level of groundwaters occurs under communities of Alhagi pseudalhagi + ephemerals (Fig. 1A), located on the banks of channels. The volume of water in the channels also affects the fluctuations in groundwater level under these communities. Groundwaters lie somewhat deeper under communities with the edificator Lagonychium farctum (Banks et Sol.) E. Bobr., a species with a very deep root system. Next in depth of groundwaters come communities of annual saltworts: Petrosimonia brachiata + Salsola crassa + Suaeda confusa + Artemisia meyeriana and ephemerals. Deepest of all are groundwaters under the A. meyeriana + Salsola dendroides community and under annual saltworts, Salsola crassa M.B. with Petrosimonia brachiata (Pall.) Bge.

With further development of the vegetation and evaporation from the soil surface, groundwaters under all the communities become deeper and do not cease this deepening process until November or December. During this period of the year, the vegetation of the Kura-Araksinsk lowland completes its development and ceases to transpire. Rainfall occurs and the rain replenishes the groundwaters so that the latter again begin to rise. To illustrate the changes in depth of groundwaters, data are shown on the total expenditure of water by the vegetation and soil during the growth period of the same year for the same communities. As shown in Fig. 1B, the maximal total consumption falls in May and June, after which the water consumption by transpiration and evaporation begins to decrease as the depth of the water table falls, the minimum being reached in July and August, to be followed by an increase on account of rainfall. The figures show that the curves for lowering of the water table and water consumption by total evaporation (evapotranspiration) are coincident. Maximum evapotrans-

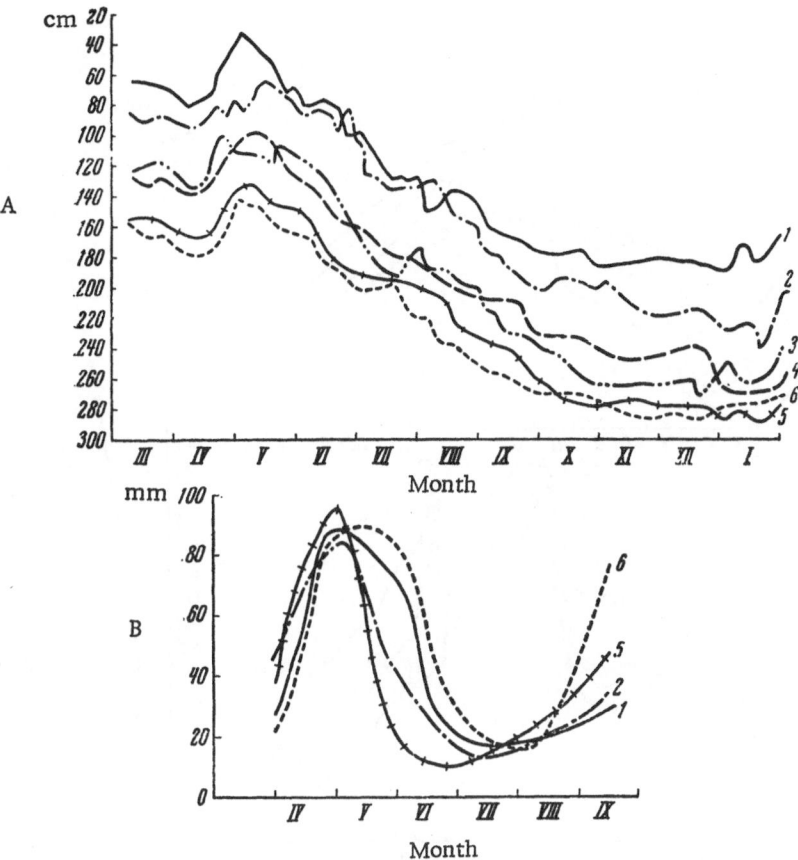

Fig. 1. Changes in the level of groundwaters and total loss (by evapotranspiration) under different communities, composed of phreatophytes
and trichohydrophytes, during the period from March 1931 to January
1932. Mugan' steppe, Kura-Araksinsk lowland. A) Changes in the
level of groundwaters (in cm) away from soil surface under the following communities: 1) Alhagi pseudalhagi and ephemerals; 2) Lagonychium farctum + ephemerals; 3) Petrosimonia brachiata + Salsola
crassa + Suaeda confusa; 4) Artemisia meyeriana and ephemerals; 5)
A. meyeriana + Salsola dendroides; 6) Salsola crassa + Petrosimonia
brachiata and ephemerals; B) changes in total evaporation (in mm)
under the same communities during the period from April to
September 1931.

piration is noted in spring, when the groundwater lies closest to the surface. The high consumption of moisture
in spring, attributable to the groundwater absorption by plants, to water from the capillary fringe being used up,
and to moisture evaporation from the soil surface, leads to the beginning of a drop in the level of the groundwaters. When the supply of irrigation water ceases, the groundwater is subjected primarily to the influence of
the vegetation. As the water table falls and the availability of groundwaters to the roots is reduced, evapotranspiration decreases, but the quantity being evaporated through soil and plant continues to drop until such
time as the vegetation completes its cycle of development and the plants finally cease to utilize groundwater.
When this happens all the atmospheric waters filter through the soil, reach the groundwaters, and raise their
level. During the summer period, the groundwater level drops from 30 to 190 cm and from 140 to 280 cm —in
total, by 160 cm.

In 1945-46, parallel observations were made at the same point in the Mugan' steppe on changes in the
level of groundwaters during the year in transpiration of plants during their developmental season. The observations were made from July 1945 to November 1946. The vegetative communities could be divided into two

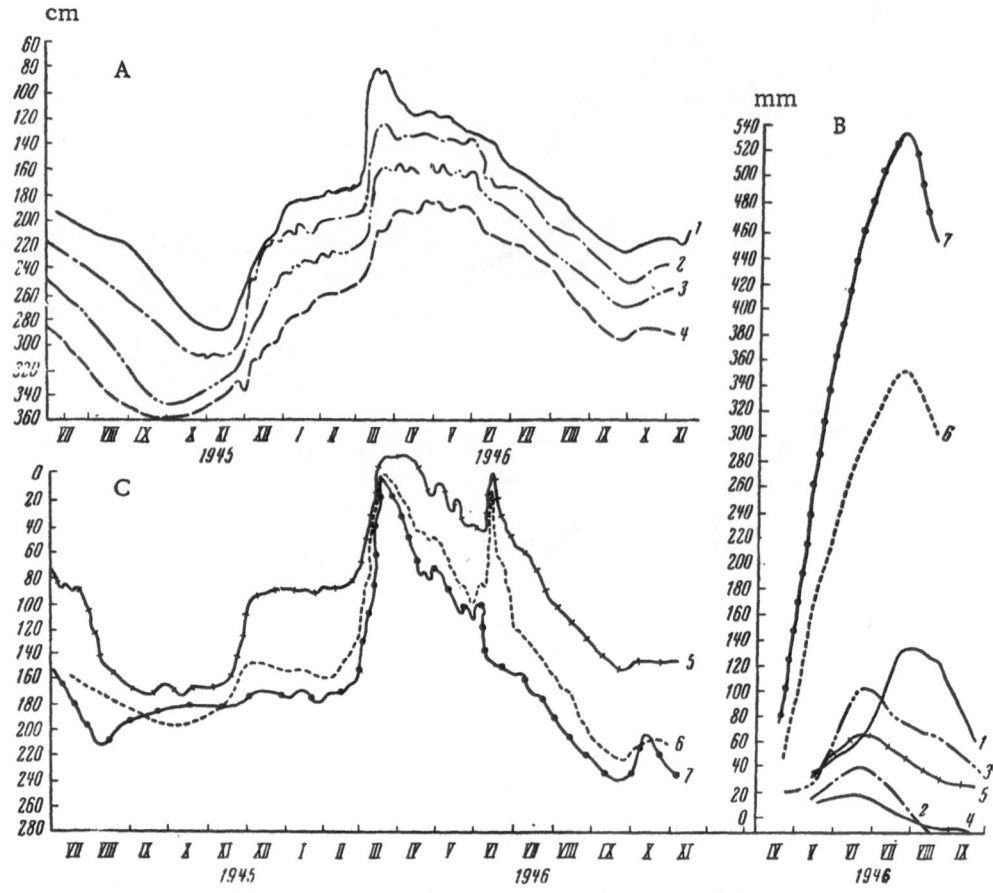

Fig. 2. Changes in the level of groundwaters and consumption of water by transpiration, under communities composed of phreatophytes and trichohydrophytes, from July 1945 to November 1946. Mugan' steppe, Kura-Araksinsk lowland. A) Changes in the level of groundwaters (in cm) under communities of the first group, under plakor: 1) Halostachys caspica + Limonium scoparium + Aeluropus repens + Tamarix ramosissima; 2) Salsola crassa + Petrosimonia brachiata; 3) Salsola dendroides + Artemisia meyeriana + Petrosimonia brachiata; 4) Petrosimonia brachiata (overgrowth). B) Changes in consumption of water in transpiration (mm) from April to September 1946 by the following communities: 1) Halostachys caspica + Limonium scoparium + Aeluropus repens + Tamarix ramosissima; 2) Salsola crassa + Petrosimonia brachiata; 3) Salsola dendroides + Artemisia meyeriana + Petrosimonia brachiata; 4) Petrosimonia brachiata (overgrowth); 5) Salicornia europaea + Aeluropus repens; 6) Glycyrrhiza glabra + Cynodon dactylon + Atriplex tatarica.*C) Changes in the level of groundwaters (in cm) under communities of the second group in the lowland: 5) Salicornia europaea + Aeluropus repens; 6) Glycyrrhiza glabra + Cynodon dactylon; 7) Tamarix ramosissima + Glycyrrhiza glabra + Atriplex tatarica.

groups according to habitat: the first group of communities was located on a plakor plain, and the second was around the vast depression. The first group comprises communities of annual saltworts with occasional bushes of Tamarix ramosissima Ldb. and Alhagi pseudalhagi (M.B.) Desv. The second community consists principally of two edificators, Artemisia meyeriana Bess. and Salsola dendroides Pall., with a significant admixture of Petrosimonia brachiata (Pall.) Bge. The third community of this group is located in artificially formed low lands on chloride-sulfate solonchak; it consists of Halostachys caspica + Limonium scoparium + Aeluropus repens.

* B7 not given in Russian original. — Tr.

Figure 2 shows the fluctuations in groundwater level under all the communities of the second group from July 1945 to November 1946. All our curves are similar in shape. The greatest volume of groundwaters is observed in mid-March and early April 1946. The level then falls gradually until September-October, when the lowest position is reached in both years. As in 1931, the high spring water table is due to the autumn, winter, and spring rainfall and the lack of any water consumption through transpiration in the winter period. Changes in the level of groundwaters in relation to transpirational consumption are shown in Fig. 2B. Water consumption is seen to be at its maximum in the community represented by the curve numbered 7. The greatest consumption occurs in July and August, when the plants produce their maximal mass. The fall in water table is associated with increased transpiration, the maximal drop in water level in this case being 140 cm.

The next community in terms of quantity of moisture utilized is Salsola dendroides + Artemisia meyeriana + Petrosimonia brachiata; the maximal consumption falls in June and July, with a gradual drop from then until September. It is apparent that the drop in water table is associated with increased transpiration in June and July and is quite substantial in August and September. The level falls by 110 cm. Groundwaters are deeply bedded under communities with the edificator Salsola crassa M.B. The maximal consumption falls in June and July, with a decrease toward August and September. The level falls by 110 cm. The least water consumption for transpiration is observed in the community made of Salsola dendroides and Petrosimonia brachiata. The maximum consumption falls in June, with a decrease from then till the following month. The water table falls by 100 cm. The above demonstrates that the higher the water consumption in transpiration, the greater is the deepening of the water-table level.

The same pattern is observed in the second group of communities (Fig. 2C). The Salicornia europaea + Suaeda confusa occupies the bottom of the depression, while the Glycyrrhiza glabra + Cynodon dactylon community inhabits the margin. Groundwaters are somewhat lower. Groundwaters are at a still lower layer under Tamarix ramosissima + Glycyrrhiza glabra + Atriplex tatarica community.

Transpiration is very high in plants of the margin group. The water utilized by these plants is weakly mineralized and is available to the root systems of the plants. The maximal consumption falls in June, July, and August (Fig. 2C). This high consumption is responsible for a significant lowering (by 210-230 cm) in the water-table level. The water consumption by annual saltworts is rather low. Its maximum occurs in June, with a gradual decrease from then till September, while the groundwater level drops at a slow rate and to a lesser extent, by 140 cm. As may be seen, the level of transpirational consumption of water by the plants controls to a significant degree the range of spring and autumn fluctuations in the groundwater level. This is illustrated in Table 1.

TABLE 1. The Transpirational Consumption of Water by Phreatophyte and Trichohydrophyte Plants of Different Communities and the Decrease in Depth of the Water Table During the Growing Period (Mugan', 1946)

| Community | Water consumption by transpiration, mm | Decrease in depth of water table, cm |
|---|---|---|
| Petrosimonia brachiata . . . . . . . . . . . . . . | 115 | 100 |
| Salsola crassa + Petrosimonia brachiata . . . . . | 157 | 110 |
| Salsola dendroides + Artemisia meyeriana + Petrosimonia brachiata . . . . . . . . . . . . . . | 372 | 110 |
| Salicornia europaea + Aeluropus repens . . . . . | 344 | 140 |
| Halostachys caspica + Limonium scoparium + Aeluropus repens + Tamarix ramosissima* . . | 953 | 140 |
| Glycyrrhiza glabra + Cyndon dactylon . . . . . . | 1311 | 210 |
| Tamarix ramosissima + Glycyrrhiza glabra + Atriplex tatarica. . . . . . . . . . . . . . . | 1372 | 230 |

*Tamarix ramosissima Ldb. was present as a single specimen only in the community, and, therefore, the water consumption by this plant was reduced by a factor of three.

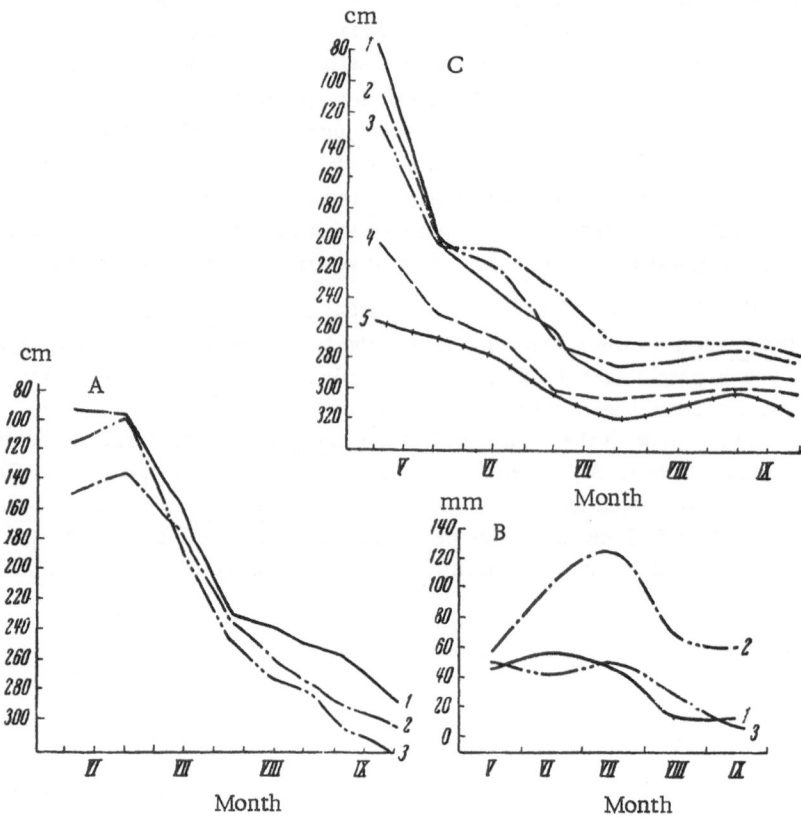

Fig. 3. Changes in the level of groundwaters and consumption of water by transpiration under different communities, composed of phreatophytes and trichohydrophytes, during the period from May to September 1954-1955. Nogaisk steppe, Northern Caucasus. A) Changes in the level of groundwaters (in cm) in the flood solonchak meadow under the following communities. 1) Agropyron pectiniforme + Artemisia taurica + Petrosimonia triandra; 2) Carex melanostachya + Alopecurus ventricosus; 3) Artemisia taurica + Petro-simonia brachiata B) Transpiration water consumption (in mm) by the same communi-ties in the flood solonchak meadow. Notations same as in Fig. 3A. C) Changes in the level of groundwaters (in cm) under the following different communities in the depres-sion: 1) Carex melanostachya + Alopecurus ventricosus + Phragmites communis; 2) Li-monium scoparium + Artemisia monogyma; 3) Limonium scoparium + Carex melano-stachya + Artemisia monogyna; 4) Limonium scoparium + Artemisia monogyna + Arte-misia taurica; 5) Agropyron pectiniforme + Artemisia taurica + Limonium scoparium.

All the plants, apart from Artemisia meyeriana, are phreatophytes and trichohydrophytes. The effect of transpirational water consumption on the level of groundwaters is therefore considerable in all the communities. It is evident that the higher the water consumption, the greater is the lowering of the groundwater level. As confirmation of these patterns, we shall present some data which we obtained for plants pertaining to the same groups with respect to conditions of hydrometabolism, in the Nogaisk steppe in the Northern Caucasus.

During 1954-1955, we carried out observations near the towns of Achikulak and Bazhigan, in the Nogaisk steppe of the Northern Caucasus, on the water regime of plants and on the changes in level of groundwaters and transpirational consumption of water by plants. The Nogaisk steppe is very similar to the Kura-Araksinsk low-land in origin and geomorphology. The Kura-Araksinsk lowland comprises the delta of the Kura and Arakso Rivers, while the Nogaisk steppe is the delta of the Kuma and Terek Rivers. As an example, we may cite fluc-tuations in the level of groundwaters in 1945 under three communities composed of phreatophytes and tricho-hydrophytes. The communities were located in an inundated solonchak meadow. The lowest part, with very closely lying groundwaters, was covered with the community Carex melanostachya + Alopecurus ventricosus; some Artemisia taurica + Petrosimonia brachiata occurs at a somewhat higher level, and, finally, at the most plakoric position, with deep groundwaters, the community Agropyron pectiniforme + Artemisia taurica + Petro-simonia triandra is present.

Figure 3A shows that after the spring flooding the groundwater is at a high level in May under all the communities, and then it drops as a result of the increase in transpirational activity (Fig. 3B), reaching its lowest position toward September. The transpirational consumption of water under Carex—Alopecurus community was at its maximum in June and July, with a decrease from then till September. The two other communities show a slightly lower consumption. The level of groundwaters under the first community fell by 160 cm, compared with 220 cm under the Artemisia taurica—P. brachiata community and 180 cm under the Agropyron pectiniforme—Artemisia taurica—P. triandra community.

We made observations in 1955 on five further communities in the Nogaisk steppe also in a similar flood solonchak meadow. These communities were situated on a profile perpendicular to an irrigation canal. A community of sedge together with some mixed grasses was located on the somewhat elevated banks of the canal. A community of Limonium scoparium + Artemisia monogyna with some annual halophytes occurred in the main depression area beyond the canal hilly strip. Limonium scoparium (Pall.) Klock. with Carex melanostachya M.B. and Artemisia monogyna Waldst. et Kit. were in a still lower position at the bottom of the depression. The slope opposite the floodplain was covered with L. scoparium together with Agropyron pectiniforme Roem. et Schult. and Artemisia taurica Willd. An Agropyron pectiniforme + Artemisia taurica + Limonium scoparium community was located at the plakor site, on the driest solonchak and along the outlet from the floodplain.

Figure 3C presents curves which show the fluctuations in groundwaters during the growth period of the vegetation, while Table 2 shows the consumption of water by the communities during the same period and the drop in the water-table level. It should be mentioned that, as in the previous case, groundwaters under all the communities deepened substantially from spring to autumn. Water consumption in nearly all the communities was at a maximum in July and August, when the groundwater level was at its lowest position. All the species at this site utilize ground and capillary water and affect the movement of groundwaters.

It may be concluded from these results that shrub, semishrub, and grassy trichohydrophytic and phreatophytic vegetation directly affect changes in the level of groundwaters during the developmental season of the plants, under conditions where the groundwaters are shallow-lying. A relationship was noted between the amount of water used up and the decrease in depth of the water table.

In the case of ombrophytic vegetation, which relies on moisture from precipitation, no such fluctuating changes in the level of groundwaters under vegetation during the growing season are observed.

TABLE 2. Transpirational Consumption of Water by Ombrophytes of Different Communities and the Decrease in Depth of the Water Table During the Growing Period

| Community | Water consumption by transpiration, mm | Decrease in depth, of water table, cm |
|---|---|---|
| Agropyron pectiniforme + Artemisia taurica + Limonium scoparium. . . . . . . . . . . . | 25 | 60 |
| Limonium scoparium + Artemisia monogyna + A. taurica. . . . . . . . . . . . . . . . | 65 | 100 |
| Limonium scoparium + Artemisia monogyna . . | 50 | 160 |
| Limonium scoparium + Carex melanostachya + Artemisia monogyna . . . . . . . . . . | 85 | 170 |
| Carex melanostachya + Alopecurus ventricosus + Phragmites communis. . . . . . . . . . . | 100 | 210 |

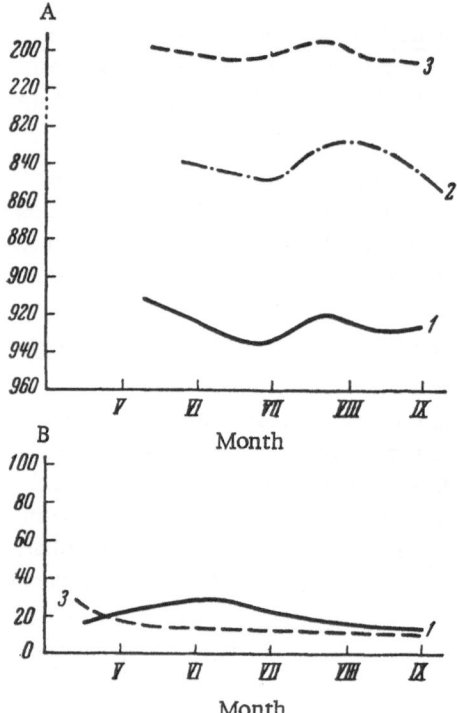

A

B

Month

Month

Fig. 4. Changes in the level of groundwaters and the transpirational consumption of water under communities, composed of umbrophytes, in a plakor valley, Nogaisk steppe, Northern Caucasus, during the period from June to October 1954. A) Changes in the level of groundwater (in cm) under the following communities: 1) Stipa capillata + Agropyron desertorum + Festuca sulcata; 2) Stipa joannis + A. desertorum + F. sulcata; 3) Kochia prostrata + Salsola dendroides + Salsola crassa; B) changes in the transpirational water consumption (in mm) with the following communities: 1) Stipa capillata + A. desertorum + F. sulcata; 2) Artemisia taurica + K. prostrata.

Direct observations on the changes in level of deep-lying groundwaters were made in the Nogaisk steppe under steppe and desert vegetation of the ombrophytic type. The species constituting the communities were growing in areas where the water-table depths ranged up to 9 m.

Two of the communities, Stipa capillata + Agropyron desertorum + Festuca sulcata, and Stipa ioannis + Agropyron desertorum + Festuca sulcata, were located on pale-chestnut soils in a sandy plakor plain. Nine of the communities observed were in Bazhigan district and were variants of the Kochia prostrata—Poa—Artemisia associations. Thus, three communities were grouped around the Bazhigan meteorological station on plakor: Artemisia taurica + Kochia prostrata + Poa bulbosa, with the admixture of Salsola laricina Pall., was situated on an undulated plain, while K. prostrata + Agropyron desertorum + Artemisia taurica was on the flattened summit of a ridge, and A. taurica + K. prostrata occupied a flattish mound. The remaining six communities were situated along the landscape from the above-mentioned group of communities to the Bazhigan plain. The entire massif through which the profile ran is a plain covered with wormwood. Two communities on undulating plain were Poa bulbosa—Kochia prostrata—Artemisia lercheana in composition. One of these comprises A. lercheana + Kochia prostrata + Poa bulbosa, together with the inclusion of Agropyron desertorum (Fisch.) Schult. and Festuca sulcata Hack. The supplementary species in the other are Salsola laricina Pall. and S. crassa M.B. The next two communities are composed basically of Agropyron desertorum + K. prostrata + Stipa: one consists of K. prostrata + Agropyron desertorum + S. capillata with the incorporation of Artemisia lercheana Web., while the second incorporates the saltworts Salsola dendroides Pall. and S. crassa M.B. Finally, the last two communities are Kochia—Salsola in type: one consists of K. prostrata + S. crassa + Petrosimonia triandra + Limonium scoparium, and the other of K. prostrata + S. dendroides + S. crassa + L. scoparium.

During the plant-growth season, the water-table level changes very slightly. The level of groundwaters rises under the first communities in July and August. The depth of the water table (8-9 m) and the short roots of the vegetation completely exclude the expenditure of groundwaters by transpiration.

Figure 4B shows the transpiration for two of these communities (S. capillata + A. desertorum + F. sulcata and Artemisia taurica + Kochia prostrata). Water consumption is very slight and its source is entirely precipitational moisture. It is greatest in spring and drops off toward September. The water-table level shows a very slight lowering toward September, and this may be associated with plant interception of summer precipitation and the failure of this moisture to reach the groundwaters.

We did not, of course, analyze all the elements which affect the level of groundwaters, such as the ebb and flow of the water. However, the examples cited do clarify the influence of vegetation on the movement of water in relation to time.

The clearest relationship appeared in the case of shallow-lying waters under phreatophytic and trichohydrophytic vegetation through the transpiration process. The groundwaters begin to get deeper with the commencement of vegetational development and then rise with the fading-out of growth activity. This was especially clear in the Mugan' example in 1946. The lack of influence of ombrophytes on the movement of groundwaters was evident from the Nogaisk steppe.

Summary

The author investigated the groundwater and, at the same time, the transpirational consumption of water by plants and plant communities in the Kura-Araksinsk lowland and the Nogaisk steppe (Caucasus). The phreatophytes and trichohydrophytes in desert and steppe regions evaporate great amounts of water and therefore affect the level and the degree of mineralization of groundwater. Evapotranspiration reaches its maximum in spring, when the groundwaters come closest to the surface. The greatest amounts of water are used by the Halostachys caspica community (in the Mugan' steppe) and the Carex melanostachya—Alopecurus ventricosus—Phragmites communis community (in the Nogaisk steppe).

A direct relationship was observed between the quantity of water used by plants of these ecological groups and the fall of the groundwater level. Shallow-rooted plants exert no marked influence on water consumption.

Literature Cited

Beideman, I. N. 1949. "The role of the vegetation cover in the water-salt regime of soils." Pochvovedenie, No. 7.
Beideman, I. N., and Filenko, R. A. 1959. "Basic hydrogeological surveys in geobotanical research." In book: Field Geobotany, Vol. 1. Moscow—Leningrad, Izd. Akad. Nauk SSSR.
Viktorov, S. V., Vostokova, E. A., and Vyshivkin, D. D. 1962. Introduction to Indicator Geobotany, Izd. MGU.
Vostokova, E. A. 1961. Geobotanical Methods of Searching for Subsurface Waters in the Dry Areas of the USSR. Moscow, Gosgeoltekhizdat.
Znamenskii, A. A. 1938. "The vegetation cover and fluctuations in the level of groundwaters." Pochvovedenie, No. 9.
Kitredge, J. 1951. The Effect of Forests on the Climate, Soil, and the Water Regime [Russian translation]. IL.
Molchanov, A. A. 1952. Hydrogeological Role of Pine Forests on Sandy Soils. Moscow—Leningrad, Izd. Akad. Nauk SSSR.
Ototskii, P. V. 1899. "On the influence of forests on groundwaters." Pochvovedenie, No. 2.
Ototskii, P. V. 1905. "Groundwaters: their origin, existence and distribution." Chap. 2 in book: Groundwaters and Forests Predominant in the Plains of Median Latitudes. SPb.
Ototskii, P. V. 1914. "Representational scheme of the bedding of groundwaters in the plain of European Russia." (Supplement to Russian translation of K. Keil'gak's book, "Subsurface waters and sources" section.) SPb.
Meintzer, O. E. 1927. "Plants as indicators of groundwaters." U.S. Geol. Survey, Water-Supply, Paper No. 577.
Robinson, T. W. 1957. "Phreatophytes." U.S. Geol. Survey, Water-Supply, Paper No. 1423.
White, W. H. 1931. "Groundwater supply of Mimbres Valley, N. Mex." U.S. Geol. Survey, Water-Supply, Paper No. 637-13.
White, W. N. 1932. "A method of estimating groundwater supplies based on discharge by plants and evaporation from soil." U.S. Geol. Survey, Water-Supply, Paper No. 659-A.

# THE TRANSPIRATION RATE AND WATER CONSUMPTION OF PLANT EDIFICATORS OF THE PRINCIPAL COMMUNITIES AND THE WATER REGIME OF THE DESERT STEPPES OF CENTRAL KAZAKHSTAN

## Zh. Zh. Zhatkanbaev

There has so far been little study of the water regime of soils and plants in the desert steppe of Central Kazakhstan, which is of great practical and scientific significance for the agriculture of the republic.

In view of the importance of this problem, the joint expedition of the USSR Academy of Sciences to Central Kazakhstan included within its program of experimental work a study of the water regime of soils and plants as a part of the main problem, "Biological complexes of new developmental regions, and their rational utilization and amelioration," assigned to the expedition.

Field studies were carried out under the direction of L. N. Beideman in May-September, 1958 and 1959, on the territory of the USSR Academy of Sciences Botanical Institute station (Zhanarkinsk district, Karaganda Province).

The climate of the region is characterized by very marked continentality: the hot summer is followed by a severe winter, the distribution of rainfall being irregular. The greatest quantity of rain usually falls in spring and autumn, and the least amount in winter.

The two years of our research on this aspect were different. The first year, 1958, seemed more favorable for plant development: the spring was warm, and the dry period began in July, whereas in 1959 the spring was dry and cold and the drought period began as early as May.

The main soil type on the territory of the station is light chestnut, but in addition solonetz and solonchak occur quite frequently.

A light-chestnut soil, which we studied in a profile under an Artemisia incana—Stipa lessingiana community, was characterized by the following: thickness of horizon A, 5-11 cm; B, 11-23 cm; C, 23-100 cm. Effervescence with 10% HCl was noted from a depth of 23 cm. Gypsum lies below the 100-cm level. Analyses of soil samples were carried out in the soil ecology laboratory of the Botanical Institute of the USSR Academy of Sciences and showed that the dominant anion was $SO_4''$, the content of which was minimal in the upper horizon (0.11 meq), but gradually increased down to a depth of 100 cm (17.93 meq); there was less $HCO_3'$ (0.051-0.35 meq), and still less Cl'. The dominant cation was $Ca''$, the content of which was 0.09-0.46 meq in the upper horizon, and reached a maximum at a depth of 100 cm (15.85 meq), falling subsequently to 0.26 meq.

There was little solid deposit in the upper horizon (from 0.01 to 0.06%) but it increases with depth, reaching a maximum at 140 cm (1.27%). As regards the mechanical structure under the Artemisia incana—Stipa lessingiana community, light loams are the dominant type.

Somewhat different data were obtained in a profile of the same light-chestnut soil, taken under an Artemisia incana—S. capillata community. The thickness of horizon A was equal to 6-24 cm, that of horizon B was 24-35 cm, and that of horizon C was 35-95 cm, while effervescence with 10% HCl was observed from 6 cm; accumulation of gypsum was noted from a depth of 75 cm.

The analysis of an aqueous extract from samples of this soil showed that the maximum content of anions and cations rises at a higher rate than under the other community, the amount of $SO_4''$ at a depth of 100 cm being equal to 19.0 meq, i.e., greater than the corresponding figure under the A. incana—Stipa lessingiana community. The content of the $Ca''$ cation reaches 17.5 meq at a depth of 75 cm. There is little solid deposit in the top horizon (from 0.13 to 0.30%), but its content increases with depth, reaching a maximum (2.36%) at 140 cm. The humus content in the top horizon reaches 2.05%, expressed as a percentage of the dry soil, and is 0.02% at a depth of 140 cm.

As regards mechanical structure, the soil here is a medium and heavy loam. The soils under the two communities thus differ in both chemical and mechanical composition, the content of anions and cations being greater in the soil under the A. incana—S. capillata community.

Let us now examine the changes in moisture content under these communities.

Two layers were differentiated under the A. incana—S. lessingiana community in June and July of both 1958 and 1959, one layer (0-40 cm) being desiccated, with the moisture content 4.41-7.19% of the dry soil weight, and the other (40-140 cm) being more moist, with a moisture content fluctuating from 11.7 to 14.97%. In August, and particularly in September, the soil in the entire profile (to 140 cm) was already desiccated. The moisture content fell to the wilting point.

This pattern of behavior can largely be explained by the high air temperature during the spring and summer seasons and the type of rainfall distribution.

Moisture in the soil under the A. incana—S. capillata community was distributed in a somewhat different manner.

In this case, shallow desiccation of the soil (from 0 to 5 cm) was observed in June, July, and August 1958. The soil layers down to a depth of 60 cm began to dry up only in September.

Soil desiccation under the same community in 1959 began in May and June in the 0-20-cm layer, in July in the layer down to 40 cm, and in August down to 140 cm. The moisture content in the soil was somewhat higher than under the A. incana—S. lessingiana community in both 1958 and 1959.

Observations were made on soil temperature, in the same layers as those in which moisture was determined, from the surface to a depth of 140 cm. The data obtained show that the temperature of the soil surface in desert steppes fluctuates from 20 to 30°C during the daytime in summer, reaching a maximum of 35°C, while at night it falls to 5-11°C. The temperature shows a rather strong vertical fall from the surface to a depth of 140 cm. During the day, only the 0-40-cm soil layer is actively heated up; it is also warmed up lower than 40 cm but only at a very slow rate; at the bottom, there is established a virtually constant temperature, which changes only slightly in relation to plant growth seasons. The soil temperature at all depths increases from May to August, after which a decrease begins in September.

The vegetation cover of the district studied pertains to the Eastern Kazakhstan subprovince, and the Artemisia—Festuca sulcata—Stipa lessingiana steppe zone of the Aral-Ulutau region.

On the territory of the desert-steppe station of the Botanical Institute, two zonal steppe communities are most widely distributed: A. incana—S. lessingiana (the northern zonal type of the desert steppes) and A. incana—S. capillata (the southern zonal type of the desert steppes), together with A. pauciflora, Atriplex cana, Anabasis salsa, and Nanophyton erinaceum associations. We studied two communities: A. incana—S. lessingiana and A. incana—S. capillata. The edificator in the A. incana—S. lessingiana community is the xerophytic, solid-turf grass Stipa lessingiana Trin. et Rupr. and the semibush type, Artemisia lercheana Web. In addition, Festuca sulcata Hack., Stipa sareptana Becker., S. kirghisorum P. Smirn., and other species are found. The edificators in the A. incana—S. capillata community are S. sareptana and A. lercheana. The composition of this community also includes S. lessingiana, F. sulcata, Kochia prostrata (L.) Schrad., and other species.

From the phenological point of view, these species develop differently. For example, flowering in S. lessingiana and F. sulcata ensues at the end of May and the beginning of June. The flowering phase in S. sareptana occurs at the end of June or beginning of July, and in K. prostrata at the end of July and beginning of August. A distinguishing feature of the last two communities is the duration of their floral budding period.

It should be noted that the phenological development of the different species is closely associated with environmental factors, namely air temperature and humidity, soil moisture, and other factors, as noted also by other investigators.

For calculating the expenditure of water by the plants during the growth period, we also studied the increase in green mass.

The greatest quantity of transpiring leaf mass was noted in Artemisia, which gave 17-22 centners per hectare in the two communities in June 1958; S. sareptana gave 18.9 c/ha, and S. lessingiana 11.0 c/ha. F. sulcata and K. prostrata gave the minimal transpiring leaf mass (from 4.5 to 2.6 c/ha).

Correspondingly, the maximum in May 1959 was also provided by Artemisia (from 6.4 to 11.2 c/ha in the two communities), followed by S. sareptana (9.1 c/ha), while S. lessingiana, F. sulcata, and K. prostrata provided the lowest amounts (from 1.8 to 2 c/ha). The greater transpiration of the leaf mass in 1958 can obviously be explained by the fact that the spring of 1959 was drier and colder, and this also affected plant growth and development. A comparison of the crude leaf mass in the two communities indicated that the quantity was greater in the A. incana—S. capillata community than in the A. incana—S. lessingiana community.

Before proceeding to discuss the results of the observations on transpiration rate in the different plants and communities, it is necessary to examine the climatic and soil conditions under which transpiration progressed.

In the A. incana—S. capillata community, the plant roots absorbing moisture obtained a more concentrated salt solution than in the A. incana—S. lessingiana community. The quantity of moisture in the soil under A. incana—S. capillata community also exceeded that under the A. incana—S. lessingiana community. Water destruction proceeded in the heated soil, although at a varied rate; the upper soil layers had undergone more heat and were therefore more desiccated. The degree of soil heating was similar in both communities. The relationship of transpiration rate with temperature and other climatic indices is described below.

The transpiration rate of all the plants fluctuated strongly during the course of a day. Thus, the transpiration rate in S. lessingiana ranged from 0.320 to 0.720 g/g/hr during the course of the day on June 1, 1958, from 0.24 to 0.780 g/g/hr on June 11, and from 0.291 to 0.592 g/g/hr on June 21. The transpiration rate of S. lessingiana in June 1959 differed sharply from that in June 1958. The mean daily values were 0.710 to 0.790 g/g/hr. The maximum increase in transpiration rate during the day on all dates in 1959 did not coincide with the increases in transpiration rate in 1958. However, two daily maxima in transpiration rate were observed in both observational years: one was in the hours preceding midday, and the other was in the post-midday period.

A high rate of transpiration was observed in S. lessingiana in July of both years. The average diurnal figures in 1958 were as follows (in g/g/hr): from 0.260 to 1.330 on July 1; from 0.220 to 0.660 on July 11; and from 0.600 to 1.800 on July 21. The corresponding figures in 1959 were 0.600-1.020 on July 1 and 0.800-1.440 on July 11.

The transpiration rate of S. lessingiana in August and September fell by 0.300-0.460 g/g/hr as compared with transpiration in June and July. The transpiration rate in this plant therefore drops from June to September, the maximum rate of transpiration occurring in July.

The transpiration rate in F. sulcata also fluctuates. The mean values in June 1958 ranged from 0.584 to 0.876 g/g/hr; and in July 1958 from 0.451 to 0.943. Similar values with some fluctuations were obtained in June and July 1959.

Some slackening in transpiration rate was noted in F. sulcata in August and September 1958 and 1959 (0.240-0.300 g/g/hr).

A comparative analysis of data on the daily patterns of transpiration in S. lessingiana and F. sulcata shows that the transpiration-rate levels in the two species are different: the level is somewhat higher in F. sulcata than in S. lessingiana, although the diurnal course of changes in transpiration rate are very similar.

The transpiration rate of Artemisia lercheana was substantially higher than in the grasses and fluctuated markedly; its value, expressed in g/g/hr, was as follows: from 0.564 to 1.440 on June 1, 1958, and from 0.720 to 1.320 on June 11, 1958. The mean daily figures were 0.618 on June 1 and 1.400 on June 10.

Transpirational Consumption of Water by Different Plant Species and Communities During the Growth Period (in mm)

| Species | 1958 | | | | | 1959 | | | | | |
|---|---|---|---|---|---|---|---|---|---|---|---|
| | June | July | Aug. | Sept. | total for 4 months | May | June | July | Aug. | Sept. | total for 5 months |
| Wormwood — feather grass community | | | | | | | | | | | |
| *Stipa lessingiana* Trin. et Rupr. | 18.6 | 7.3 | 3.5 | 0.6 | 30.0 | 1.4 | 4.3 | 3.2 | 1.3 | 0.3 | 10.5 |
| *Festuca sulcata* Hack. | 4.7 | 5.0 | 2.9 | 0.5 | 13.1 | 4.2 | 5.7 | 2.8 | 2.3 | 0.7 | 15.7 |
| *Artemisia lercheana* Web. | 49.7 | 20.7 | 11.5 | 1.2 | 83.1 | 6.5 | 19.2 | 11.2 | 6.4 | 2.0 | 45.3 |
| *Kochia prostrata* (L.) Schrad. | 3.5 | 2.5 | 1.8 | 0.3 | 8.1 | 0.6 | 1.0 | 1.3 | 1.1 | 0.3 | 4.3 |
| Total for community edificators | 76.5 | 35.5 | 19.7 | 2.6 | 134.3 | 12.7 | 30.2 | 18.5 | 11.1 | 3.3 | 75.8 |
| *Artemisia lercheana* — feather grass community | | | | | | | | | | | |
| *Stipa sareptana* Becker. | 36.2 | 27.5 | 15.3 | 0.8 | 79.8 | 6.3 | 11.4 | 5.1 | 3.0 | 1.2 | 27.0 |
| *Artemisia lercheana* Web. | 55.0 | 27.9 | 14.1 | 2.3 | 99.3 | 14.0 | 30.7 | 5.8 | 5.5 | 3.0 | 59.0 |
| *Kochia prostrata* (L.) Schrad. | 2.7 | 5.9 | 2.4 | 1.3 | 12.3 | 0.8 | 4.7 | 2.8 | 1.7 | 0.7 | 10.7 |
| Total for community edificators | 93.9 | 61.3 | 31.8 | 4.4 | 191.4 | 21.1 | 46.8 | 13.7 | 10.2 | 4.9 | 96.7 |

The transpiration values on August 1 of both years were similar and their patterns were more or less uniform. The transpiration rate in mid-August 1958 (August 11) was substantially higher than in 1959, but the daily patterns were very similar. At the end of August, on the other hand, the transpiration rate was substantially higher in 1959, although the patterns in the two years were again similar. The mean transpiration rates, in g/g/hr per day, were 0.788 at the beginning of August, 1.340 in mid-August, and 0.449 at the end of August. A marked abatement in transpiration was noted in September 1958, the mean daily value falling to 0.340. In September 1959, on the other hand, Artemisia lercheana transpired very actively during the day. This can apparently be explained by the fact that the A. lercheana plants were severely depressed during May, when there was no rainfall, but when there was a significant increase in rainfall young shoots appeared and the plant again began to develop vegetatively.

There is thus no doubt that the transpiration rate during this period is linked both with the growth and development of the plant itself, and with the amount of rainfall.

The daily course of transpiration rate in K. prostrata was close to that of A. lercheana and also fluctuated markedly. It varied from 0.568 to 0.726 at the beginning of June 1958 and 1959, from 0.541 to 0.704 in July and from 0.516 to 0.605 g/g/hr in August and September of both years.

The daily patterns of transpiration rate in Stipa sareptana, Artemisia lercheana and K. prostrata underwent analogous fluctuations (A. incana—S. capillata community).

Analyses of the curves for the daily pattern in transpiration rate showed that all the species usually have 2-3 increases, which depend on the simultaneous change of many ecological factors, particularly relative humidity, saturation deficit and, to a lesser degree, air temperature.

The daily transpiration-rate pattern in all the species corresponds basically to the saturation-deficit pattern. With high relative humidity and a small air saturation deficit, the transpiration rate falls in all species, while it increases where there is low relative humidity and a high air saturation deficit.

The seasonal patterns of plant transpiration rate in 1958 and 1959 (from May to September) in the different species within the A. incana—S. lessingiana and A. incana—S. capillata communities are also subjected to changes, and show increases or decreases depending on the level of soil moisture.

The seasonal pattern of transpiration in all the species corresponds basically, as does the daily pattern, to the atmospheric saturation deficit and the soil moisture. The rate of transpiration is somewhat higher in the A. incana—S. lessingiana community than in the A. incana—S. capillata community. Where there is adequate soil moisture and a high air temperature, the transpiration rate increases, whereas it decreases in dry soil in spite of the fact that the air temperature may be high.

Transpirational consumption of water during the growing season (from May to September) is shown in the table, from which it is apparent that the nature of the changes in water consumption by plants and plant communities during the growing season fluctuates with the year of observation. Thus, it was greater in 1958 than in 1959. After increasing irregularly from June to August, it then falls sharply in September.

The maximal rate of transpirational consumption of water was observed in 1958 in A. lercheana, S. Sareptana, and S. lessingiana, and the lowest rate in F. sulcata and K. prostrata. This can apparently be explained by the fact that the quantity of above-ground herbage produced by these species is less than in the others; the rates of transpiration of F. sulcata and K. prostrata show only slight differences.

Summary

The author studied transpiration in Artemisia lercheana, Stipa lessingiana, S. sareptana, Festuca sulcata, and Kochia prostrata in the steppes of Central Kaẑakhstan. The highest total water consumption rate was noted in A. lercheana, the rate being much lower in K. prostrata. The rates of water consumption are closely related to the developmental rhythm of the plants, which is dependent to a large extent on the amount of rainfall during the period of vegetative growth.

# THE PRESENT POSITION OF THE PROBLEM
# OF THE UTILIZATION OF THE INDICATOR ROLE
# OF THE VEGETATION COVER OF BOGS IN RELATION
# TO THE STRUCTURE AND PROPERTIES OF PEAT DEPOSITS

## M. S. Boch

The resolution of the problem of using the vegetation covers of bogs as indicators of the structure and properties of peat deposits opens wide prospects in practical work on peat deposits. An understanding of the interrelationship patterns between bog vegetation and peat would greatly simplify work which is at present cumbersome, lengthy, and expensive.

It has been calculated (Yanushevskii, 1961) that the cost of reconnaissance surveys of bogs, carried out by the indicator method, is 30-50% less than the cost of a survey conducted according to technical methods now in use; in addition, the latter take much longer.

It is known that a peat deposit is genetically linked with the bog vegetation and its composition therefore depends on the nature of the latter. Changes in bog vegetation with time (expressed in the peat deposit by stratification of different peats) show regular patterns in bog massifs of different types. Thus, if we know the course of these changes, it is possible to say which vegetational groupings were dominant in the past, on the basis of the modern vegetation, i.e., what type of peat deposit is present in the particular ease.

The first studies on this problem were carried out in Finland by Prof. Lukkala in the 1920's (Lukkala, 1920). He established that certain types[*] of bog are underlain to a depth of 1 m with peats of definite botanical composition and a definite degree of decomposition. Lukkala, however, also considered that the types of peat underlying some bog communities cannot be determined.

As early as the 1930's, some papers appeared in the USSR which were closely related to the problem: of primary interest were the studies of Z. F. Ruoff (1934) and N. Ya. Kats (1934), in which the problem of the interrelationship of bog vegetation and peat was dealt with in detail. Of particular value was the fact that these workers studied the structure of deposits in mosaic communities and in complexes, and established that complexes of different types of deposits correspond to some vegetational complexes. As Z. F. Ruoff writes, "columns of mochezhina peats alternate here with silt columns." This point is exceptionally important in practice, since the investigation of deposits on bog sites with complex vegetation frequently becomes procedurally incorrect and provides an inaccurate picture of the deposit at the particular site. Whereas N. Ya. Kats and Z. F. Ruoff discuss the structure of deposits under oligotrophic communities, the work of D. A. Begak, S. N. Tyuremnov, et al. (1934) is the first in the Soviet literature in which we find data on the structure of deposits under a number of definite mesotrophic and eurotrophic bog communities. Among the authors who have worked specifically on the problem of using bog vegetation as an indicator of the structure and properties of peat deposits, we might name T. G. Abramova (1947, 1951, 1954) who, on the basis of data on the bogs of Leningrad province, deals with the structure and properties of deposits under many oligotrophic and mesotrophic communities and presents a series of principles governing the problem; E. A. Shirokovskaya (1947), who made a detailed study of the structure of the top layers of deposits under different phytocenoses; and V. V. Yanushevskii (1961), who worked out a procedure for studying peat deposits involving the large-scale application of aerial photography.

---

[*] "Types" according to the usage of A. Cajander (1913) or, as we understand the term, formations or groups of associations.

With this procedure, bog phytocenoses are identified by deciphering aerial photographs, and predictions are then made as to the properties of the deposits under these cenoses. This technique places the indicator work with which we are concerned on a higher level and opens up vast new prospects. Among the works specifically dealing with this aspect we might cite our own contributions between 1958 and 1960 (Boch, 1958, 1958a, 1959, 1960).

All the papers cited are concerned with the use of plant communities as indicators of peat deposits. There does, however, exist another trend in bog research, namely when individual plants are used as indicators of peat properties. Thus, the paper by Kh. Kh. Trass (1955) enumerates bog plants which are indicators of types of peat, of particular levels of peat decomposition, or of specific pH values or mineralization of peats. Interesting studies were carried out by M. Salmi (1956); he employed individual bog plants as indicators of rare minerals contained in the soil.

Much factual material on the structure and properties of peat under the most varied types of communities may also be obtained from papers by G. A. Blagoveshchenskii (1936), E. A. Galkina (1941, 1946, 1959), V. D. Lopatin (1947, 1954), L. A. Gorshkov (1952), N. I. P'yavchenko (1953), L. D. Bogdanovskaya-Gienéf (1936, 1949, 1956), R. P. Tikhonova (1952, 1955), A. A. Grebenshchikova (1956), E. A. Romanova (1960), N. G. Solonevich (1960), and others. These papers are devoted to various problems of bog research and can serve as an excellent source of data of interest to us.

Some of these works even contain special chapters devoted to the indicator role of vegetation with respect to peat. Finally, factural material can also be obtained from works such as the marsh evaluation surveys of different provinces of the USSR, and also the major reports on bog studies, compiled by S. N. Tyuremnov (1940, 1949, 1953), N. Ya. Kats (1941, 1948), and others, which present varied stratigraphic profiles and information on the vegetation cover corresponding to the particular types of bog site. The following conclusions may be enunciated on the basis of a survey of all the available data.

1. Peat deposits of similar structure and properties are located on bogs of the same types, in the same geographical region, and under uniform phytocenoses; the vegetation cover can consequently be used as an indicator of these deposits.

2. It is, however, frequently impossible in practice to determine the structure of a peat deposit on the basis of the properties of a single vegetation cover; it is necessary to know the type of bog massif in which the given community is located and its natural characteristics, since a deposit can be different on massifs of different types under identical communities. Deposits under similar vegetational groupings on bogs in different geographical regions may also be different. It must not be forgotten that massifs pertaining to different groups have undergone distinct developmental pathways, and this is reflected in the structure of their peat deposits, and though identical communities may appear on them at a particular stage of development, the peat under the two communities remains different. The adaptation of the indicator method for research on peat deposits should therefore be carried out concurrently with investigations on types of bog massifs in different districts.

3. In using plant communities as indicators of peat deposits, it is necessary to take into account in what part of the massif these communities are located, what proportion of the area they occupy in a complex with other communities, and so on.

4. Within certain vegetational formations, peat is as a rule uniform under communities of different associations in the particular formation. This is, for example, true in the case of formations of Spagneta fusci and S. magellanici and for various sedge and grassy formations. In other formations, for example Sphagneta baltici and S. dusenii, peat deposits (or at least the upper layers) are different under communities of different associations (in this case we are of course referring only to communities which are located in similar conditions in massifs of a single type).

5. Very clear patterns in peat structure are evident under bog sites occupied by community complexes. Eutrophic and eumesotrophic* complexes are usually underlain by floodplain deposits, whose form can be de-

_____

*Eumesotrophic complex—a complex in which one of the elements consists of mesotrophic vegetation occupying hummocks, ridges, etc., while the other is eutrophic and occupies the depression of the microrelief.

termined from the vegetation of the depressions. Mesotrophic and particularly oligomesotrophic complexes are underlain by transitional deposits. Peat deposits under ridges and hillocks, to a depth of not more than 0.5 m, differ from those of the adjacent depressions. A complex of deposits 0.5 m or more in thickness usually occurs at sites with oligotrophic complexes.

The vegetation cover can serve as an index not only of the structure of a peat deposit (particular attention is paid to this aspect in the literature) but also other properties of the deposit such as the degree of decomposition, the humidity, and the mineralization. It is true that the absolute levels of these indices can only be estimated approximately on the basis of vegetation, but at any rate their degree of increase from one layer to the next under particular communities and also the differences in these indices under ridges and mochezhinas can be determined. Thus, various data has shown that peat under ridges is less moist and has a lower ash content than under mochezhinas in some districts, while in other districts the reverse was true.

In conclusion, we should say that a relatively large amount of factual material has quite clearly demonstrated the great indicator significance of the vegetation cover of bogs in relation to their peat deposits. The further development of the problem is retarded, however, due to several reasons:

1. Investigations specifically devoted to the study of the interrelationships between the vegetation cover of bogs and peat deposits are very few in number.

2. Factual material which has been studied by the workers enumerated above deals in most cases with oligotrophic bogs of the central and northwestern peat-bog regions. There are virtually no data on the structure of deposits under various types of communities of bogs in the tundra and forest tundra, or in Siberia, the Far East, and other regions. There is much less material on mesotrophic and eutrophic communities than on oligotrophic types.

3. The majority of workers do not mention on what types of bog massifs their work was carried out and rarely refer to the habitat conditions, on the massif, of those vegetational groupings whose indicator role is under study. Since special investigations have shown that peat deposits under identical communities may be different in bogs of different types and in different geographical areas, material presented without this information cannot be used for comparative purposes.

4. No reference is made in most investigations to the type of deposit which underlies a particular vegetational grouping: sometimes there is merely a mention that peats of one or another type predominate. This approach may also include a description of those vegetational groupings under which the deposits under investigation were located.

All these factors indicate that although there is now sufficient data for using vegetation as an indicator of peat, much still remains to be done in this field.

## Summary

An analysis of data in the literature and the author's own observations have shown that peat deposits found under identical phytocenoses or complexes of phytocenoses have a similar structure in bogs of similar types, situated in the same geographical area. However, on bogs of different types, deposits may differ even under similar communities. They may also differ under the same phytocenosis in different geographical regions. The vegetative cover is indicative not only of the structure of deposits but may also be used to provide a rough estimate of the degree of decomposition, humidity, and ash content of the peat. However, the whole complex of problems related to indicational work on bogs has been considered mainly in regard to oligotrophic bogs, with almost no attention to mesotrophic and eutrophic bogs.

## Literature Cited

Abramova, T. G. 1947. "The vegetation cover as an indicator of some properties of the upper layers of a peat deposit." Vestn. LGU, No. 5.

Abramova, T. G. 1951. "Materials on the problem of the association between the vegetational cover of an upland bog and some properties of the upper layers of its peat deposit." Uchenye zap. LGU, seriya biol. Vol. 30. Geobotanika.

Abramova, T. G. 1954. "On the association between the vegetational cover of bogs and the structure of the upper layers of a peat deposit." Uchenye zap. LGU, seriya biol., Vol. 34. Geobotanika.

Begak, D. A., Tyurmenov, S. N., et al. 1934. "Technological research on the 'Orshinskii mokh' peat bog." Tr. Inst. torfa, 14.

Blagoveshchenskii, G. A. 1936. "Evolution of the vegetational cover of the '1007 km' bog massif at the Loukhi station." Tr. Bot. Inst. Akad. Nauk SSSR, seriya III. Geobotanika, No. 3.

Bogdanovskaya-Geinéf, L. D. 1936. "The formation and development of ridges and mochezhinas on bogs." Sov. botanika, No. 6.

Bogdanovskaya-Gienéf, I. D. 1949. "Types of upland bogs in the USSR." Tr. 2-go Vses. geogr. S"ezda, III.

Bogdanovskaya-Gienéf, L. D. 1956. "On some regressive phenomena in upland bogs." In collection: To Academician V. N. Sukachev on his 75th birthday. Moscow—Leningrad, Izd. Akad. Nauk SSSR.

Boch, M. S. 1958. "A contribution to the problem of utilizing the vegetation cover as an indicator of the structure of a peat deposit." Vestn. LGU, seriya biol., No. 3.

Boch, M. S. 1958a. "The vegetation cover and its association with the peat deposit of bog massifs of different types." Bot. zhur., Vol. 43, No. 7.

Boch, M. S. 1959. The Vegetation Cover as an Indicator of the Structure of a Peat Deposit. Author's Abstract of Candidate's Dissertation, Leningrad.

Boch, M. S. 1960. "The indicator role of the vegetational communities of bogs in relation to the structure of peat deposits (as exemplified in some bog communities in Leningrad province, Karelia, and White Russia)." In book: Regional Conference on Problems of Geobotanical Research on Bogs in the Northwest of the USSR. Tartu (Summaries of Papers).

Galkina, E. A. 1941. The Development of Bog Massifs of the Central Karelian Types and the Indicator Properties of Their Vegetation Cover. Author's Abstract of Candidate's Dissertation.

Galkina, E. A. 1946. "Bog landscapes and the principles of their classification." In book: Collection of Scientific Investigations by the Botanical Institute of the USSR Academy of Sciences, Conducted in Leningrad in 1941-1943. Moscow—Leningrad, Izd. Akad. Nauk SSSR.

Galkina, E. A. 1959. "Bog landscapes of Karelia and principles of their classification." Tr. Karel'sk. fil. Akad. Nauk SSSR, No. XV.

Grebenshchikova, A. A. 1956. "Slightly decomposed peat deposits (their genesis, structural characters, and peat properties)." In book: Collection of Articles on the Study of Peat Formations. Moscow.

Gorshkov, L. A. 1952. "Microstratigraphy of the upper layers of a peat deposit." Torfyanaya prom., No. 8.

Kats, N. Ya. 1936. "A contribution to the study of the structure and procedure for investigating a sub-peat layer." Torfyanoe delo, No. 1.

Kats, N. Ya. 1941. Bogs and Peats. Moscow, Uchpedgiz.

Kats, N. Ya. 1948. Types of Bog in the USSR and Western Europe and Their Geographical Distribution. Moscow, Geografgiz.

Lopatin, V. D. 1947. "Principal conclusions from a study of the Tēsovsk bog massif." Vestn. LGU, No. 2.

Lopatin, V. D. 1954. "A flat bog (its peat deposit and bog features)." Uchenye zap. LGU, seriya geogr., Vol. 9.

P'yavchenko, N. L. 1953. "A contribution to understanding the nature of ridge-mochezhina bog complexes of the Karelian type." Tr. Inst. lesa Akad. Nauk SSSR, XIII.

Romanova, E. A. 1960. "On the association between vegetation, the upper layers of a peat deposit, and the water regime of upland bogs of the Northwest." Tr. GGI, Vol. 89.

Ruoff, Z. F. 1934. "The morphology and age of layers in the upper stratum of sphagnum peat in the Central Russian bogs." Tr. Inst. torfa, Vol. 14.

Solonevich, N. G. 1960. "The vegetational cover and the structure of the 'Bor' of the Shirinsk bog system." Tr. Bot. Inst. Akad. Nauk SSSR, seriya III. Geobotanika, No. 12.

Tikhonova, R. P. 1952. A Contribution to the Problem of Using the Indicator Properties of the Vegetational Cover for the Agricultural Development of the Bog Massifs of the Karelo-Finnish SSR. Author's Abstract of Candidate's Dissertation. Leningrad.

Tikhonova, R. P. 1955. "Natural features of bog massifs of drainage basins in Central Karelia." Tr. Karel'sk. fil. Akad. Nauk SSR, III.

Trass, Kh. Kh. 1955. "The Flora and Vegetation of Lowland Bogs in Western Estonia." Author's Abstract of Candidate's Dissertation.

Tyuremnov, S. N. 1940. Peat Formation. Moscow, Gostopizdat.

Tyuremnov, S. N. 1949. Peat Formations and Exploration for Them. Moscow—Leningrad, Gosenergoizdat.

Tyuremnov, S. N., and Vinogradova, E. A. 1953. "Geomorphological classification of peat formations." Tr. Mosk. torf. Inst., No. II.

Shirokovskaya, E. A. 1947. "Interrelationship between the vegetation cover and the surface layer of a peat deposit." Torfyanaya prom., No. 8.

Yanushevskii, V. V. 1961. Methods of Carrying Out Peat Prospecting Investigations with the Use of Aerial Survey Material. Author's Abstract of Candidate's Dissertation.

Cajander, A. K. 1913. "Studien über die Moore Finnlands." Acta forest. fennica, No. 2.

Lukkalo, O. I. 1920. "Studien über das Verhältnis zwischen dem Moortypus und dem Oberflachentorf der Moore." Acta forest. fennica, Vol. 16, No. 3.

Salmi, M. 1956. "Peat and bog plants as indicators of ore materials in Vihanti Ore field in western Finland." Bull. Commiss. géol. Finlande, No. 175.

# THE INDICATOR SIGNIFICANCE
# OF THE VEGETATION COVER OF THE BOGS
# OF LENINGRAD PROVINCE

## T. G. Abramova

The role of the most recent vegetation cover of bogs as an indicator of the structure and properties of peat deposits has not yet received adequate study, in spite of the current significance of this problem for the peat industry and for agriculture. There has also been little attention paid to procedures for studying correlations between vegetation cover and peat deposits.

There are few papers directly concerned with the indicator role of bog vegetation. They include for example, the contributions of Abramova (1947, 1951, 1954) begun in 1945 at the suggestion and under the direction of I. D. Bogdanovskaya-Gienéf, and papers by M. S. Boch (1958), V. D. Lopatin (1947), E. A. Shirokovskaya (1947), dealing with the bogs of Leningrad province, and by R. P. Tikhonova (1952) and M. S. Boch (1958, 1958a, 1959), dealing with bogs in the Karelian ASSR. The Finnish bog expert O. Lukkala (1920), who worked on the bogs of Finland, is among the foreign workers on the problem of the interrelationship of vegetation with peat.

The reason for the difficulty in the further successful development of the indicator problem in bog research is the distinction between the botanical composition of peats and the species composition of the phytocenoses which deposit them. I. D. Bogdanovskaya-Gienéf (1945) notes that "the botanical composition of peats always differs to some extent from the composition of the maternal phytocenoses." In his view this depends on the following, taking into account the procedural imperfections: 1) the difference in the resistance of different plant species to decomposition processes, which leads to the accumulation of residues of some plants and the disappearance of others and thus to the formation of peats of residual composition, and 2) the penetration of plant roots from later phytocenoses into earlier deposited peats, which markedly changes the initial botanical composition of the peat and leads to the formation of peats with so-called secondary botanical composition.

As regards papers concerned with these interesting aspects of the problem which are of importance for indicator research, one can cite, apart from I. D. Bogdanovskaya-Gienéf's contribution (1945), only a short article by E. A. Shirokovskaya (1947) and S. N. Tyuremnov's book (1949), which mention that the features of the peat-formation process distort the natural character of the association between the modern bog vegetation and the structure of the deposit, and this renders difficult the final solution of this problem.

Different authors vary in their evaluation of the depth at which the role of the present-day vegetation cover as an indicator is evident. Our view (Abramova, 1951, 1954) is that the interrelationship of vegetation with peat deposits is clearly expressed down to a depth of 2-3 m, sometimes 4 m in deep deposits, and through the entire depth of the deposit in the case of small types of peat deposit. The manifestation of the indicator role of the vegetation cover in the upper layer only can apparently be explained by the widespread occurrence of convergence of plant associations in bogs (Shennikov, 1929; Shennikov and Golubeva, 1930) and consequently convergence of peat deposits also (Bogdanovskaya-Gienéf, 1949a). This phenomenon of convergence limits the possibility of determining the properties of peat deposits on the basis of vegetational groupings.

In the present communication, new data are put forward on the indicational significance of the vegetation cover in relation to the structure and degree of decomposition of peat in bogs of the northwestern part of Leningrad province (Karelian isthmus) and a comparison is made with previous conclusions by the author (Abramova,

1947-1954) on the bogs of the central part of the province ("Gladkoe" bog massif, Tosnensk district) and the eastern part ("Shirinskie mkhi" bog system, Kirishsk district).

The usual procedure of carrying out field geobotanical investigational work was applied in the various years of our study of the vegetational cover and the peat deposits of bogs. When boring through peat deposits in association complexes, profile sections were prepared concurrently in different elements of the microrelief. A relatively large number (3-4) of parallel boreholes in different elements of a complex of associations is required for a correct understanding of the history of the development of each complex.

## Indicator Significance of the Vegetation Cover of Bogs

### A. Oligotrophic Bogs

#### 1. Northwestern Part of Leningrad Province

Oligotrophic bogs predominate in the Karelian isthmus, comprising 64% of the area of all bogs. Two types of bogs are encountered. The first are large (1000-2500 ha), convex (bog convexity 3.5-5.5 m), watershed peat bogs of the Russian-Baltic type (Tsinzerling, 1938) or the west Russian type of group of typical upland bogs of the Fuscum-bog subgroup (Bogdanovskaya-Gienéf, 1949), with ridge-mochezhina and ridge-lake-mochezhina complexes, developed in watershed drainage basins. The second type are predominantly small (50-150 ha), hummocky-hilly pine-cotton grass or cloudberry-shrub bogs (sometimes with an abundance of heather) of the Reisermoor group of types (A. Cajander, 1913), and are associated with enclosed basins of kame relief and with terraced lacustrine-glacier plains on the coasts of the Bay of Finland.

#### a. Association Complexes of Both Types of Bog

Two types of ridge-mochezhina complexes can be distinguished in convex oligotrophic bogs: Sphagneta fusci + Sphagneta baltici, a more recent type with secondary (according to Bogdanovskaya-Gienéf, 1936) mochezhinas; and S. fusci—Cladinae + S. cuspidatii or S. dusenii with deep primary mochezhinas. Both types of complexes are dominated by large- or small-shrub and cloudberry associations of S. fuci, frequently with lichens, and with Pinus silvestris f.f. willkommii Sukacz., and litwinowii Sukacz. In secondary associations, cotton-grass and butterbur associations of Sphagneta baltici develop; the more flooded types have associations of S. cuspidatii or dusenii with Scheuchzeria palustris L. and Rhynchospora alba (L.) Vahl. Narrow strips of a Sphagnum fuscum complex are sometimes situated on the slopes of bogs between ridge-mochezhina complexes and the margins of swamps (Bogdanovskaya-Gienéf, 1928). This complex is characterized by the slight breaking-up of the surface and the predominance of shrub and cloudberry associations of S. fusci. The S. fuscum (Schpr.) Klinggr. complex frequently also occupies broad sites in marshes of the Reiserling types. Mosaic pine-shrub associations* (Pineta fruticulosa-sphagnosa) associations with pine, f.f. litwinowii and uliginosa Abol. are located on the better-drained sections of convex bogs. The margins of convex bogs are occupied by swamp-hummock complexes, with the oligotrophic Sphagneta angustifolia or baltici + S. fusci or magellanici, or the oligomesotrophic S. apiculati + S. magellanici.

#### b. Structure of Turf Deposit

As regards structure, oligotrophic deposits of convex peat bogs and marshes of the Reisermoor group are represented by the following types: complex oligotrophic magellanicum, oligotrophic Eriophorum, and fuscum (Tyuremnov, 1949).

The complex oligotrophic type of structure is widely distributed in convex oligotrophic bogs and develops only under complexes with a very definite microrelief, i.e., under ridge-mochezhina and ridge-lacustrine-

---

*Mosaic associations are those in which different sinuses of shrub-grass and mossy layers occur in a specific plot and are under the influence of a particular edificator, in this case pine (Abramova, 1951).

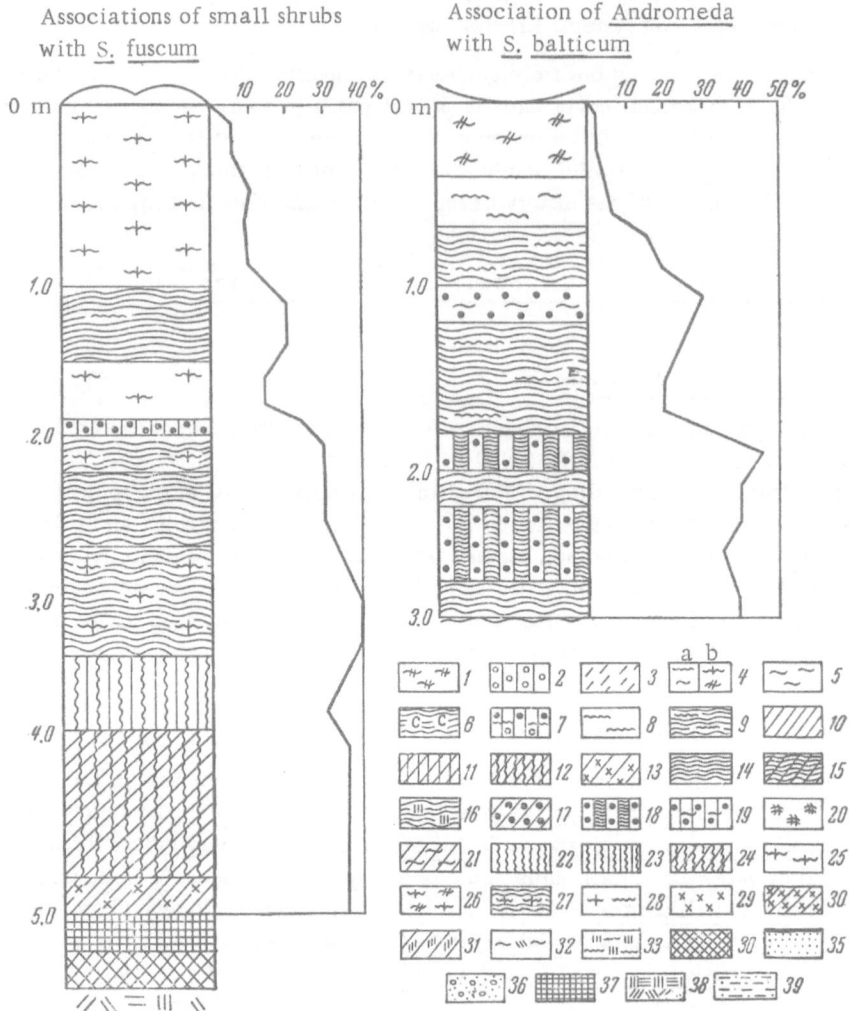

Fig. 1. Ridge-mochezhina small-shrub complex of Sphagneta fusci + S. baltici. Complex oligotrophic type of deposit structure. Shallow-mochezhina variety. 1st variant. Left—structure and degree of decomposition (in %). Ridge deposits, from right—same, adjacent mochezhina.

Symbols for peats and soils for Figs. 1-14: 1) balticum peat; 2) birch peat; 3) gipnovyi; 4) complex oligotrophic (a and b); 5) cuspidatum peat; 6) cuspidatum vaginatum peat; 7) forest transitional peat; 8) magellanicum peat; 9) magellanicum vaginatum peat; 10) surge; 11) sedge—woody; 12) sedge—reed; 13) sedge—Equisetum peat; 14) Eriophorum; 15) Eriophorum mesotrophic; 16) Eriophorum—Scheuchzeria peat; 17) pine eutrophic; 18) pine—Eriophorum; 19) pine—sphagnum peat; 20) sphagnum eutrophic; 21) sphagnum mesotrophic peat; 22) reed; 23) reed—woody; 24) reed—sedge peat; 25) fuscum peat; 26) fuscum—balticum peat; 27) fuscum—vaginatum peat; 28) fuscum—magellanicum peat; 29) Equisetum; 30) Equisetum—sedge peat; 31) Scheuchzeria eutrophic peat; 32) Scheuchzeria—cuspidatum peat; 33) Scheuchzeria oligotrophic peat; 34) argillaceous sapropel; 35) sand; 36) sand with pebbles and gravel; 37) sapropel; 38) loam; 39) sandy loam.

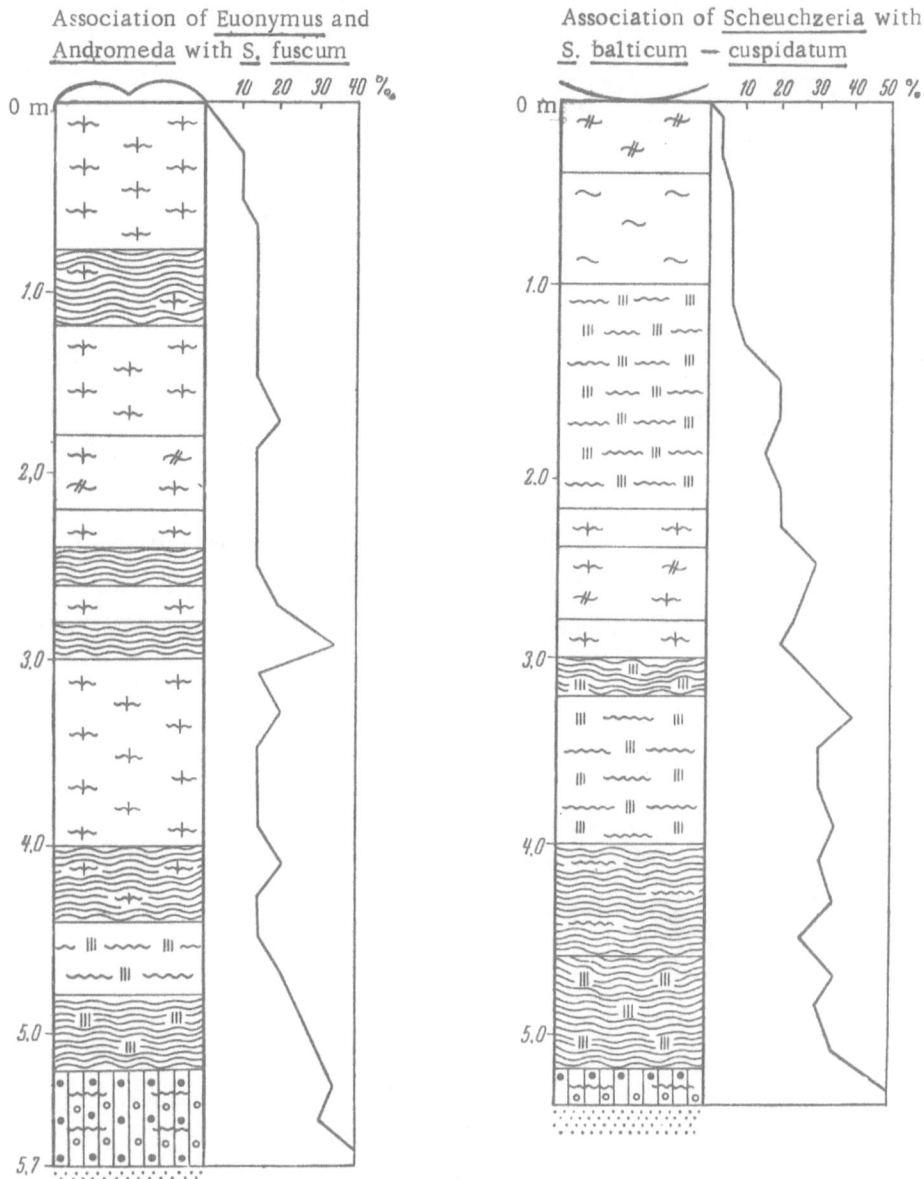

Association of Euonymus and Andromeda with S. fuscum

Association of Scheuchzeria with S. balticum — cuspidatum

Fig. 2. Ridge-mochezhina shrub-Rhynchospora-Scheuchzeria complex of Sphagneta fusci + Sphagneta baltici—cuspidati. Complex oligotrophic type of deposit structure. Deep-mochezhina variety. Structure of ridge (left) and adjacent mochezhina (right). Notations same as in Fig. 1.

mochezhina complexes of both types. Within the complex oligotrophic type, two varieties can be differentiated —shallow mochezhina and deep mochezhina, which differ in the thickness of the "peat complex." The characteristics of the structure of both varieties are determined by the degree of differentiation of the microrelief in the maternal complex and the "age" of the complex.

The shallow mochezhina variety is characterized by a shallow complex of peats (from 0.5 to 2 m) and is represented by two variants, differing in the thickness of the "peat complex." Figure 1 shows the characteristics of the structure of one of the variants, encountered under associations of the ridge-mochezhina small-shrub complex Sphagneta fuscii + S. baltici in a convex part of a bog massif. The complex of peats reflects a depth of 0.5 m.

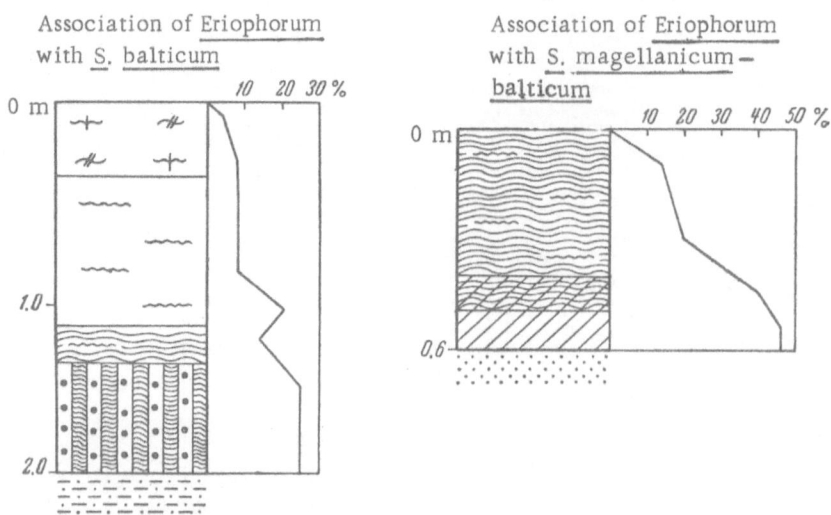

Fig. 3. **Magellanicum** type of deposit structure. Shallow-deposit variety.
Notations same as in Fig. 1.

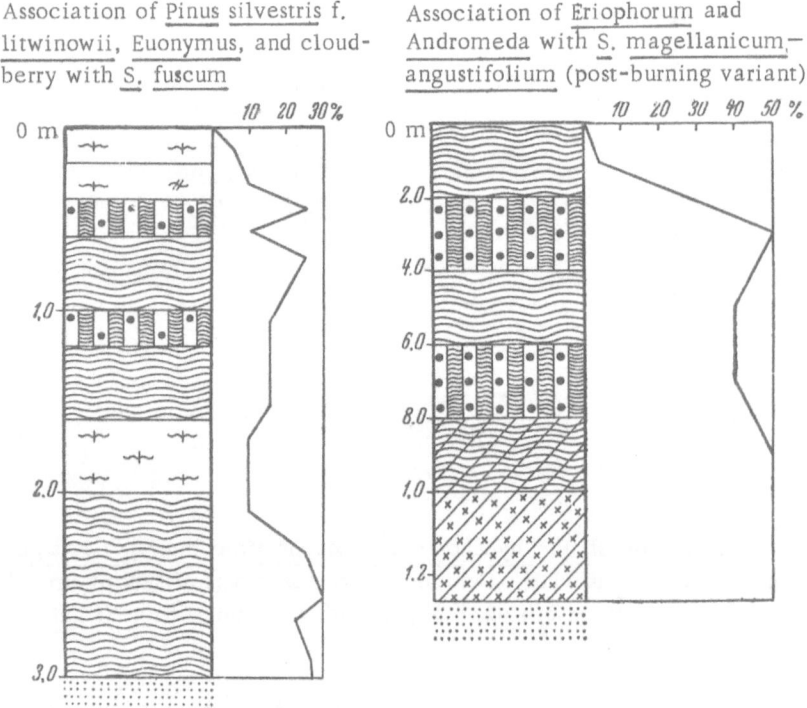

Fig. 4. Oligotrophic **Eriophorum** type of deposit structure. Notations
same as in Fig. 1.

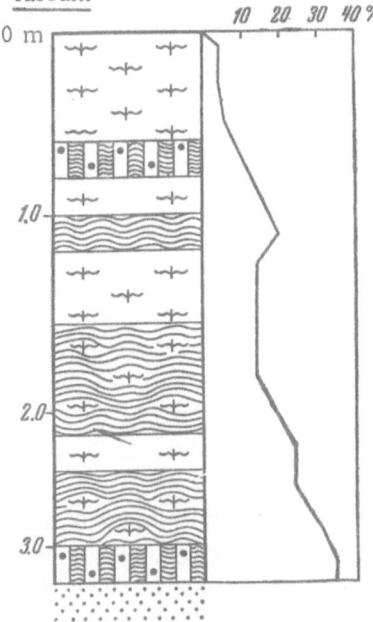

Association of Euonymus and Andromeda with S. angustifolium—fuscum

Fig. 5. Fuscum type of deposit structure. Shallow-deposit variety. Notations same as in Fig. 1.

The deep mochezhina variety is found more frequently and is characterized by a deep (up to 3.20-4.0 m) peat complex. This variety is represented by a single variant with a complex of peats 4.0 m in depth. From 0 to 4.0 m, the deposit comprises an alternation of vertical strata of ridge and mochezhina peats (Fig. 2). This is evidence of the prolonged existence of clearly expressed complexity in the area of the present-day ridge-mochezhina complex.

The magellanicum type of deposit structure is found only on the margins of convex upland bogs and bogs of the Reisermoor group and is represented by a shallow-peat variety (the deposit thickness does not exceed 2 m). This type of structure (Fig. 3) develops under associations of complexes of marginal Eriophorum swamps with Sphagneta baltici or S. angustifolii and of hillocks with S. magellanici and S. fusci.

The oligotrophic Eriophorum type of deposit structure is encountered more rarely and is also represented by a shallow-deposit variety. This type of structure (Fig. 4) is found under mosaic pine-shrub associations (Pineta fruticulosa—sphagnosa), associated with drained slopes or the margins of convex oligotrophic bogs. Mosaic associations, subjected to repeated burnings are frequently replaced by small hummocky-hilly shrub—Eriophorum complexes with dry, charred pines, Pinus silvestris f. litwinowii and an extensive overgrowth of pine and birch. The structure of deposits in variants subsequent to burning (Fig. 4) is similar to that under the initial mosaic associations.

The fuscum type of deposit structure is represented by a single shallow-deposit variety (thickness of deposit does not exceed 3.20 m) (Fig. 5) and is observed under Sphagnum fuscum complexes (with predominance of heather and cloudberry associations) in small bogs of the Reisermoor type, located on the coasts of the Bay of Finland.

In the northwestern part of Leningrad province (Karelian isthmus), the fuscum type of deposit structure is encountered mainly in the southern part of the isthmus in small bogs of the Reisermoor group along the northern coast of the Bay of Finland. Fuscum peat also plays a significant role in deposits under ridge associations of ridge-mochezhina and ridge-lacustrine-mochezhina complexes of convex upland bogs. The fuscum type of structure is also typical for convex upland bogs on the southern coast of the Bay of Finland (Kingisepp district, Leningrad province) (Chernova-Lepilova, 1928). According to a verbal communication of L. D. Bogdanovskaya-Gienéf, deposits of fuscum peat predominate in almost all the oligotrophic peat bogs in the Russian Baltic zone. Fuscum peat is widely distributed in oligotrophic bogs in Estonia (Veber, Kurm, Laasimer, et al., 1957). The presence of the fuscum type of structure or the predominance of fuscum peat in deposits of oligotrophic bogs in western districts of Leningrad province (Karelian isthmus, Kingisepp district) and the Estonian SSR can probably be explained by the fact that these districts have a coastal climate with a rather cool summer (moderating influence of the Baltic Sea). These conditions are less favorable for the thermophilic Sphagnum magellanicum Brid. and promote its replacement by S. fuscum. The magellanicum type of deposit structure is typical for the rather more continental climate of the central ("Gladkoe" bog) and especially the eastern ("Shirinskie mkhi" bog massif) districts of Leningrad province.

2. Central Part of Leningrad Province

For purposes of comparison, let us now turn to the bogs in the central part of Leningrad province (Tosnensk district), to the oligotrophic, strongly convex (convexity in the northwestern part reaches 7 m) "Gladkoe" bog massif of Tsinzerling's Russian type (1938). A description of this bog has been published (Abramova, 1951) and we shall therefore not dwell long on the topic. It is sufficient to mention that the vegetation cover of the massif is characterized by considerable uniformity and the dominance of oligotrophic complexes (ridge-moch-

Association of Empetrum
nigrum with S. fuscum

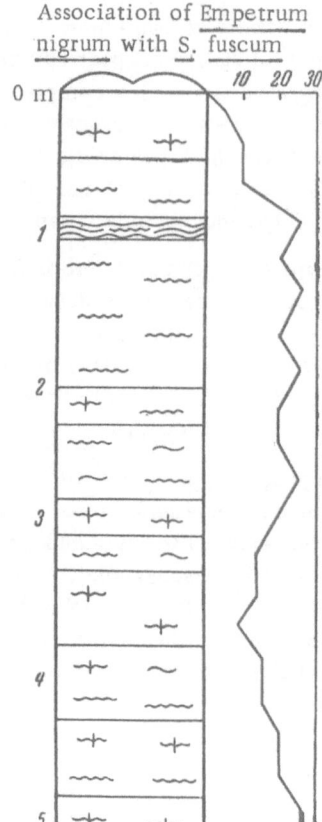

Association of Eriophorum
with S. balticum

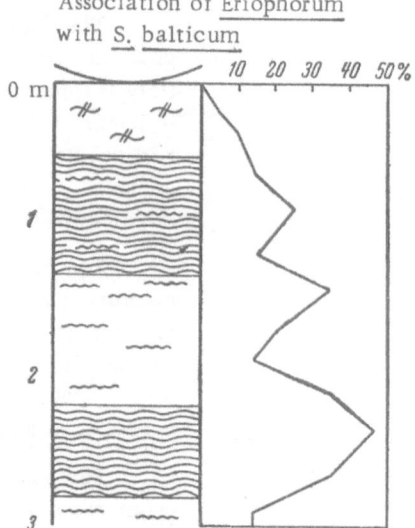

Fig. 6. Ridge-**mochezhina** shrub—Eriophorum complex of Sphag-neta fusci + S. baltici. Complex oligotrophic type of structure. Shallow-mochezhina variety. Structure of ridge (left) and adjacent mochezhina (right). **Notations** same as in Fig. 1.

Association of Pinus silvestris f. litwi-nowii and large shrubs with S. fuscum

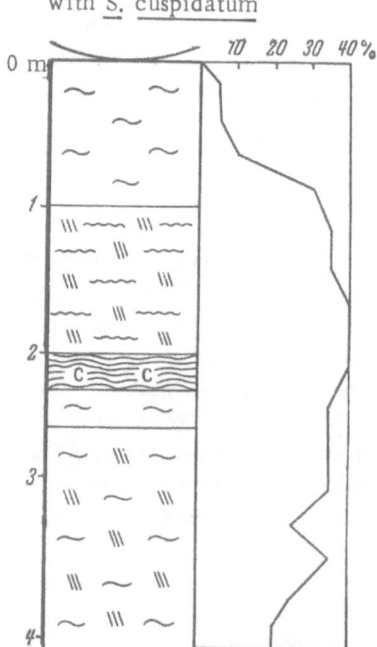

Association of Scheuchzeria
with S. cuspidatum

Fig. 7. Ridge-mochezhina pine—shrub—Scheuchzeria complex of Sphagneta fusci + S. cuspidati. Complex oligotrophic type of structure. Deep-mochezhina variety. Structure of ridge (left) and the adjacent primary mochezhina (right). Notations same as in Fig. 1.

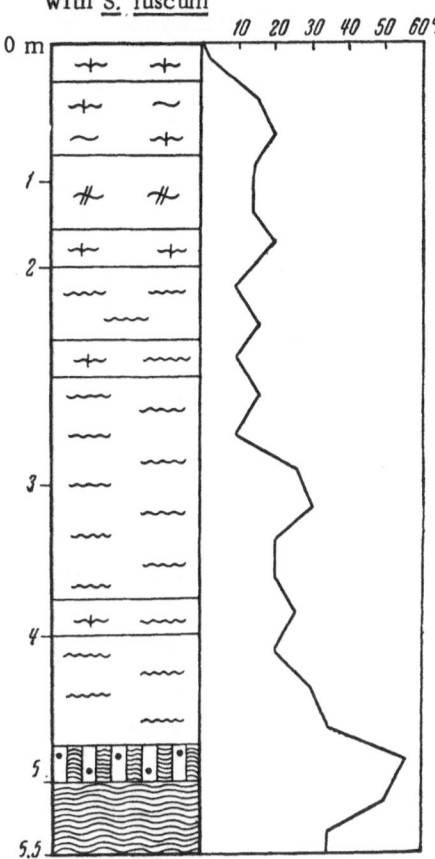

Association of <u>Pinus silvestris</u> f.
<u>litwinowii</u> and <u>Empetrum nigrum</u>
with <u>S. fuscum</u>

Fig. 8. <u>Magellanicum</u> type of deposit
structure. Deep-deposit variety. Notations same as in Fig. 1.

ezhina, <u>Sphagnum fuscum</u> complex, and others), which occupy
the top and most convex central parts. Well-drained sites show
strips of mosaic associations (<u>Pineta fruticulosa</u>—<u>sphagnosa</u>) with
pine, <u>Pinus silvestris</u> f. <u>uliginosa</u>; oligomesotrophic, predominantly forest-free grass-sphagnum swamps develop on the periphery.

The "Gladkoe" peat deposit is also uniform and simple in
structure: the oligotrophic layer of 3-3.5 m is composed in the
main of <u>magellanicum</u> and <u>fuscum</u> peats. Complex oligotrophic
and <u>magellanicum</u> types of deposit structure are widely distributed. It is characteristic that the <u>fuscum</u> type of structure
is absent.

The complex oligotrophic type of structure develops under
ridge-mochezhina complexes:

1. <u>Sphagneta fusci</u> + <u>S. baltici</u>. The shallow-mochezhina
variety; the peat complex is expressed in a depth of 0.5 m (Fig. 6).

2. <u>Sphagneta fusci</u> + <u>S. cuspidati</u>. The deep-mochezhina
variety; the peat complex is expressed in a depth of 4 m (Fig. 7).

Complexes of peats of similar composition are formed
under similar associations of ridges and mochezhinas in ridge-
mochezhina complexes of the "Gladkoe" bog and the convex
bogs of the Karelian isthmus. A distinctive feature is the predominance of <u>fuscum</u> and <u>fuscum</u>—<u>vaginatum</u> peats rather than
<u>magellanicum</u> peat in deposits under ridges in bogs of the isth-
mus. <u>Eriophorum</u> and pine—<u>eriophorum</u> peats occupy a significant place in deposits under mochezhinas in the isthmus
bogs. The <u>magellanicum</u> type of deposit structure in the "Gladkoe" bog, but not in the bogs of the Karelian isthmus, is represented by two varieties, deep-deposit and shallow-deposit. This
type of structure develops under different complexes than under
isthmus bogs. The deep-deposit variety is associated with small-
shrub associations of the <u>Sphagnum fuscum</u> complex in convex
central parts of the bog. The thickness of the <u>magellanicum</u> peat is 5 m, the total depth of the deposit being
6-6.85 m (Fig. 8). The shallow-deposit variety, in which the thickness of the deposit varies from 2 to 3.5 m,
is found under small-shrub associations of the <u>Sphagnum fuscum</u> complex on the slopes of the bog and its marginal areas (Fig. 9), under mosaic pine-shrub associations (Fig. 10), and under peripheral oligotrophic hummocky
swamps of <u>Sphagneta angustifolii</u> + <u>Sphagneta magellanici</u> (Abramova, 1951).

The <u>fuscum</u> type of structure of peat deposits is absent in the eastern part of Leningrad province (Kirishsk
district), in the oligotrophic, sharply convex (Galkina, 1946) "Bor" bog massif which belongs to the Shirinsk bog
system, according to the data of M. S. Boch (1958) and N. G. Solonevich (1960). In the complex oligotrophic
type of structure under associations of the ridge-mochezhina complex <u>Sphagneta fusci</u> + <u>S. cuspidati</u>, <u>magel-
lanicum</u> deposits occur under both elements of the microrelief, while the <u>fuscum</u> type is rarer and is found only
in ridges (Boch, 1958). The <u>magellanicum</u> type, shallow-deposit variety, is found under mosaic pine-shrub
associations with <u>Pinus silvestris</u> f. <u>uliginosa</u>, occurring on the drained edges of the "Shirinskie mkhi" mesotrophic bog system (Abramova, 1954).

B. Mesotrophic and Eutrophic Bogs

1. Northwestern Part of Leningrad Province

The indicator role of the vegetation cover was traced in a large (more than 1000 ha in area) eumesotrophic, well-irrigated, forest-less bog, of the water-carrying discharge basin type (Galkina, 1959).

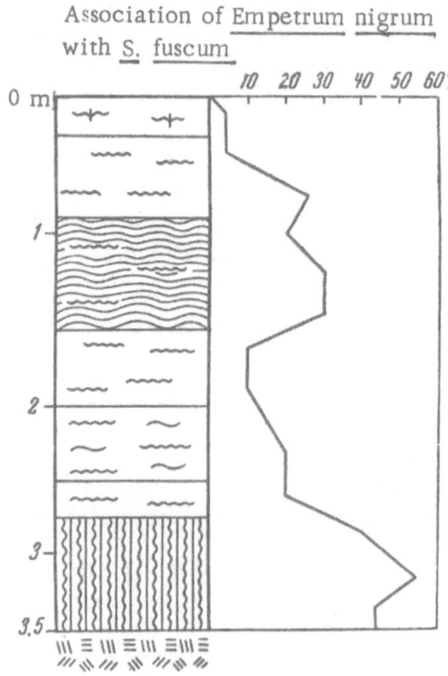

Association of Empetrum nigrum
with S. fuscum

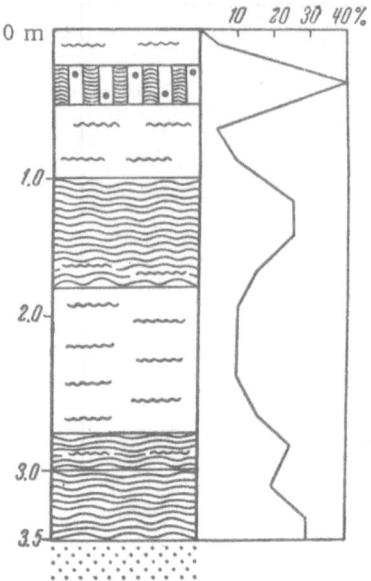

Fig. 9. Magellanicum type of deposit
structure. Narrow-deposit variety. Notations same as in Fig. 1.

Association of Pinus silvestris f.
uliginosa and Ledum with S.
angustifolium—magellanicum

Fig. 10. Magellanicum type of deposit structure under mosaic pine-
shrub associations. Narrow-deposit
variety. Notations same as in Fig. 1.

Association of hairy 100-fruit sedge
and Menyanthes with hypnum and
S. apiculatum

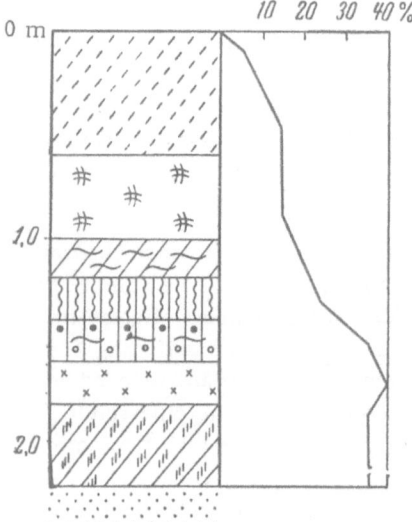

Fig. 11. Mesoeutrophic multilayer
deposit. Forest-swamp type of structure. Shallow-deposit variety. Notations same as in Fig. 1.

## a. Vegetation Cover of Bog

Numerous rock mineral "islands," overgrown with forest, are
scattered through the bog. These islands separate the bog into distinct sectors, causing differences in the water and mineral regimes
of the sectors and producing currents of mineralized waters around
the islands. The bog is dominated by sedge—buckbean, buckbean—Scheuchzeria, and Eriophorum—sedge, swampy groupings of
Sphagneta apiculati, S. papillosi, S. centrale, and other species,
characterized by streaky and spotty distribution, caused by the
presence of the "islands" and the passage of "buried" streams
through the deposit (Bogdanovskaya-Gienéf, 1955).

Hillocks with shrub groupings of Sphagneta papillosi, S.
magellanici, and S. fusci are spread on swamp areas. A borehole
in one of the hillocks with an association of Cassandra and Sphagneta angustifolii—fusci showed that fuscum peat in this case forms
only the upper layer (0.6 m), which lies on a peat stratum of papillosum and Scheuchzeria—mesotrophic peats 3.5 m in thickness.
This is evidence of the "secondary nature" or recent origin of oligotrophic hillocks in the swamp.

## b. Structure of Deposit

The deposits under swamp associations are mesoeutrophic or
eutrophic, multilayer, and have the forest-swamp type of structure,
and are deep (5.5-8 m with saprolep). The multiple-layer character

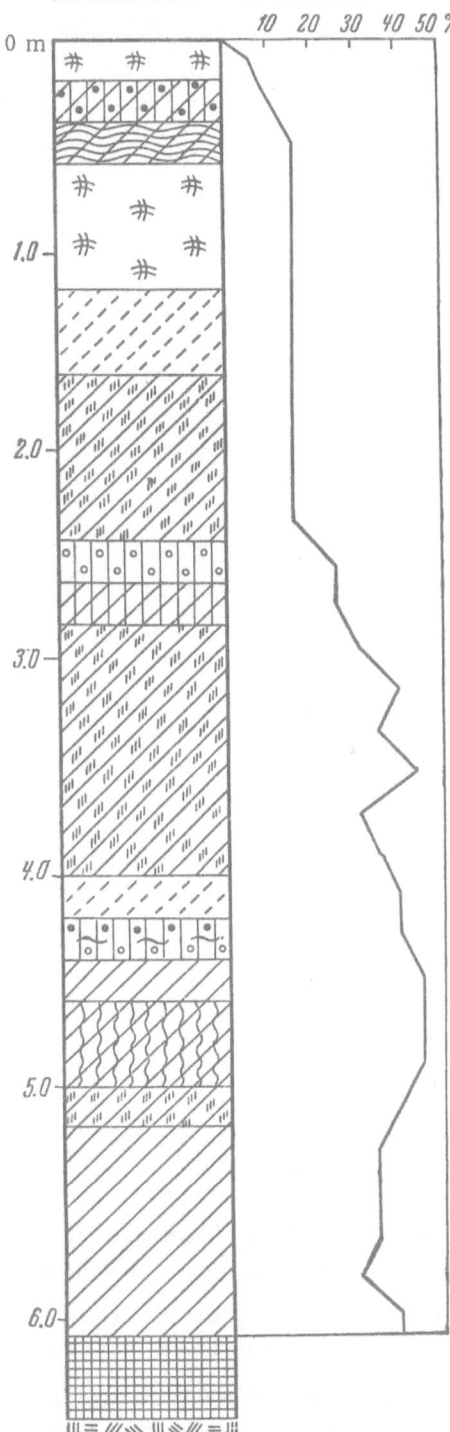

Fig. 12. Eutrophic multilayer deposit.
Forest-swamp type of structure. Deep-
deposit variety. Notations same as Fig. 1.

of the deposit and the alternation within its entire depth of moderately (15-25%) and well (35-45%) decomposed eutrophic and
mesotrophic peats of the woody, woody-grassy, and grassy (with
possibly some localities having the herbage-residue group of
sphagnum peats, as described by L. D. Bogdanovskaya-Gienéf,
1945), and mossy groups duplicate the variegation and mosaicism
in the distribution of plant associations on the surface of the bog
and are evidence of the existence of similar fragmentation in the
distribution of cenoses during the entire history of bog development.

Two varieties can be differentiated in the deposit structure
for the forest-swamp structure type: a narrow-deposit variety
(0.4-2.5 m in thickness), which occurs on the margins of the bog
and in forest-free zones on the mineral islands (Fig. 11), and a
deep-deposit variety (6-7 m in thickness), which is associated
with the central part of the bog (Fig. 12).

## 2. Eastern Part of Leningrad Province

In the east of the province (Kirishsk district), an investigation was made on a mesotrophic bog of the drainage basin type,
constituting the extreme eastern part of the "Shirinskie mkhi"
complex bog system. A mesotrophic deposit of the swamp subtype, 1.25-5.0 m in thickness, has developed under swamp sedges,
and Menyanthes and Menythes—Eriophorum associations of Sphagneta apiculati, which are characterized by streaky, spotty distribution, as in the case of the eumesotrophic bog in the Karelian
isthmus. It was multilayer only in the central, deeper part of the
bog and through its entire depth is composed mainly of mesotrophic, moderately (15-20%) and well-decomposed (30-40%)
peats of the moss and grass-moss groups and the grass-residue
groups of sphagnum peats (Bogdanovskaya-Gienéf, 1945).

Thus, the indicator role of vegetation groupings on mesotrophic and eumesotrophic bogs of a particular type (drainage
basins) in the northwestern and eastern parts of Leningrad province
is similar. The lesser degree of variegation in the vegetation
cover and the simpler structure of peat deposits in the bog in the
eastern part of the province can apparently be explained by the
absence of mineral islands in the area.

The interrelationship between the present-day vegetation
cover and the structure of the deposits in eutrophic bogs in the
Karelian isthmus was studied in a bog of the terrace peat-bog
type (Tyuremnov, 1949). found in low terraces in the coastal zone
of the Bay of Finland.

The vegetation cover of the bog develops in the form of
swamp, grassy, and willow-grass associations, arranged in stripe,
which descend perpendicularly along the slope of the bottom of
the bog in the direction of the coastline. Sphagnum mosses are
found in the form of individual patches.

Deposits of the swamp subtype of the reed-swamp (Tyuremnov, 1949) (Fig. 13) and the horsetail-swamp
(Fig. 14) types of structure are characterized by a simple structure and small width (from 0.6 to 2.80 m). The
absence of mosaicism in the distribution of cenoses on the surface of the bog indicates the relative uniformity

Association of Calamagrostis
neglecta and Lysimachia with
S. girgensohnii

Association of reed and
horsetail

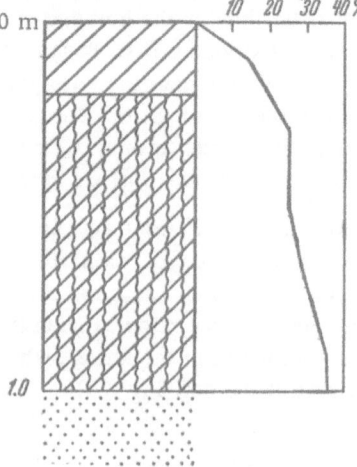

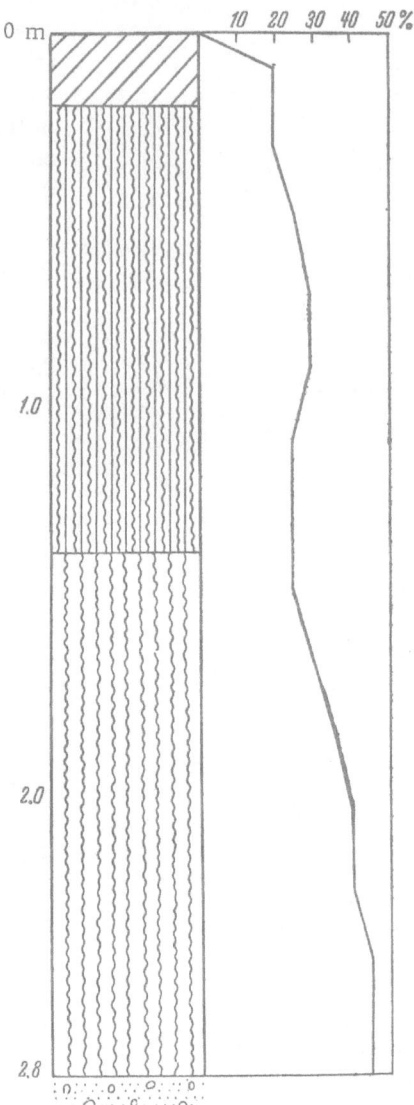

Fig. 13. Eutrophic deposit.
Reed-swamp type of struc-
ture. Notations same as
in Fig. 1.

and the simplicity of the deposit structure, which consists of a rather sparse assortment of moderately (20-25%) and predominantly well-decomposed (30-45%) peats of the grassy group (reed, sedge, and horsetail).

Of the properties of the peat deposit, the author studied the degree of decomposition of the peat, mainly the upper 3-meter layer of the deposit, i.e., the part where the indicator role of the vegetation cover is evident.

Under the ridge-mochezhina complexes of the "Gladkoe" bog, the average degree of peat decomposition under mochezhina associations usually exceeded that under ridges. This difference was especially significant (up to 10%) in the top 3-meter layer of the deposit.

Previous reference to this difference in the degree of decomposition of mochezhina and ridge peats has been made by L. D. Bogdanovskaya-Gienéf (1936) for the bogs of the Polistov bog system (northwestern bog zone, Tyuremnov, 1949). At the same time, L. A. Gorshkov (1952) noted quite the opposite in ridge-mochezhina complexes of upland bogs in Vladimir province (central peat-bog zone, Tyuremnov, 1949)—a 10% higher degree of peat decomposition under ridges than under mochezhinas.

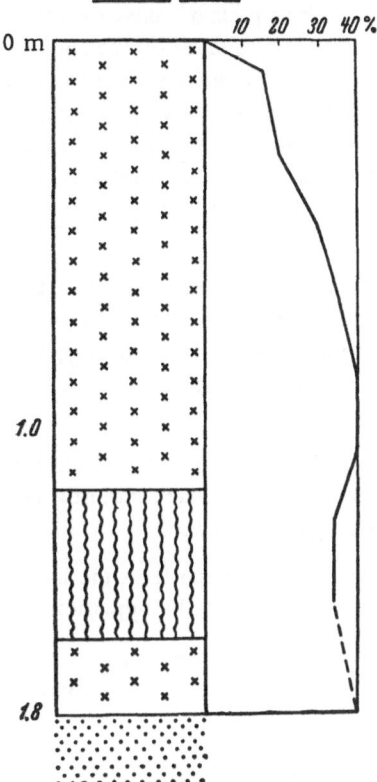

Association of horsetail,
Calamagrostis neglecta
and Cicuta virosa

Fig. 14. Eutrophic deposit. Horsetail-
swamp type of structure. Notations
same as in Fig. 1.

In determinations of the degree of peat decomposition in deposits of ridge-mochezhina and ridge-lacustrine complexes in the bogs of the Karelian isthmus, we did not find the type of clear pattern noted in the "Gladkoe" (Abramova, 1951, 1954) and mentioned by L. D. Bogdanovskaya-Gienéf (1936). Neither the position of the complex on the surface of the bog nor its age showed any influence on the fluctuations in the degree of peat decomposition. In one case, the degree of peat decomposition under ridge and mochezhina was equivalent, in another case decomposition in the peat complex under a ridge exceeded that under a mochezhina by 9%; in a third case, the degree of decomposition of peat under a mochezhina exceeded that under a ridge by 3%. Data on the degree of peat decomposition in the complex oligotrophic deposits in the bogs of the Karelian isthmus were thus very variable, and the determination of the degree of peat decomposition on the basis of the vegetation cover and the degree of surface disintegration of the bogs appears highly difficult.

Summary

1. The results of investigations between 1953 and 1955 on bogs in the northwestern part of Leningrad province (Karelian isthmus) confirm the existence of a definite relationship of the present-day vegetation cover of bogs and the character of their surfaces with the structure of the upper (3-4 m) layers of peat deposits, as was established by studies of the present author (Abramova, 1947-1954) on the bogs of the central and eastern parts of Leningrad province and by work of other bog researchers on peat bogs in various geographical regions.

2. The present-day vegetation cover can serve as an indicator of the structure of the upper layers of a peat deposit within the limits of associations and large taxonomic units of the vegetation cover, i.e., formations and, more rarely, groups of associations.

3. An oligotrophic vegetation grouping corresponds to the oligotrophic or mixed types of structure (Tyuremnov, 1949), while mesotrophic groupings are associated with mesotrophic, mesoeutrophic, or eutrophic deposits, and eutrophic associations develop on eutrophic deposits.

4. Clearly expressed complexity of the vegetation cover and marked disintegration of the surface of convex oligotrophic bogs (ridge-mochezhina and ridge-lacustrine-mochezhina complexes) of the water-carrying drainage basin type are evidence of complexity in the structure of the deposit (complex oligotrophic type of structure with shallow- and deep-mochezhina varieties). The thickness of the 'peat complex" depends on the age of the complex and its microrelief properties.

5. In complexes with only slight disintegration of the surface in the same types of bogs, differences in the structure of deposits under associations of different elements of the microrelief are in most cases absent or are evident only in the topmost layer (0.5-0.6 m) of the deposit.

6. Mosaic pine-shrub associations (Pineta fruticolosa—sphagnosa with Pinus silvestris f.f. litwinowii and uliginosa, which are linked with drained bog sites, develop on the shallow-deposit variety of magellanicum deposit (in the more eastern districts of Leningrad province) or on a deposit of the oligotrophic Eriophorum type of structure (in the northwestern part of the province).

7. Associations of the Sphagnum fuscum (Schpr.) Klingr. complex in bogs of the Reisermoor group in terraced lacustrine-glacier plains in the northwestern part of Leningrad province develop on a shallow (3 m) deposit with the fuscum type of structure.

77

Oligotrophic complexes in convex upland bogs in the southwestern part of Leningrad province and in Estonia develop on the deep-deposit (4.5-9.0 m) variety of fuscum deposit.

8. "Incompletely segmented" phytocenoses of Sphagneta fusci on the marginal parts of convex oligotrophic bogs indicate the presence of magellanicum peat underneath, with a thin surface layer of "young" fuscum peat. The species composition of associations of the Sphagnum fuscum complex also serves to some extent as an index of the thickness of the fuscum-peat layer of bogs in the more eastern districts of Leningrad province. Small-shrub associations of Empetrum nigrum L. and Andromeda polifolia L. develop on a magellanicum deposit with a weakly developed (0.25-0.75 m) fuscum layer; large-shrub associations of Ledum palustre L., Chamaedaphne calyculata (L.) Moench, and Calluna vulgaris (L.) Hill are frequently found in magellanicum peat with a thick (up to 3.5 m) fuscum layer.

9. Mosaicism and fragmentation in the distribution of grassy-sphagnum swamp associations of mesotrophic and eumesotrophic bogs (water-carrying drainage basin type of bogs) with abundant mineral islands (which, by changing the water and mineral regimen of the bog, cause or at least increase the variegation of its vegetation cover) are evidence of the multilayer nature and structural variegation of mesotrophic, mesoeutrophic or eutrophic deposits.

10. The relative uniformity in the distribution of swamp grass associations on eutrophic bogs of the terrace peat-bog type in the northwestern part of Leningrad province is evidence of the existence underneath of a eutrophic deposit of simple structure and of uniform species composition in its component peats.

11. The assessment of the degree of peat decomposition in the upper layers of the deposit on the basis of the nature of the vegetation cover and the degree of bog microrelief differentiation can only be hypothetical.

12. Data on the indicator role of the vegetation cover in relation to the structure of peat deposits are of significance for bogs in nearby geographical districts and for specific types of bog massifs and bog landscapes, associated with certain bog regions.

The results of investigations by the author during 1945-1955 and by other workers are significant for bogs of the following types: oligotrophic bogs of the Russian and Russo-Baltic types, as recognized by Yu. D. Tsinzerling (1938) or the west Russian or central Russian types of L. D. Bogdanovskaya-Gienéf (1949), or bog meso-landscapes with the central-oligotrophic pattern of development (E. A. Galkina, 1946), and bogs of the Reisermoor group (Cajander, 1913) in the Northwest and Baltic peat-bog regions (Tyuremnov, 1949). In addition, the author's data are also of significance for mesotrophic and eumesotrophic bogs similar to the type of mesotrophic, open, Herbosphagneta bogs (Tsinzerling, 1932).

13. Conclusions reached by the author and other researchers facilitate more accurate drilling and analytical work in the study of peat deposits for industrial and agricultural uses.

14. A final decision on the indicator problem in bog research requires further detailed study of the degree of preservation of plant residues in the peat, and the rate of the processes of decomposition of the various peat-forming plant species.

## Summary

The present-day vegetation cover of bogs can be regarded as an indicator of the structure of the upper layers of peat deposits. Oligotrophic vegetation groups reveal a corresponding deposit of the oligotrophic or mixed-type pattern, while the mesotrophic and eutrophic groups have definite deposit patterns also. A mosaic and fragmented arrangement in the distribution of mesotrophic and eutrophic bog associations is evidence of a multilayer and variegated deposit structure. The uniform appearance of vegetation in eutrophic bogs is evidence of the simplicity of the deposit structure. The rate of peat decomposition can not really be estimated accurately on the basis of the vegetation cover. The indicator role of plant communities in bog areas can be of some value only within the range of areas which are very similar in respect of natural conditions and within a specific type of bog.

## Literature Cited

Abramova, T. G. 1947. "The vegetation cover as an indicator of the properties of the upper layers of a peat deposit." Vestn. LGU, No. 5.

Abramova, T. G. 1951. "Materials on the problem of the relationship between the vegetation cover of an upland bog and some properties of the upper layers of its peat deposit." Uchenye zap. LGU, seriya biol., Vol. 30, No. 143.

Abramova, T. G. 1954. "On the association between the vegetation cover of a bog and the structure of the upper layers of its peat deposit." Uchenye zap. LGU, seriya biol., Vol. 34, No. 167.

Bogdanovskaya-Gienéf, I. D. 1928. "The vegetation cover of upland bogs of the Russian Baltic zone." Petergofsk. estestv.-nauchn. inst., No. 5.

Bogdanovskaya-Gienéf, I. D. 1936. "The formation and development of ridges and mochezhinas in bogs." Sov. botanika, No. 6.

Bogdanovskaya-Gienéf, I. D. 1945. "Principles of the genetic classification of peats." Uchenye zap. LGU, seriya biol., Vol. 15, No. 75.

Bogdanovskaya-Gienéf, I. D. 1949. "Types of upland bogs in the USSR." Tr. 2-go Vses. geogr. s"ezda.

Bogdanovskaya-Gienéf, I. D. 1949a. "On the principles of the classification of bog massifs and on the types of bogs in Karelia." In collection: Natural Resources, History, and Culture of the Karelo-Finnish SSR, Vol. 2. Petrozavodsk.

Bogdanovskaya-Gienéf, I. D. 1955. "Bog streams." Uchenye zap. LGU, seriya geogr. nauk, Vol. 10, No. 199.

Boch, M. S. 1958. "A contribution to the problem of the utilization of the vegetation cover as an indicator of the structure of peat deposits." Vestn. LGU, seriya biol., Vol. 1, No. 3.

Boch, M. S. 1958a. "The vegetation cover and its relationship with peat deposits in bog massifs of different types." Bot. zhur., Vol. 18, No. 4.

Boch, M. S. 1959. "A contribution to the problem of the structure of peat deposits in bogs of central Karelia." In collection: Peat Bogs of Karelia. Tr. Karel'sk. fil. Akad. Nauk SSSR, Vol. XV. Petrozavodsk.

Veber, K., Kurm, Kh., Laasimer, L., Raudsepp, A., and Truu, A. 1957. Peat Stocks in the Estonian SSR. Collection of Articles Dealing with the Study of Peat Resources, No. 2. Moscow.

Galkina, E. A. 1946. "Bog landscapes and principles of their classification." In book: Collection of Scientific Work by the Botanical Institute of the USSR Academy of Sciences, Carried Out in Leningrad in 1941-1943. Leningrad, Izd. Akad. Nauk SSSR.

Galkina, E. A. 1959. "Bog landscapes of Karelia and the principles of their classification." In collection: Peat Bogs of Karelia. Tr. Karel'sk. fil. Akad. Nauk SSSR, No. XV. Petrozavodsk.

Gorshkov, L. A. 1952. "Microstratigraphy of the upper layers of an upland peat deposit." Torfyanaya prom., No. 8.

Lopatin, V. D. 1947. "Principal conclusions from a geobotanical study of the Tĕsovsk bog massif." Vestn. LGU, No. 2.

Solonevich, N. G. 1960. "The vegetation cover and the structure of the 'Bor' bog of the Shirinskii bog system." Tr. BIN im. Komarova Akad. Nauk SSSR, Geobotanika, seriya III, No. 12. Moscow—Leningrad, Izd. Akad. Nauk SSSR.

Tikhonova, R. P. 1952. A Contribution to the Problem of Using the Indicator Properties of the Vegetation Cover for the Agricultural Development of the Bog Massifs of the Karelo-Finnish SSR. Author's Abstract of Candidate's Dissertation, Leningrad.

Tyuremnov, S. N. 1949. Peat Deposits and the Exploration. Moscow—Leningrad, Gosenergoizdat.

Tsinzerling, Yu. D. 1932. "Geography of the vegetation cover of the Northwestern section of the European part of the USSR." Tr. Geomorfol. Inst., No. 4. Moscow—Leningrad, Izd. Akad. Nauk SSSR.

Tsinzerling, Yu. D. 1938. "The vegetation of bogs." In collection: Vegetation of the USSR, 1. Moscow—Leningrad, Izd. Akad. Nauk SSSR.

Chernova-Lepilova, G. K. 1928. "Upland peat bogs of the Kurovits plateau." Tr. Petergofsk. estestv.-nauchn. inst., No. 5.

Shennikov, A. P. 1929. "On convergence among plant associations." In book: Essays on Phytosociology and Phytogeography. Leningrad, Izd. Akad. Nauk SSSR.

Shennikov, A. P., and Golubeva, M. M. 1930. "The vegetation of the bog section of the Arkhangel'sk bog experiment field station." Tr. Ar'khangel'skogo bolotn. opytn. polya. Arkhangel'sk.

Shirokovskaya, E. A. 1947. "The interrelationship between the vegetation cover and the surface layer of a peat deposit." Torfyanaya prom., No. 8.

Cajander, A. K. 1923. "Studien über die Moore Finlands." Fennia, Vol. 35, No. 5.

Lukkala, O. J. 1920. "Studien über das verhältnis zwischen dem Moortypus und dem Oberflächtendorf der Moore." Acta forest. fennica, Vol. 16, No. 3.

# THE VEGETATION COVER AS AN INDICATOR
# OF GROUNDWATER LEVELS IN UPLAND BOGS

## E. A. Romanova

The first observations on groundwater levels in bogs were made by A. D. Dubakh (1936). More complete systematic data on groundwaters in the forest zone were provided by K. E. Ivanov (1957), who established definite general principles for the fluctuation of groundwater levels in different bog microlandscapes.

For the present article, data obtained from our studies on upland bog landscapes in the Northwest natural geographical region (Baltic area, Karelian isthmus, Volkhovsk-Il'mensk lowland) were utilized. To provide a picture of the groundwater level regime, we analyzed observations at eight bog stations and posts of the Hydrometeorological Service over five years (1951-1955), of which one year (1951) had low moisture while another (1953 or 1954) had much water. In addition to the station observations, we also used data on the groundwater levels in the summer period (July-August), obtained during expeditions to the region.

The position of the groundwater levels in convex upland bogs and its changes with time are determined by several factors, of which the most important are the air temperature, the quantity and annual distribution of rainfall on the surface of the bog, and the type of bog landscape (system or massif, acutely convex or sloping-convex bog massif, etc.). These factors influence the level of transpiration, water loss by the active layer, and rate of water runoff, on which the groundwater level is directly dependent.

There does, however, exist a general pattern of change in levels during the year, controlled by the annual cycle of moisture supply and consumption in the bog. There are spring and autumn maxima and summer and winter minima in the groundwater level, and also occasional increases in level with heavy rainfall or decreases in the presence of a temporary lack of rainfall during the warm period of the year.

The spring maximum level coincides with maximal melting of snow, usually in April, when the groundwaters stand 10-20 cm below the average surface of the upper elements of the bog microrelief and frequently drench its lower parts. The subsequent drop in levels, caused by water being drained away and an increase in evaporation from the surface of the bog, leads to the summer minimum, with the groundwater level 40-70 cm from the top of the bog surface. The autumn increase in groundwater level is associated with a reduction in evaporation and partly with an increase in precipitation. The subsequent drop in level in winter, caused by the water drainage in the absence of any atmospheric water supply to the frozen top of the bog, leads to the winter minimum in February-March (40-60 cm).

It is quite clear that, in spite of the variable pattern in groundwater levels during the year and the variable range of fluctuations, there does occur some seasonal synchrony in fluctuations in different bog microlandscapes of a particular bog massif. The observed synchrony can be explained by the proximity of the groundwaters to the surface of the bog and is the result of the fact that the level of groundwaters reacts rapidly to all the changes in moisture supply and consumption through the surface of the bog. In upland bogs, filtration runoff occurs mainly in a thin surface layer, 30-70 cm thick, of the peat deposit. Its lower limit represents the average minimal groundwater level.

It is of particular importance that the dynamics of groundwater levels in similar bog microlandscapes of different wet bog massifs in a particular natural geographical district are practically identical. It is this fact which permits the use of the vegetation cover as an indicator of the groundwater level in upland bogs of the

TABLE 1. Average Depths of Groundwaters in Different Microlandscapes of Northwestern Upland Bogs, According to the Data Provided by Bog Stations and Posts for 1951-1955. (Of these years, 1951 was a low-moisture year, while 1953-1955 were high-moisture years)

| Type of vegetation | Bog microlandscapes | Position on microrelief | Depths of groundwaters (cm) | | | | | | | | | No. of water-measuring drill holes tested |
| --- | --- | --- | --- | --- | --- | --- | --- | --- | --- | --- | --- | --- |
| | | | average yearly | | | average monthly, April | | | average monthly, August | | | |
| | | | for five years | in high-moisture year | in low-moisture year | for five years | in high-moisture year | in low-moisture year | for five years | in high-moisture year | in low-moisture year | |
| Forest | Pine-shrub (Pinus f. uliginosa, 8-12 m in height) | Well-drained margins and slopes | -45 | -38 | -54 | -35 | -28 | -45 | -52 | -30 | -72 | 3 |
| Moss-forest | Pine-shrub-sphagnum (Pinus f. uliginosa, 6-8 m in height) | The same | -32 | -26 | -42 | -24 | -21 | -28 | -39 | -22 | -55 | 3 |
| | Sphagnum-shrub-Eriophorum-pine (Pinus f. litwinowii Sukacz, 4-6 m in height) | Slopes | -27 | -25 | -30 | -20 | -18 | -22 | -29 | -24 | -36 | 5 |
| Moss | Sphagnum fuscum (Schpr.) | Slopes, margins | -26 | -22 | -35 | -15 | -12 | -16 | -34 | -24 | -44 | 2 |
| | Klingr.—Eriophorum—shrub (rarely afforested with Pinus f. willkommii Sukacz, 1-3 m in height) | | | | | | | | | | | |
| | Sphagnum angustifolium C. Jensen—Eriophorum | Margins | -12 | – 6 | -24 | 0 | + 4 | – 3 | -22 | – 7 | -33 | 3 |
| Grass-moss | Eriophorum—S. angustifolium, S. apiculatum H. Lindb. | | -10 | – 3 | -16 | + 1 | + 1 | 0 | -20 | -10 | -25 | 3 |
| Ridge-mochezhina | S. fuscum—shrub (rarely afforested with Pinus f. litwinowii) | Slopes / Ridges | -31 | -27 | -40 | -22 | -23 | -22 | -37 | -25 | -48 | 6 |
| | S. angustifolium—Eriophorum | Mochezhinas | (-10) | (-6) | (-19) | (-1) | (-2) | (-1) | (-10) | (-4) | (-27) | 3 |
| | S. fuscum—shrub—Eriophorum, with sparse Pinus f. willkommii | Ridges | -24 | -19 | -33 | -17 | -22 | -18 | -29 | -19 | -43 | |
| | S. balticum Russ.—Eriophorum | Mochezhinas | (- 7) | (- 2) | (-16) | (0) | (- 5) | (- 1) | (-12) | (- 2) | (-26) | 5 |
| | S. fuscum—shrub—Eriophorum, rarely afforested with Pinus f. willkommii | Ridges | -26 | -22 | -34 | -16 | -12 | -17 | -30 | -22 | -38 | |
| | S. balticum—Scheuchzeria | Mochezhinas | (- 8) | (- 4) | (-16) | (+ 2) | (+ 6) | (+ 1) | (-12) | (- 4) | (-20) | 3 |
| | S. fuscum—shrub, rarely afforested with Pinus f. willkommii | Ridges | -30 | -26 | -37 | -24 | -24 | -24 | -33 | -21 | -49 | |
| | S. dusenii C. Jensen, S. cuspidatum Ehrh.—Scheuchzeria | Mochezhinas | (- 5) | (-11) | (-12) | (+ 1) | (+ 1) | (+ 1) | (- 8) | (+ 4) | (-24) | |

TABLE 2. Depths of Summer Groundwaters in Different Bog Macrolandscapes and for Different Dominant Sphagnum-Moss Species*
(from data of the author's expedition research)*

| Type of vegetation | Bog microlandscapes | Depth of groundwaters from the surface of the bog (cm) | | | | | | Number of field measurements |
| | | Dominant species of sphagnum moss | | | | | On the microlandscape as a whole | |
| | | Sph. fuscum | Sph. magellanicum | Sph. angustifolium | Sph. balticum | Sph. dusenii | | |
|---|---|---|---|---|---|---|---|---|
| Forest | Pine-shrub-sphagnum (Pinus f. uliginosa) | 40–45† / 41 | 32–39 / 35 | — | — | — | 33–45 / 37 | 8 |
| Moss-forest | Sphagnum-Eriophorum-shrub-pine (Pinus f. litwinowii) | 30–39 / 34 | 22–24 / 23 | 20–21 / 20 | — | — | 20–39 / 27 | 30 |
| Moss | Sphagnum-Eriophorum-large shrub, rarely afforested with Pinus f. willkommii | 25–27 / 26 | 20–27 / 23 | 18–19 / 18 | — | — | 18–27 / 24 | 10 |
| | Sphagnum-Eriophorum-large shrub, with an abundance of deadwood. | 21–22 / 21 | — | — | — | — | 21–22 / 21 | 29 |
| Ridge-mochezhina complexes | On ridges | | | | | | | |
| | Sphagnum-small shrub-Eriophorum, afforested with pine | 32–33 / 32 | — | — | — | — | 32–33 / 32 | 3 |
| | Sphagnum-shrub-Eriophorum, afforested with pine | 28–35 / 31 | — | — | — | — | 28–35 / 31 | 4 |
| | Sphagnum-shrub-Eriophorum | 25–28 / 27 | 21–24 / 22 | — | — | — | 21–28 / 26 | 21 |
| | In mochezhinas | | | | | | | |
| | Sphagnum-Eriophorum. | 22–27 / 25 | — | — | — | — | 22–27 / 25 | 11 |
| | Sphagnum-Scheuchzeria | — | — | 11–19 / 15 | 9–11 / 10 | — | 9–19 / 13 | 28 |
| For sphagnum-moss species as a whole | | — | — | — | 2–8 / 5 | (+2)–(+8) / +5 | (+8)–(–8) / 0 | 20 |
| Number of field measurements | | 21–45 / 29 | 20–39 / 26 | 11–21 / 16 | 2–11 / 7 | (+2)–(+8) / +5 | (+8)–(–45) / 22 | — |
| | | 79 | 32 | 20 | 26 | 7 | — | 164 |

* Depths of groundwaters measured from the surface of the bog (respectively, elevations for ridges and depressions for mochezhinas) during the periods July 17–August 29, 1948, and July 21–August 11, 1949.

† Average monthly depths.

region, and, in particular, the average level of groundwaters over a period of several years in bog microlandscapes, without reference to the type of bog massif (acutely convex, sloping convex, and so on).

The results of analyses of station observations on the level of groundwaters in different bog microlandscapes on eight upland bogs are correlated and presented in Table 1. The depths of groundwaters were calculated from the average characteristic surface of the bog microlandscape using microlevel data in a water-measuring drill hole. To provide the basic information on the level of groundwaters, we selected the average yearly and average monthly observational data for April and August, both for the whole five-year period 1951-1955, and for the characteristic low-water (1951) and high-water (1953 or 1954) years.

For each type of bog microlandscape, the depths of groundwaters were determined by taking averages from not less than 3-5 water-measuring drill holes, not only in one but in several different bog massifs. As a rule, the divergence in groundwater depths did not exceed 2-3 cm for different data from the water-measuring drill holes taken into account when calculating averages.

Let us survey the level of groundwaters in the most widely distributed types of bog microlandscapes with forest, moss-forest, moss, and grassy-moss types of vegetation, and also ridge-mochezhina complexes.

The average annual depths (for five years) of groundwaters are lowest (32-45 cm) in bog microlandscapes with the forest type of vegetation, and are slightly less deep (27-33 cm) in the case of the moss-forest type; moderate values, although with a rather wide range (12-31 cm) occur in the case of groundwaters in bog microlandscapes with moss-type vegetation, and the highest levels (5-10 cm) occur with the grass-moss type. In bog microlandscapes of ridge-mochezhina complexes, the average mean depths of groundwaters in ridges correspond to the higher values of the depths in microlandscapes with the moss type of vegetation (24-31 cm). More complete data are given in another paper by the present author (Romanova, 1960).

Turning to the average monthly depths of groundwaters in different bog microlandscapes in April and August, attention should be focused on the clear correlation between these depths themselves, as well as between them and the average annual groundwater depths in the same microlandscapes. There is a direct correlation between these depths, although the level of groundwaters in August is 12 cm lower than that in April.

A comparison of edificator species with the depths of groundwaters in bog microlandscapes reveals a correspondence between them. In forest and moss-forest microlandscapes of upland bogs, the growth form of pine and the height of trees are good indicators of the depth of groundwaters. The sphagnum-moss edificator species is a good indicator of levels of groundwater in moss and grass-moss bog microlandscapes.

Expeditionary researches on upland bogs of the Northwest were carried out by us in the summers of 1948 and 1949, more or less during the same part of the summer (July-August).

The depths of groundwaters in different bog microlandscapes were measured in temporary 1.5-m-deep water-measuring drill holes, which were prepared in the course of working with a peat drill-borer in the bogs. The analyzed measurement data for similar types of bog microlandscapes and forms of sphagnum moss were quite similar, as follows: the different depths did not as a rule vary by more than 3-4 cm. The results of treatment of these data are given in Table 2, which are highly representative, the more so since numerous measurements were made in the main types of bog microlandscape and species of sphagnum moss.

If we analyze the data provided, we note that a substantial difference in depths of groundwaters occurs both in different types of bog microlandscapes and for different edificator species of sphagnum moss. However, the edificator species of sphagnum moss provides a clear indication of the groundwater level for each type of bog microlandscape. Thus, the type of bog microlandscape together with a designated edificator species of sphagnum moss constitute a good indication of the groundwater level, and the data for each respective group, determined by the type of bog microlandscape and the edificator sphagnum-moss species, are very constant (the difference between the lower and higher groundwater depths is usually no more than 5 cm for each group).

A comparison of the data in Tables 1 and 2 shows that there is agreement between them. It is that the depths of groundwaters for all similar types of bog microlandscapes and species of sphagnum moss were, according to the data station investigations for August 1951-1955. The difference between these depths was only 2-6 cm.

Summary

1. Microlandscapes of high bogs can be arranged in the following order according to their degree of inundation (in ascending order): forest, moss-forest, moss, and grass-moss; ridge-mochezhina bog microlandscapes correspond to moss on the ridges and to grass-moss in the mochezhinas. The depths of groundwaters may vary considerably in different mochezhinas, being higher in sphagnum-cotton grass mochezhinas and lower in sphagnum—Scheuchzeria mochezhinas.

2. It has also been confirmed that the major sphagnum-moss edificators of upland bogs fall into the following order, according to increasing inundation of habitat: S. fuscum (Schpr.) Klingr., S. magellanicum Bridel., S. angustifolium C. Jensen, S. balticum Russ., S. dusenii C. Jensen, S. cuspidatum Ehrh.

3. It can also be affirmed that measurements of the depths of groundwaters in upland bogs, even if carried out during short-term expeditions, are highly representative and useful for comparing different types of swamp microlandscapes.

Literature Cited

Dubakh, A. D. 1936. Essays on the Hydrology of Bogs. Leningrad, Gidrometeoizdat.
Ivanov, K. E. 1957. Principles of the Hydrogeology of Bogs in the Forest Zone. Leningrad, Gidrometeoizdat.
Romanova, E. A. 1960. On the relationship between vegetation, the upper layer of peat deposits, and the water regime of upland bogs in the Northwest. Tr. GGI, No. 89.

# EXPERIENCE IN THE USE OF GEOBOTANICAL METHODS
# IN THE EXPLORATION OF PEAT BOGS IN SIBERIA

## G. G. Yasnopol'skaya

This paper is the result of a preliminary examination of material obtained by aerovisual and land studies of the southeast part of the Vasyugan bogs, carried out during the summer of 1960.

Basic to the peat-exploration work was the use of the indicator properties of the vegetation cover, as this led to the possibility of studying the vast, rather inaccessible bog massifs of Siberia by means of aviation. The procedure utilized in the expedition consisted of a preliminary mapping of the bog vegetation on a large-scale topographical basis, with subsequent terrestrial study of the most characteristic sectors which could be differentiated.

Eighteen aerovisual surveys were made in 1960 on an area of 800,000 ha in the Vasyugan massif, and 150 plots were marked out for land work. The use of a helicopter provided the possibility of landing at any given point.

The author made a botanical analysis of 500 peat samples, a hydrochemical analysis of 19 water samples, and tested 100 soil samples for effervescence with HCl. On the basis of these data and the use of data from earlier land itineraries covering a total distance of 350 km, we compiled a geobotanical map and a cartographic representation of peat deposits showing the different types and forms of deposit and these indicate that the Vasugan massif is very complex.

Bogs of first- and second-order watersheds, which have different pathways of bog formation, fuse into a single vast massif, located on the border of two zones, taiga and deciduous forest, although differences in the degree of salinization and soil alkalinity produce conditions for the formation of different types of bogs, even within the limits of one watershed.

The part of the Vasyugan bog massif toward the center of the wide flat watershed is of greatest interest.

Substantial areas on the southern slopes of the latter are occupied by sedge-hypnum bogs. It is typical that in second-order watersheds these bogs have tonguelike streaks which stretch out for some kilometers, being replaced further by upland pine-sphagnum and ridge-mochezhina bogs.

A distinctive feature of the sedge-hypnum bogs is the absence of a tree layer. Trees (usually Betula pubescens Ehrh. and occasionally Pinus) are found as individual specimens or are associated with the uncommon, elongated ridges, which have a length of one to several kilometers, a width of 2-5 m, and a height of 10-30 m. The ridges are characterized by a profusely developed layer of shrubs of Chamaedaphne calyculata (L.) Moench and Ledum palustre L. and a moss layer of Tomenthypnum nitens (Schreb.) Loeske, Pleurozium schreberi (Willd.) Mitt., Sphagnum warnstorfii Russ., and S. subbicolor Hampe.

The space between ridges has a completely flat surface. The grass layer is formed by Carex lasiocarpa Ehrh., C. limosa L., C. chordorrhiza Ehrh., C. inflata Huds., C. diandra Schrank, C. omskiana Meinsh., and Equisetum palustre L., with Comarum palustre L., Eriophorum angustifolium Roth., Scheuchzeria palustris L., Cicuta virosa L., Menyanthes trifoliata L., and other species present in addition. The moss cover is composed of Drepanocladus vernicosus (Lindb.) Warnst., Calliergon stramineum (Brid.) Kindb., Meesia triquetra (L.) Aongst., and Helodium blandowii Warnst. Patches of Betula nana L. and Salix lapponum L. bushes are noted on the bog. A comparison of the maps shows that the contours of the sedge-hypnum bogs (Fig. 1) are fully repeated in Fig. 2, coinciding in this case with the contours of sectors with a 1-2-m-thick lowland sedge deposit. Numerous analyses of peats showed great uniformity in their composition in different points of the massif. The pre-

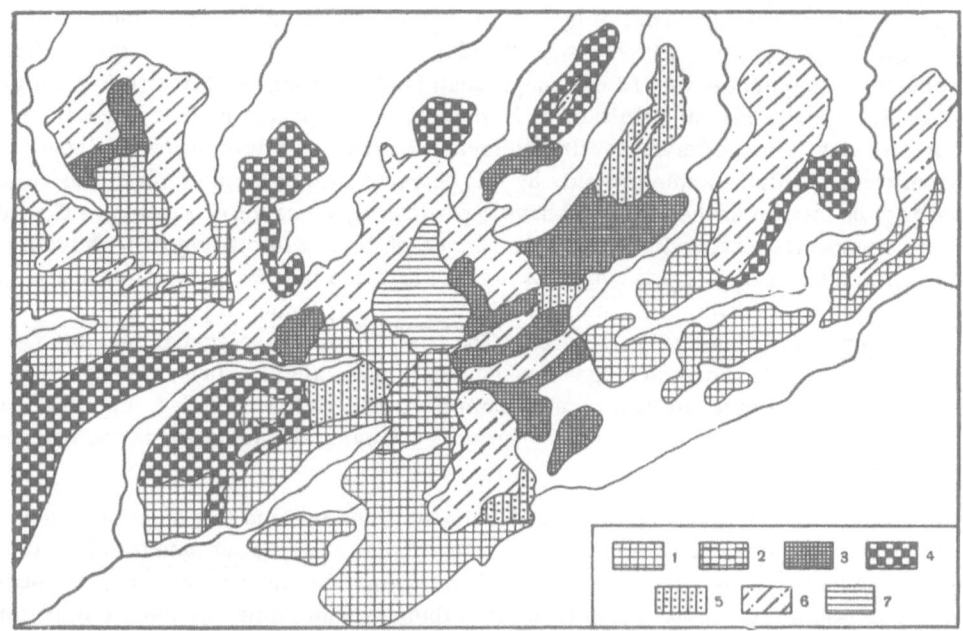

Fig. 1. Map scheme of the bog vegetation in the southeastern part of Vasyugan: 1) sedge—hypnum bogs; 2) same, polygonal; 3) sedge-sphagnum complexes; 4) pine-shrub-sphagnum; 5) pine-sphagnum; 6) ridge-mochezhina; 7) mochezhina-lacustrine bogs.

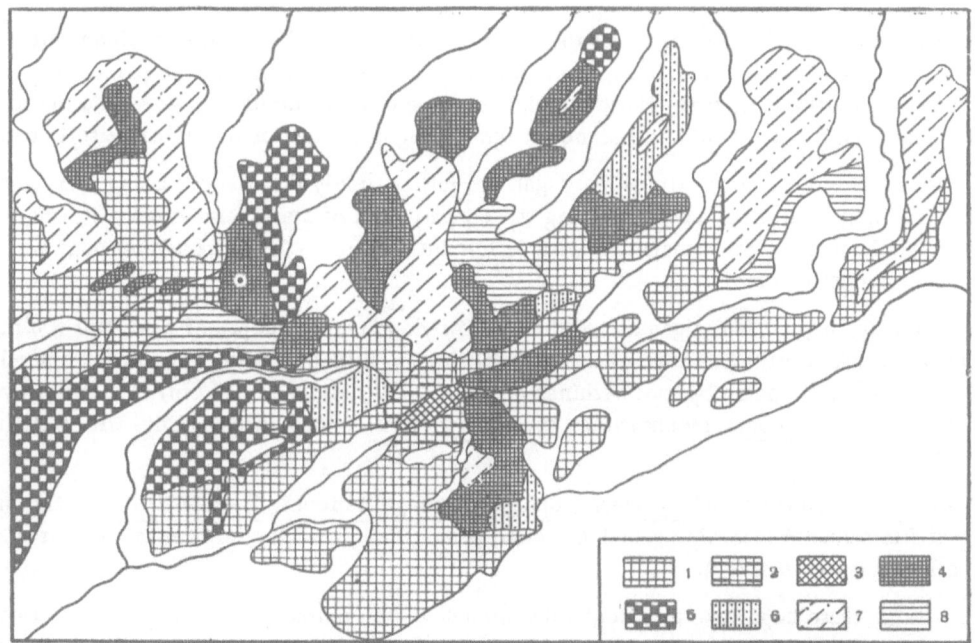

Fig. 2. Map scheme of the types of peat deposit in the southeastern part of Vasyugan: 1) sedge peat; 2) sedge-hypnum; 3) hypnum; 4) transitional swamp; 5) transitional forest-swamp; 6) fuscum peat; 7) complex upland; 8) mixed.

dominance of Carex lasiocarpa Ehrh. is characteristic, as is the presence of C. inflata Huds., C. omskiana Meinsh., and C. paradoxa Willd. Remnants of hypnous mosses are found almost everywhere but in very small amounts. The degree of peat decomposition reaches 25-30%, while the ash content fluctuates from 5 to 8%. The presence of salts of calcium (4-6 mg/liter), magnesium (1-2 mg/liter), and iron (0.5-0.6 mg/liter) is noted in the water of these bogs; the content of sulfates is 15 mg/liter. The peat deposit is underlain everywhere with a bluish-grey or dark clay, which effervesces with hydrochloric acid; the origin of lowland bogs in watershed areas of Vasyugan can probably be explained only by this characteristic of the maternal rocks at the bottom of the bog. Those bogs on the rim of the watershed háve a particularly characteristic nature. The ridges interlock together and form, as it were, a network, dividing the surface of the bog into polygons with more or less rounded corners, 200-500 m in size. The polygons consist of highly inundated mochezhinas with a sparse grassy layer; the moss cover is formed by the hydrophilic mosses: Campylium polygamum Bryhn., Scorpisium scorpioides (L.) Limpr., and Mnium cinclidoides (Blytt.) Huben. The peat deposits of polygonal bogs are 3-4 m in depth and are composed of slightly decomposed sedge—hypnum peat, underlain by a layer of sedge peat. At the sources of the river Ot', the deposits are purely of hypnum peat, underlain with sedge—hypnum. The displacement of gramineous plants by mosses is evidence of the constant water accumulation in the bog and the exceptionally slight runoff.

Also on the top of the watershed and in direct proximity to the lowland bog described, there is an upland mochezhina-lacustrine sphagnum bog with infrequent, flat cushions of dying Sphagnum fuscum (Schimp.) Klingr. and S. papillosum Lindb. and occasional forest-free ridges. The mochezhina areas are occupied by S. balticum Russ. and S. papillosum Lindb.; the grass layer is thinnish, and formed by clumps of cotton grass, Rhynchospora alba (L.) Vahl., and Scheuchzeria palustris L. The peat in the upper horizons is very thin and consists of residues of those mochezhina Sphagnum species occurring on the surface; a complex upland peat is bedded in the lower horizons, while in the bottom layers there is generally a sphagnum peat intermingled with a little sedge material. The thickness of the deposits reaches 6 m in some places. The peat is very slightly decomposed and low in ash content. The underlying soils do not effervesce with hydrochloric acid; the water contains very slight amounts of calcium salts (3 mg/liter) and magnesium salts (0.5 mg/liter) and traces of sulfates.

These two types of bog, developing under the same conditions in a flat, almost drainage-free watershed, have similar features: considerable hydration, the absence of a wood layer, a thinnish grass cover and the profuse development of hydrophilic mosses. However, the presence of carbonated soils governed the formation of lowland bog with hypnum mosses, while an upland sphagnum bog developed under alkaline soil conditions.

Upland bogs in the investigated part of Vasyugan are rather widely distributed, being focused on the northern slope of the main watershed and occupying the central parts of the second-order watersheds. Ridge-mochezhina bogs predominate; pine-sphagnum (with S. fuscum) and pine-shrub-sphagnum bogs are drawn to the more well-drained margins of the massif.

Ridge-mochezhina bogs in Vasyugan are of the shallow-deposit type (depth of peat, 2.0-3.5 m) and, depending on their age, indicate the presence of transitional swamp, mixed swamp, or complex upland deposits. At earlier stages, ridges are denoted by mochezhinas of small area (height, 10-15 cm) in a moss cover of S. magellanicum Brid. and S. fuscum. Mochezhinas have Eriophorum—sphagnum groupings with S. balticum and S. dusenii C. Jensen.

Carex lasiocarpa, C. inflata, Scheuchzeria, Sphagnum magellanicum, S. angustifolium, and other species occur in the peat filament. Frequently, a lowland sedge peat is bedded in the bottom layers. At some points, the soils show effervescence with HCl.

At a later stage, these bogs have a typical ridge-mochezhina complex. The ridges are up to 40 cm in height and are long and narrow; Sphagnum fuscum forms a dense sod, and Pleurozium schreberi (Willd.) Mitt., Dicranum rugosum Brid., and lichens are found. Mochezhinas are voluminous, markedly inundated, with cotton-grass, Scheuchzeria, Sphagnum papillosum, and S. balticum growing on the periphery, and Rhynchospora alba (L.) Vahl., Carex limosa, and S. balticum at the center. The peat deposit in the lower layers consists of transitional sphagnum and sphagnum-sedge peats. In the upper horizons under ridges, fuscum peat is located, with a separate transitional interlayer of magellanicum peat, while under mochezhinas, sphagnum-mochezhina and complex-upland peats occur. The soils do not show effervescence when HCl is applied.

At many points in the bottom layers of the peat, tree residues (birch, rarely spruce) are found, although they are not observed in sedge-hypnum bogs. It is evident that in the period preceding bog formation, southern and northern slopes differed to some extent in the character of soils and vegetation. The southern slopes were occupied by gramineous communities; they are at present occupied by eutrophic bogs. On the northern slopes, the grass-bog stage was of less duration and was rapidly replaced by an upland sphagnum bog. The process of transition of a lowland bog to the oligotrophic phase of development is still observed at the present time.

On the eastern slopes of Vasyugan, we noted peculiar sedge-sphagnum complex bogs, where eutrophic vegetation groupings alternate with oligotrophic groupings.

On the surface of the sedge-hypnum bogs there appear cushions of S. magellanicum with inclusion of Aulacomnium palustre (L.) Schwaegr., covered with Cassandra and dwarf Convolvulus, having a diameter from one to several meters, and dispersed very thinly or else forming an almost solid moss cover.

Carex lasiocarpa, C. inflata, C. limosa, Eriophorum vaginatum L., Equisetum palustre, Menyanthes trifoliata, Scheuchzeria palustris, Sphagnum obtusum Warnst., Drepanocladus vernicosus (Lindb.) Warnst., and Calliergon stramineum (Dicks). Kindb. grow in areas between hillocks.

The peat layer is composed of a sedge lowland peat. Under the sphagnum beds, there begin to form peat deposits of a transitional type with S. magellanicum; the greater the sphagnum bed on the bog surface, the greater is the layer of transitional peat in the upper horizons of the deposit. There is no doubt that ridge-mochezhina bogs of northern slopes have attained the stage in development of a transitional bog with magellanicum beds, evidence being the interlayer of magellanicum peat under fuscum peat in ridges.

Thus, sedge-hypnum lowland, sedge-sphagnum transitional and ridge-mochezhina upland bogs constitute elements of a single ecological series of bog formation in broad, flat watersheds under the conditions of the deciduous-tree zone.

The territorial distribution of individual elements of this series relative to one another is determined by the distributional pattern of mineral earths.

Pine-sphagnum bogs with S. fuscum, which have deposits composed of fuscum peat throughout their entire depth (2.0-2.5 m), occupy a somewhat isolated position; they are developed in the most well-drained sites with very barren alkaline soils.

Pine-shrub-sphagnum bogs with S. magellanicum, having a shallow deposit, are made up of forest-swamp peats and arise in bogged forest areas.

In the compilation of the map of peat deposits, all the bog communities described were used as indicators of a specific type of deposit structure.

Groupings of sedge-hypnum bogs indicate the presence of a 2-m-thick sedge deposit; sedge-hypnum deposits 3-4 m in depth lie under polygonal sedge-hypnum bogs; sedge-sphagnum complex bogs at different stages of development have a 2-m sedge lowland or transitional swamp deposit 2.0-2.5 m in thickness; ridge-mochezhina upland bogs have a transitional, mixed, or upland deposit, depending on their age, with an average thickness of 3 m.

A comparatively deep deposit (3-6 m) is characteristic for mochezhina-lacustrine bogs, the layer of upland complex bog accounting for more than half of this depth.

A pine-sphagnum grouping with S. fuscum is an indication of a 2.5-m thick fuscum deposit, while a pine-shrub-sphagnum grouping indicates transitional forest-swamp and swamp deposits.

The identification of deposit sectors on the basis of data of land surveys without the use of a geobotanical map leads to errors in delimiting borders and to an incorrect determination of resources. Only an analysis of a geobotanical map in conjunction with data on the stratigraphy of peat deposits permits the determination of the establishment of the type of bog-forming process, and thus the discovery of the principles of identifying peat-deposit sectors.

Summary

    The author took part in investigations of the Vasyugan peat bog (West Siberian lowlands) and, on the basis of aerovisual observations and land work, has evolved the following hypotheses regarding the indicational value of plant communities:  groups of sedge-hypnum bogs are indicative of sedge deposits up to 2 m in thickness; sedge-sphagnum bogs are characteristic of sedge-moss and transitional swamp deposits; bogs with small lakes and shallow depressions filled with hydrophilic Sphagnum species (mochezhinas) usually have deposits of 3-6 m, the upland peat layer accounting for about half of the total depth; finally, pine-sphagnum groupings are an indicator of swamp or forest-swamp deposits about 2.5 m in thickness.

# THE ROLE OF THE GEOBOTANICAL METHOD
# IN THE OVERALL COMPLEX
# OF ENGINEERING GEOLOGY RESEARCH

## I. S. Komarov, A. V. Sadov, and L. N. Tagunova

The present state of our country's economic development is characterized by the extensive spread of all forms of construction. In different parts of the country, large heating and hydroelectric plants are being put up, new industrial enterprises are being constructed as well as towns and workers' settlements, railroads and highways are being built, and the gas and petroleum pipeline networks are under extension. One of the most important features of contemporary construction is the scientific trend toward siting construction work in different regions of the country on the basis of complex regional planning.

Regional planning represents the future economic geography of individual economic regions. Its tasks comprise the allocation of industrial and energy works on the territory of the economic regions, the settlement of workers in convenient and healthy conditions, the organization of transport and other communication systems, the creation of zones for the recreational activities of the population, and so on. The projects of regional planning should provide for the maximal utilization of local raw materials, energy, water, and other natural resources.

Many of these new industrial regions are located in remote, little inhabited, rather inaccessible parts of the Soviet Union.

The successful working out of the projects of regional planning can be achieved on the basis of the comprehensive consideration of the features of the natural geographical conditions of territories, in such a manner that all the projected works conform in the best possible way with natural conditions.

Only under these circumstances will we get the most appropriate locations for the allocation of industrial works, towns and settlements, zones of large-scale recreation, and so on.

One of the most important divisions of this work is the study of the engineering-geology conditions of the territory, which should result in a map of the engineering-geology zonation of the territory with the demarcation of the following sectors:

a) those suitable for industrial construction and multistorey town building without preliminary engineering preparatory work;

b) those of limited suitability for building, requiring a more or less substantial amount of engineering preparation, the use of complicated and expensive artificial foundations, or the use of special constructions, foundations and building (sectors with poorly suited soils or the close proximity of the groundwater level, or affected by active physical or geological processes, etc;

c) those unsuitable for construction (peat bogs, areas with the large-scale development of active land-slip, karst, and other processes, inundated territories, and so on).

Based on the above composition of engineering-geology zonation, and the requirements affecting the natural foundations of the object of large-scale construction (industrial, civilian, or highway), the following data should be the basis of zonation: a) geological-lithological profiles of deposits to a depth of 10-12 m; b) level of groundwaters (within the limits of the depth mentioned above); c) occurrence and distribution borders of physical and geological phenomena of danger to construction (karst, land-slip, frost deformations, etc.); d) the properties of rocks as the basis of engineering construction.

In the regions where perennially frozen rocks are distributed, this list should be supplemented by data on the depth of bedding of the top of the frozen rocks, the location and limits of steatites, the presence of subterranean ice, and various other specific features of this territory.

Engineering-geology zonation of a territory, even with some elements of schematization, and ready for the stage of regional planning, is a highly complicated problem, mainly because of the vast size of the areas requiring study and the limitations on the periods when work can be carried out.

In carrying out engineering-geological investigations by normal methods, a survey of territories of such size must of necessity be stretched out over many years, and the planners do not obtain all the required primary data in good time. This position leads in practice to the frequent use in regional planning of material on too small a scale, of a fragmentary nature and of little significance. This can ultimately lead to the compilation of inadequately based regional planning schemes and the necessity for their further treatment and conversion.

On the basis of these foundations, we must search for rapid methods of engineering-geology investigations, based on the latest achievements of geological science in this field.

In the present communication, we do not propose to describe in detail the methods worked out, as this is not our task. We shall mention only that aerial methods—aerial survey and aerovisual inspection—occupy an important place in the investigational complex. The principal application of these methods in districts with a more or less thick cover of loose tertiary deposits is to provide information on the landscape as a regular and interrelated combination of natural elements, governed by the overall developmental history of different sectors of the earth's crust.

This most important aspect of modern geographical science opens great possibilities before engineering geology since it permits the establishment of some elements of the natural conditions which have definite significance for engineering geology but are concealed from direct observations and mapping, on the basis of other elements whose determination and mapping may be carried out by various simple methods. In the case of interest to us, the former group includes such elements as the composition of deposits, the depth of the water table or the top of the perennially frozen rocks, etc., while the latter group includes structure of the relief (macro-, meso-, and microrelief forms) and the character of the vegetation cover (species composition, structure, distribution of plant communities, and so on).

Data in the literature and the experimental work of VSEGINGEO show that the study of the structure of the relief by geomorphological methods and of the vegetation cover by geobotanical methods through the use of readily established direct features (outcrops of main rocks, sources of groundwaters) and data on the economic utilization of the territory not only provide landscape zonation of the territory which in itself is of considerable interest for regional planning, but also afford an engineering-geological interpretation to landscape zonation, i.e., convert it essentially to engineering geology zonation—the basis for the accurate prediction of conditions for the construction of buildings and installations of the territory under study.

The study of the structure of the relief by geomorphological methods has long been incorporated into practical programs, utilized in engineering geology. Recommendations on this problem can be found in many reference books and handbooks on engineering geology. We shall therefore not deal in detail with this problem. As regards geobotanical methods, they have still not gained wide acceptance. At the same time, they are in many cases not inferior to geomorphological methods in effectiveness.

Data on the relationship of the vegetation cover with the composition of deposits and the level of groundwaters are given in numerous papers by Soviet and foreign specialists for many landscape zones (Abramova, 1947; Viktorov, 1955, 1956, 1959; Viktorov and Vostokova, 1959; Vinogradov, 1958; Vostokova, 1955 and 1956; Vysotskii, 1904; Ososkov, 1909-1912; Pogrebnyak, 1944; Priklonskii, 1935; Benninghoff, 1953; Clements, 1920; Meintzer, 1927; Sykora, 1959). Determinations have now been made of plant indicators which afford a sufficiently accurate indication of the composition of surface deposits and the depth of the water table. The depths for which a more or less accurate prognosis can be made fluctuate from 1-3 m in the tundra zone to 5-10 m in the forest zone and 15-20 m, sometimes more, in the semidesert and desert zones.

Problems Solved with the Aid of the Geobotanical Technique in Engineering-Geology Site Evaluation

| Geological structure | Hydrogeological conditions | Physical-geological processes |
|---|---|---|
| 1. Composition of surface deposits to a depth of 1-3 to 15-30 m | 1. Locations of subsurface water outlets to the surface | 1. Locations and nature of gravitational processes (cave-ins, avalanches, talus shifts) |
| 2. Thickness of top layer of surface deposits (sands on clays, etc.) | 2. Position of the groundwater level | 2. Presence, distribution limits, age, and degree of intensity of land-slip processes |
| 3. Position of tectonic disturbances of a rupturing nature | 3. Presence and borders of leakage water | 3. Locations of subsidence and suffosion phenomena, shifts of the earth's surface above mining works; covered forms of karst |
| | 4. Type and degree of salinization of groundwaters | 4. Distribution limits and depth of bedding of perennially frozen rocks, position and range of steatites, location and intensity of congelation processes (thermokarst, solifluction, nadels, and so on) |
| | | 5. Type of bog formation, thickness of peat deposits, water-supply regime, and so on |
| | | 6. Type and degree of soil salinization |
| | | 7. Locations and intensity of abrasion and erosion processes |
| | | 8. Location and intensity of aeolian processes |
| | | 9. Character of recent tectonic movements |

Many workers have also noted the possibility of using geobotanical methods for establishing the position, distribution limits. and degree of intensity of many physical-geological phenomena such as, for example, land-slips, subsidence and suffosion* phenomena, soil salinization, bog formation, and several others (Abramova, 1954; Bobrovskii, 1957; Viktorov, 1955; Vyshivkin, 1955; Grebenshchikova, 1939; Shirokovskaya, 1947).

Geobotanical methods have recently been used in studying permafrost (the depth of the top of the frozen rocks, the presence and range of steatites, etc.) and the associated congelation phenomena (thermokarst, solifluction, etc.) (Andreev, 1957; Petrusevich, 1954; Benninghoff, 1953).

An attempt is made (see table) to schematize the problems of engineering-geology site evaluation, in the solution of which geobotanical methods may be (and have been) applied. This table must be regarded as far from exhaustive.

The experience of VSEGINGEO shows that an engineering-geology survey by the rapid "landscape method" can be carried out particularly rapidly and successfully with the MI-4 helicopter, which can be used both for aerial survey and for aerovisual observations.

* "Suffosion"—concerns the extraction of fine elastic particles from unconsolidated sediments, leading to settling and the formation of depressions or "saucers."

To verify conclusions on the relationships between vegetation and other elements of the landscape, to study soil properties and to explain other processes, the method of key plots may be used, which has found wide application in landscape investigations. This is effected by identifying typical plots within the districts differentiated on the map, and carrying out all further detailed investigations on these plots. Geophysical, excavation, and drilling work can be set up in the key plots; rock samples are collected, and the required laboratory investigations are carried out.

Experience shows that here too the use of the MI-4 helicopter is of considerable assistance, as it can be used for transporting the necessary equipment and permits landing within the key plots.

We have surveyed above the possibilities of applying geobotanical methods at the very earliest stage of planning—in the compilation of regional planning schemes.

In the subsequent stages—in the regional planning of industrial zones—the tasks of engineering-geology research remain as before. Only the scale of the surveys increases. As before, the survey can be carried out with the use of aerial methods with subsequent verification on key plots on land. Geobotanical methods completely retain their significance.

In more detailed investigations, i.e., in the stages of project assignment and the planning of technical work, the significance of landscape methods, including geobotanical methods, decreases markedly, since the study of individual construction sectors requires a detailed study of the profile of deposits, involving a considerable amount of excavation, drilling, and laboratory investigations. The only exceptions are communication lines (highways, gas and petroleum pipelines, and electricity lines), where the landscape method retains its significance to a substantial extent in the project assignment stage.

It should, however, be noted that the work of recent years has opened clear prospects for the utilization of geobotanical methods not only in detailed investigations but also in the period of construction and building works. In this case, one important characteristic of vegetation is used, namely, its capacity to react sensitively to changes in ecological conditions, particularly changes in soil and ground moisture content. Observations show that changes in the conditions of the water supply to plants are reflected not only in the disappearance of some species and the appearance of others, but also in the morphology of existing forms. Thus, for example, changes in hydrogeological conditions frequently disrupt the entire pattern of growth and development of tree forms (symptoms of depression appear, the rate of increment of annual rings changes, and so on).

Trees also react to disturbances in the stability of land-slip masses (landslides, talus movements, etc.), this being reflected in twisted growth, the appearance of trunk eccentricities and other symptoms.

Changes in the form of plants or the above-mentioned modifications in the morphology of existing plants can be used in observations on the dynamics of some physical-geological and engineering-geological processes such as drainage or raising of the water table in territories, the development of land-slip upheavals, and so on.

The features enumerated permitted G. L. Galazii (1960) to suggest an interesting procedure for assessing the age of land-slips and certain other relief features.

One should not, however, overestimate the value of geobotanical methods, since, like any other approach, they are characterized by certain deficiencies and limitations. The main limitation is undoubtedly the fact that geobotanical methods can only be successfully applied where the natural vegetation still remains in a more or less unchanged condition. The application of the geobotanical method seems less effective, however, in densely populated inhabited locations, where the natural landscape has been replaced by a cultivated landscape and the principal areas are occupied by plowed fields, gardens, and orchards. The greatest success from the application of the rapid methods of engineering-geology work, involving the use of geobotany, should therefore be expected in the remote regions of Siberia, the Far East, and Central Asia, where the natural landscape has as yet been little disturbed by the economic activity of man. It is just these territories, little studied from the geological aspect, which represent the regions having the most urgent need for the application of rapid methods of engineering-geology research.

It is necessary in conclusion to mention the need for preparing special handbooks on geobotanical methods, as these would greatly facilitate the use of these methods by a wide circle of specialists in engineering geology and hydrogeology, who have no botanical education and, in this field, rely mainly on their natural powers of observation. Particularly valuable assistance would be afforded by albums of indicator plants, compiled for different landscape zones.

Specimens of such handbooks for certain regions of the USSR (deserts of Central Asia, Central region) have now been compiled by VSEGINGEO under the direction of S. V. Viktorov (1959) and E. A. Vostokova (1955). Thanks to their clarity, these albums are exceptionally suitable for practical work by geologists, and should be published more rapidly. Further work along these lines should be regarded as highly useful and promising.

## Summary

Geobotanical investigations can be widely applied in engineering-geology work conducted on a small or medium scale. Data obtained by the interpretation of aerial photographs of indicator communities permit the compilation of indicator maps reflecting the distribution of different soils, the deposition of subsoils of varied mechanical composition, and the general hydrogeological conditions. Geobotanical investigations are especially important in the first stages of engineering-geology studies of a territory and can therefore be used in regions of new settlement in the Northeast and North of our country.

## Literature Cited

Abramova, T. G. 1947. "The vegetation cover as an indicator of the upper layers of a peat deposit." Vestn. LGU, No. 5.

Abramova, T. G. 1954. "On the association between the vegetation cover of a bog and the structure of the upper layers of its peat deposit." Uchenye zap. LGU, seriya biol., Vol. 34.

Andreev, V. N. 1957. "The deciphering of different types of tundra by aerial photography and their aerovisual characterization on the basis of frost fissures." Geogr. sb., No. 7.

Bobrovskii, R. V. 1953. "On the effect of the Rybinsk Reservoir on the forest of the Darwin Reserve"; Chap. 1. Changes in the nature of the banks of the reservoir. In book: The Rybinsk Reservoir. Izd. MOIP.

Viktorov, S. V. 1955. The Use of the Geobotanical Method in Geological and Hydrogeological Investigations. Moscow, Izd. Akad. Nauk SSSR.

Viktorov, S. V. 1955a. "Geobotanical features of Karst-suffosion processes in the desert." Byull. MOIP, otd. biol., Vol. 60, No. 1.

Viktorov, S. V. 1956. "Lichens as indicators of lithological and geochemical conditions in the desert." Vestn. MGU, No. 5.

Viktorov, S. V. 1959. "On geobotanical methods in searches for water leakages in dry regions." Razvedka i okhrana nedr, No. 11.

Viktorov, S. V., Vostokova, E. A., and Voronkova, L. F. 1955. "The utilization of geobotanical features for finding tectonic disturbances." Tr. Vses. aerogeol. tresta, No. 1.

Viktorov, S. V., and Vostokova, E. A. 1959. "Experience in using aerial methods in geological observations, carried out within the complex of geological and hydrogeological research." Tr. Labor. aerofotomet. Akad. Nauk SSSR, Vol. VIII. Moscow, Izd. Akad. Nauk SSSR.

Vinogradov, B. V. 1958. "On the relationship of vegetation with groundwaters in steppe landscapes of Northern Kazakhstan and the use of the vegetation as an indicator in the hydrogeological deciphering of aerial photographs." Izv. Akad. Nauk SSSR, seriya geogr., No. 1.

Vostokova, E. A. 1955. "The application of the geobotanical method in hydrogeological research in deserts and semideserts." Tr. Vses. aerogeol. tresta, No. 1. Moscow.

Vostokova, E. A. 1956. "The use of geobotanical observations in hydrogeological research in dry regions." Sb. nauchn.-tekhn. inform. MG i ON, No. 3. Moscow.

Vysotskii, P. K. 1904. "Some geobotanical observations in the Northern Urals." Pochvovedenie, Vol. 6, No. 2.

Vyshivkin, D. D. 1955. "Procedure for the compilation of maps of ground salinization on the basis of geobotanical data." Tr. Vses. aerogeol. tresta, No. 1. Moscow.

Galazii, G. I. 1960. "On the use of the botanical method in the solution of problems of hydrogeology and engineering geology." Byull. MOIP, otd. geol., Vol. 35, No. 1.

Grebenshchikova, A. A. 1939. "A contribution to the problem of the development of bogs in karst sinkholes of Ivanovsk province." Sov. botanika, No. 1.

Ososkov, P. A. 1909-1912. "The relationship of forest vegetation to the geological composition of indigenous rocks." Lesnoi zhur., Nos. 2-3, 4-5, 8-9.

Petrusevich, M. N. 1954. Geological-Survey and Exploration Work on the Basis of Aerial Methods. Moscow, Gosgeoltekhizdat.

Pogrebnyak, L. S. 1944. Principles of Forest Typology. Kiev, Ukrgostekhpromizdat.

Priklonskii, V. A. 1935. "Vegetation and groundwaters." In collection: Hydrogeology and Engineering Geology, No. 1. Moscow—Leningrad.

Shirokovskaya, E. A. 1947. "The interrelationship between the vegetation cover and the surface layer of peat deposits." Torfyanaya prom., No. 8.

Shumilova, L. V. 1951. "Peat bogs in Tomsk province and prospects for their explicitation." In book: Summaries of Reports at the Second Scientific Conference on the Realization of the Plan for the Transformation of Tomsk Province. Izd. Tomsk. gos. univ.

Benninghoff, W. S. 1953. "Use of aerial photographs for terrain interpretation based on field mapping." Photogr. Eng., Vol. 19, No. 3.

Clements, G. 1920. "Plant indicators; the relation of plants to process and practice." Carnegie Inst. Washington Publs., No. 290.

Meintzer, O. 1927. "Plants as indicators of groundwaters." Water-Supply Paper, No. 557.

Sykora, L. 1959. Rostliny v geologickem vyzkumu. Prague.

# RELATIONSHIP OF THE VEGETATION COVER TO CLIMATIC
# AND SOIL-LITHOLOGICAL CONDITIONS
# IN THE CENTRAL URALS

## A. G. Chikishev

Up to the present time, geobotanical indicator research has evolved mainly in lowland regions, where the influence of the relief, the vertical zonation, and the exposure of slope are very slight, as a result of which the role of the parent soil-forming rocks, soil, and hydrogeological conditions emerge with particular clarity. The number of such investigations carried out in mountainous districts is still very limited, and there have been almost no studies of this type in the USSR. It thus seems of particular interest to examine the problem of the specific features of geobotanical indicator observations on mountains.

An attempt is made in the present article to discuss this problem, taking the Central Urals as an example.

The Central Urals are included in the southern and, partly, central taiga subzone, which is predominantly occupied by spruce-fir plantations. The complex lithological-structural and orographic composition of the surface, the varied climatic conditions, and the features of the historical development of the vegetation determined the great variability in the vegetation cover and the frequent changes of plant groupings at short intervals.

A characteristic feature of the Urals is the combined presence of widely differing representatives of the Asiatic and European flora, associated with the area being the meeting-point of the two continents. For example, one finds the European spruce, Picea excelsa Linc. in addition to the widely distributed Siberian spruce, P. obovata Ldb., as well as their intermediate forms. Many broad-leaved species have their eastern limit of distribution in the Urals. Only the lime and the elm cross over the Urals. As a whole, the Ural montane taiga is a highly varied plain.

The main background of the vegetation cover in the Central Urals consists of coniferous forests, among which deciduous birch-aspen plantations occur as small massifs, having arisen in place of pine forests and spruce-fir taiga as a result of forest fires and intensive felling practice. The coniferous forests of the territory surveyed are variable. Spruce-fir formations are distributed on the Western slopes and in the watershed zone, while pine forests are dominant on the Eastern slopes. This distribution of tree stands is associated with the characteristics of the climate. The location of the dark coniferous taiga on the Western slopes is determined by the significant humidity of the territory from atmospheric precipitation. On the Eastern dry slopes, spruces and firs are easily dislodged by the more xerophytic pine plantations, which go up as high as 500-600 m, after which spruce-fir forests again begin to become dominant on account of the change in climate in the direction of greater humidity.

The coniferous forests of the Central Urals are very varied from the phytocenotic respect in the northern and southern parts of the territory. The boreal, green-moss, dark conifer taiga is situated in the north, and the so-called southern taiga is in the south, its composition including plants of the broad-leaved complex. The border between these regions, according to K. N. Igoshina (1943), runs along the line Chermoz-Kizel-Pashiya-Teplaya Gora-Kushva-Verkhotur'e. The boreal taiga, represented by spruce, fir, and some cedar, is characterized by comparatively low productiveness. It is also characterized by constant semidarkness, high soil moisture content, an abundance of decaying logs, a sparse undergrowth poor in composition, a thin, low-growing grassy layer, and a strongly developed moss cover. Tree stands in the southern taiga consist of spruce and fir and are typically of higher productivity. The undergrowth is richer in composition and the grassy cover is denser and taller. The moss cover is less developed. A typical feature of the southern taiga is the presence of tree and gramineous plants of the broad-leaved complex, and also the limited distribution of bog vegetation. Deciduous species, such as Tilia cordata Mill., Ulmus scabra Mill., and Acer platanoides L., enter the composition of tree stands in the southern taiga or grow in the undergrowth in the form of shrubs. Herbage plants, Asarum europaeum L., Asperula odorata L., Dryopteris filix-mas (L.) Schott., Actaea spicata L., and Stachys silvatica L., characteristic for broad-leaved forests, are found throughout the southern taiga.

The Central Urals are characterized by well-expressed altitudinal zonation of vegetation. Attention was first drawn to this by P. N. Krylov (1878) and S. I. Korzhinskii (1893), who differentiated alpine, forest, and forest-steppe regions. Within the mountain-ridge zone, according to K. N. Igoshina (1944), four vertical vegetation strips are differentiated: forest, warped forest, coniferous belt, and mountain tundra. On the basis of the vertical distribution of dominant plant forms, V. S. Govorukhin (1960) differentiates a mountain-tundra zone, a zone of montane coniferous and mixed forests, and a zone of montane forest steppes in the Central Urals. The mountain-tundra zone is in turn subdivided into a belt of montane tundra, a belt of shrubs, a belt of montane hessian, or thin woods, and a belt of forest tundras. P. L. Gorchakovskii (1960b) identifies the following four main altitudinal belts in the Central Urals: submontane forest-steppe, forest-steppe, sub-peak, and peak.

We consider the main units of altitudinal zonation to be altitude belts and altitude sub-belts, patterns of elevation zonation being differentiated in the regional plan according to regions, provinces, and districts. Altitudinal belts are characterized by the dominance of a definite type of vegetation, while elevation sub-belts are differentiated by vegetational subtypes belonging to that type.

On the basis of the elevation distribution of the dominating types of vegetation, we distinguish the following elevation belts and sub-belts:

1. A belt of submontane forest-steppe.
2. A belt of montane light-conifer pine forests.
3. A belt of dark-conifer spruce-fir montane taiga: a) a sub-belt of montane fir forests; b) a sub-belt of montane spruce forests.
4. A belt of thin forests and montane meadows: a) a sub-belt of thin, dry, summit warped forest; b) a sub-belt of montane tall-grass meadows.
5. A montane-tundra belt.

The belt of submontane forest-steppe in the territory under study is apparent only in the Southern Zaural'e and partly in the southern districts of the Predural'e. It is characterized by a combination of forest and steppe vegetational elements. The vegetation cover consists of birches and, more rarely, pine forest islands, alternating with meadow-steppe sections. According to the data of A. N. Ponomarev (1949), the forest-steppe floristic complex of the Central Urals includes about 60 species, the great majority of which are calciphytes.

In the montane-forest belts, dark-conifer spruce-fir taiga is dominant, with a southern exposure. Among the spruce forests, dolgomoshnik* spruce and grassy spruce groves are dominant. The spruce forests are characterized by a substantial admixture of Abies sibirica Ldb., which sometimes forms separate plantations, especially in the mountain-ridge zone. Pine forests, characterized by high productivity and purity of stands, develop on the Eastern slopes and in the most low-lying part of the Ural ridge. Pine is usually supplemented by Larix sibirica Ldb., Pinus sibirica Mayr, and Betula pubescens Ehrh., the latter being especially characteristic in southern districts. Pine-whortleberry forests and pine-bilberry forests are dominant in the Central Urals. Sphagnum pine forests and dolgomoshnik pine forests are encountered in the north. Pure P. sibirica and L. sibirica woods almost never occur within the region under study. Massifs of birch and aspen forests are disseminated inside the coniferous forests, their area tending to increase gradually toward the south. Besides the forest vegetation, meadows, bogs, and rock outcrop vegetation (Fig. 1) are distributed in the montane-forest belts.

*Dolgomoshnik—Haircap moss.

Fig. 1. The Grebeshka cliff in the middle of the Chusova course (photo by the author).

At a height of 600-700 m above sea level, the forest thins out, birch begins to appear, and the forest becomes even more transformed into a herbaceous sparsely wooded terrain, intermingled with sectors of high-mountain meadows. This is the belt of thin forests and mountain meadows. In the Central Urals, it is noted only in the north of the territory, where the highest mountain massifs are located. The belt consists of thin woodland with spruce, fir, and birch, and a thick grass cover. The latter consists of mesophytes and hydrophytes, with a small admixture of mountain-tundra species. Close to the summit, in the warped spruce forest, high-elevation meadows develop, being composed of Angelica silvestris L., Archangelica officinalis Hoffm., Cirsium heterophyllum (L.) Hill., Crepis sibirica L., Thalictrum minus L., Valeriana officinalis L., and other plants.

We thus see that the belt under study is characterized by tall herbage, this being favored by the climatic conditions of the surface layer, i.e., an abundance of precipitation, high air humidity, substantial cloudiness, and an early and thick snow cover which excludes the possibility of the ground freezing.

At a height of 800-900 m, rock placers, covered with mosses and lichens, are widely distributed, and they may still retain their coniferous woodland, while further up the mountain-tundra belt begins. In the Central Urals, this belt is best expressed in the Bosega ridge, reaching to an absolute height of almost 1000 m. Mountain-tundra vegetation is also observed on the peaks of the Khariusnaya and Lyalinskii Kamen' mountains.

Thus, the distribution of vegetation in the Central Urals is most strongly affected by the climatic conditions and the associated altitudinal zonation, and also the effect of slope exposure. The vegetation cover can therefore be primarily regarded as one of the important physiognomic features by means of which the territory studied can be divided into regions which are more or less uniform in their complex of climatic factors. This is particularly apparent in differentiating altitudinal belts, of which the lower ones—mountain-forest-steppe and mountain-steppe—are distinguished mainly on the basis of the specific vegetation which is characteristic for them; the vegetation cover also plays a great role in identifying the upper belts—the thin woodlands and tundra woodlands.

This aspect of geobotanical indicator research has still been little analyzed, although there are references to the prospects of this type of indicator work on climatic conditions in papers by several geobotanists who have worked in mountainous regions, notably M. V. Kul'tiasov (1926) and P. D. Yaroshenko (1939). In one form or another, hypotheses on the differentiation of climatically uniform territories on the basis of the nature of the vegetation are also found, for example, in the work of N. K. Vysotskii (1904) and some other geographers. All this evidence confirms that the specific nature of geobotanical research in mountain regions depends in the first place on the fact that the vegetation acquires significance mainly as a climatic indicator.

Besides the climate, relief and, particularly, lithology play a considerable role in the distribution of vegetation. K. N. Igoshina (1944, 1952) mentions that forests of the northern type in intermontane depressions and on acid grounds are displaced toward the south, whereas forests of the southern type on montane slopes and carbonaceous formations go far toward the north. The relationship between the vegetation cover and mountain rocks has been the subject of attention by many investigators. Even at the end of the last century, A. Ya. Gordyagin (1895) noted the great difference in the flora of limestone and granite rock outcrops on the Tura River. After geological investigations in the area of Mount Kachkanar, N. K. Vysotskii (1904) established the association of spruce-fir forests with crystalline paleozoic schists, and of pine forests with olivine rocks. On gabbro and diorites, forests have a mixed character, with Pinus sibirica having wide distribution. The selective role of plants

Fig. 2. The Olenii cliff, located 8 km to the northwest of the village of Kash-ka. Sectors of steppe plants are found on the top and slopes (photo by author).

in relation to different mountain rocks later found some expression in papers by A. N. Ponomarev (1949), K. N. Igoshina (1960), P. L. Gorchakovskii (1960a), and N. M. Gryuner (1960). It was established that carbonaceous and basic rocks favor the preservation, under northern conditions, not only of individual southern species but sometimes even entire vegetational formations. Rich steppe sectors, according to the data of A. N. Ponomarev (1949), are found in the valley of the Chusova River on limestone rocks in the region of the Olenii, Omatnii, Duzhnii, Vysokii, Rostun, and Ermak cliffs (Fig. 2). On the Eastern slopes, an analogous sector of steppe plants was described in the nineteenth century by A. Ya. Gordyagin (1895) at the village of Elkina on the Tura River.

Within specific elevation belts, controlled predominantly by climatic factors, the vegetation may act as a local indicator of a combination of specific geomorphological and hydrogeological conditions. Thus, for example, within the mountain-forest belts, characterized by dominance of dark coniferous taiga, broad, excessively moist depressions are marked by "Sogars"—bog surfaces covered with a thick sphagnum cover and low-growing spruce, fir, and Pinus sibirica. In the southern portions of the Eastern slopes, grassy pine forests and complex pine forests with Tilia cordata are indicators of rich, moist soils. Naturally, the clearest lithological association occurs in the rock complex of the Central Urals vegetation, as it is here that the vegetation is most closely linked with the maternal soil-forming rocks.

The rock complex of the Central Urals, according to N. M. Gryuner's data (1960), carries 131 species, comprising 17% of the total number of species of higher vascular plants recorded in the region. Of these, 56 species inhabit cliff soils exclusively. These are the so-called obligate petrophylls. In the valley of the Chusova River, on the western slopes, the rock complex is represented by 96 species, of which 42 are obligate petrophylls. N. M. Gryuner found in this area 11 species growing on limestone rocks, unknown beyond the limits of the Chusova valley. These are Cotoneaster uralensis Pojark., Carex alba Scop., Epipactis pubigiposa Cranz, Anemone silvestris L., A. biarmiensis Jas., Vicia multicaulis Ldb., Saxifraga sibirica L., Euphorbia gmelinii Steud., Seseli ledebourii G. Dou., Artemisia santolinifolia Turcz., and Cryptogramma stelleri R. Br. In the watershed sector, the rock complex was reduced to 31 species, only 8 of which were obligate petrophylls. In the valley of the Tagil River, on the Eastern slopes, N. M. Gryuner described 105 species of petrophylls (41 species of obligate petrophylls), 10 of which are not found beyond the limits of this valley. They include the following: Melica altissima L., M. transsilvanica Schur., Carex obtusata L., Minuartia helmii (Fisch.) Schischk., Arenaria graminifolia Schrad., Clausia aprica Korn.-Tr., bristly cinquefoil, Campanula wolgensis R. Smien, Linaria debilis Kuprian., and Artemisia armeniaca Lam.

Consequently, as we move from the foothills to the watershed ridge, the rock complex diminishes in size, this being evidently associated with historical, geographical, and edaphic factors.

The richest flora in the Central Urals is that of limestone outcrops, which, according to N. M. Gryuner (1960), have 38 species of petrophylls, compared with 29 species recorded on serpentines, 24 species on syenites, 18 on pyroxenites, 17 on clay shales, 13 on granites, and only 5 on amphibolites.

According to the data of P. L. Gorchakovskii (1960a), rock, Arcto-mountain, and mountain-steppe species are most characteristic for limestone cliffs. The normal rock species include some endemics of the Urals, including Schiverkia kusnetzowii, Agropyron strigosum (M.B.) Boiss., Minuartia helmii, and Asplenium rutamuraria L. The group of Arcto-mountain plants includes the following relics: Dianthus repens Willd., Potentilla kuznetzowii (Turcz.) Juz., Dryas punctata Juz., and Saxifraga caespitosa L. The mountain-steppe species are represented by the following Ural endemics: Dianthus acicularis Fisch. and Scorzonera glabra Kupr.

The association of vegetation with soil and lithological conditions is highly apparent in the river valleys of the Central Urals, as already noted by us earlier (Chikishev, 1960).

Based on the above, it seems feasible to propose the following tentative scheme for employing geobotanical indicator research in mountain regions: The first stage of the investigation should be the study of the altitudinal zonation of the vegetation cover and its relationship to the exposure of the slopes of a given mountain structure with the purpose of using the vegetation as one of the indicators in the complex climatic (and, partly, pedologic) regionation of the territory; the second stage should be the study of the relationship of vegetation with specific combinations of geomorphological, soil-lithological, and hydrogeological conditions in the different altitudinal belts and a compilation of detailed indicator schemes for each of the latter.

There are great prospects for indicator geobotany in connection with the use of aerial methods, particularly on account of color and spectrozonal aerial photography, since the vegetation is converted into a most important deciphering feature.

Indicator features have recently acquired special significance for the diagnosis of plant nutrition, facilitating the determination of the correct type and dose of fertilizers.

Summary

Indicator investigations have up till now been carried out mainly in lowland regions. Attempts to apply the method in mountain areas are therefore of considerable interest. The author has attempted this, using the Central Urals as an example. His studies show that the vegetation of mountain regions is primarily an indicator of climatic conditions. Thus, under the conditions of the Central Urals, climatic altitude zones (especially the two lower ones, mountain-forest-steppe and mountain-forest) are differentiated on the basis of their specific vegetation. Within definite vertical zones, determined by climatic factors, the vegetative cover can be regarded as a local indicator of a combination of specific geomorphological and hydrogeological conditions. Thus, for example, within the mountain-forest zone with a dominance of dark coniferous taiga, vast overmoistened depressions can be clearly discerned from the character of the vegetation, and areas of rich soil can be singled out. Rock species have a very high indicator value.

Based on the above, the author suggests the following scheme of indicator research in mountain regions: stage one—study of the vertical zonation of the vegetation cover, aimed at using vegetation as an indicator in climatic (and, partly, pedological) regionation; stage two—study of the lithological indicator and hydrologic indicator value of individual communities within the vertical zones.

Literature Cited

Vysotskii, N. K. 1904. "Some botanical observations in the Northern Urals." Pochvovedenie, Vol. 6, No. 2.
Govorukhin, V. S. 1960. "Altitudinal zonation of the vegetation of the Urals." In book: Problems of Physical Geography in the Urals. Izd. MOIP.
Gordyagin, A. Ya. 1895. "The vegetation of limestone rocks on the Tura River in Perm' province." Tr. obshch. estestvoispyt. Kazansk. univ., Vol. 23, No. 2.
Gorchakovskii, P. L. 1960a. "On the preservation of relict plants in unique plant communities in the Urals." In book: Nature Conservation in the Urals, No. 1. Sverdlovsk.

Gorchakovskii, P. L. 1960b. "Experience in the botanical and geographical subdivision of the high mountains of the Urals." In book: Problems of Botany, Vol. 5. Moscow—Leningrad. Izd. Akad. Nauk SSSR.

Gryuner, N. M. 1960. "The rock flora of the pretaiga part of the Central Urals as related to petrographic and topographic conditions." Tr. Sverdlovsk. obl. kraeved. muzeya, No. 1.

Igoshina, K. N. 1943. "Relics of broad-leaved cenoses among the fir-spruce taiga of the Central Urals." Bot. zhur. SSSR, Vol. 28, No. 4.

Igoshina, K. N. 1944. "The vegetation of the Central Urals." Sov. botanika, No. 6.

Igoshina, K. N. 1952. "The vegetation of the subalps of the Central Urals." Tr. Bot. in-ta Akad. Nauk SSSR, seriya 30. Geobotanika, No. 8. Moscow—Leningrad, Izd. Akad. Nauk SSSR.

Igoshina, K. N. 1960. "Characteristics of the vegetation of some mountains in the Urals as related to the nature of mountain rocks." Bot. zhur., Vol. 45, No. 4.

Korzhinskii, S. I. 1893. "The flora of the east of European Russia in its systematic and geographic aspects." Izv. Tomsk. univ., book 5.

Krylov, P. N. 1878. "Materials towards a flora of Perm' province." Tr. obshch. estestvoispytat. Kazansk. univ., Vol. 6, No. 6.

Kul'tiasov, M. V. 1926. "Vertical vegetation zones in western Tyan'-Shan." Byull. SAGU, No. 14, 15.

Ponomarev, A. N. 1949. "On the forest-steppe floristic complex of the Northern Urals and the northern part of the Central Urals." Bot. zhur., Vol. 34, No. 4.

Chikishev, A. G. 1960. "The relationship of vegetation with soil and hydrogeological conditions on a terrace of the Chusova River." In book: Problems of Indicator Geobotany. Izd. MOIP.

Yaroshenko, P. D. 1939. "Vegetation as an indicator of soil and climatic conditions in the moist subtropics." In book: Transactions of Young Scientists of the Armenian Branch of the USSR Academy of Sciences. Erevan.

# ON SOME PROCEDURAL FEATURES OF INDICATOR RESEARCH IN THE FOREST ZONE

## N. N. Preobrazhenskaya

At the present time, geologic indicator methods are being used more and more widely in geological and engineering-geology prospecting work, in agriculture, soil surveys, etc. These methods have become most widely used in provinces with an arid climate.

In the forest zone, less indicator work has been carried out, although the fundamental hypotheses on indicator work with mountain rocks on the basis of vegetation did in fact originate from studies on the distribution of forests (Vysotskii, 1904; Karpinskii, 1840; Osokov, 1899, 1909-1912; and others).

The contemporary literature, both Soviet and foreign, already includes fairly comprehensive surveys of the problem of the indicator role of vegetation in regions with a moderate climate (Abramova, 1954; Biské, 1949; Zauér, 1959; Mazing, 1953, 1955; Pankratov, 1958; Apalia, 1957; Krause, 1957; Krüdener, 1950; Mraz, 1959; Sykora, 1959; and others). The contributions of the authors cited contain much interesting material on the use of the vegetation of the forest zone for the evaluation of different environmental conditions and there is unanimous agreement on the great scientific and practical value of the method.

We have collected material for the indication, by means of the vegetation cover, of the mechanical composition of quaternary deposits and the depth of groundwaters in the broad-leaved forest zone. The work was carried out in 1958-1959 and was methodically planned. Some data were kindly supplied to us by E. A. Vostokova, V. I. Turmanina, and N. G. Moskalenko.

Since our task is not to provide a detailed description of the factual results of our work, we propose rather to discuss in the present article some general problems in the procedure of indicator research on forest and meadow landscapes in the broad-leaved forest zone and the recognized features of this procedure, in comparison with the generally accepted methods set out in the existing handbooks (Viktorov, 1955; Vostokova, 1955; Viktorov, Vostokova, Vyshivkin, 1959).

One of the first problems of indicator research in the broad-leaved forest zone, and one of the most important in practice, is the establishment of the limits in depth to which indication of lithological types of deposits is possible.

Of primary importance in this respect is the determination of the depth of penetration of the roots of main species of the landscape. According to the data of A. G. Soldatov (1955), I. S. Matyuk (1953), and some other investigators, the root system of oak grows to a depth of 7-9 m; P. S. Pogrebnyak (1951) shows that in a young pine forest the roots of the trees reach a depth of 4-5 m. In the fissures of rocks, oak roots can even be traced as deep as 10-11 m. In the quarries of Yaroslavl' province, according to the observations of S. V. Viktorov, oak roots penetrated to a depth of 7.5-8.5 m, having pierced through layers of sandy loam and clay and still having a comparatively large diameter at that depth. Thus, direct observations on the root systems provide evidence of the possibility of indicator research being carried out in the top 10-meter layer.

The possibility of indicator research at this depth has also been confirmed by certain indirect considerations. We carried out, for procedural purposes, repeated comparisons of prognoses of the lithological structure of deposits, compiled on the basis of geobotanical data with data from geological and also geophysical studies, and this gave satisfactory agreement of the limits of phytocenoses with the limits of mechanical types of subsoils for the top 8-10 m. Within this range, the vegetation cover reflects quite clearly the presence of particular sublayers or lenses, differing from the general background of the layers located in the plot in respect of their mechanical composition. Changes in the structure of the profile at depths below 10 m did not show any effects on the type of vegetation. It should also be noted that the above refers only to unconsolidated deposits; in places where limestones, slates, and other types of rocks are located, indicator research seems possible only up to the surface of the rocks.

We thus see that indicator research into the mechanical composition of unconsolidated layers in broad-leaved forests is possible almost to the depths previously accepted by other investigators (Viktorov, 1955; Voronkova, 1955) for deserts.

The limits of indicator research of hydrogeological conditions are substantially more restricted than in arid zones. The reasons for this should be sought in the changes in the complex of zonal physicogeographical conditions and, in the first place, the different significance of the moisture factor. Favorable conditions of humidity, in some cases up to the point of excess, result in the dominance of different ecological groups of plants. The deep-rooted phreatophytes, typical for deserts, are very rare in the broad-leaved forests; an analysis which we made of the list of species which have significance as hydrologic indicators in our region showed that almost all belong to the hygrophyte group and have comparatively weakly developed root systems. Thus, the indication of groundwaters by vegetation would be feasible only to a depth of 5 m, even if we take into account the capillary fringe of the groundwaters. The widely distributed hydroindicators in our region include such plant species as Phragmites communis Trin., Menyanthes trifoliata L., Filipendula ulmaria (L.) Maxim., Caltha palustris L., a number of bog species of sedge (Carex vesicaria L., C. gracilis Curt., and others), Molinia coerulea (L.) Moench., Scirpus silvaticus L., Jancus effosus L., Comarum palustre L., Alnus glutinosa (L.) Gaertn., and others.

Turning to an examination of the type of uses of indicators in the broad-leaved forests, it should be noted that more than in other areas there is particular significance in the phytocenological principles of indication, i.e., the use of plant communities as indicators. In arid regions, where indicational research has shown notable development, the significance of individual species as indicators has been quite great. Many of these species, particularly those which have a tendency to form single-species growths (for example, Halosnemum strobilaceum M.B.), have acquired independent indicator significance and, when they are used as indicators, all the various communities formed by them are not taken into account.

Work on the distribution of broad-leaved forests convinced us that the successful utilization of lithologic indicators and hydrologic indicators in these districts is possible only through a comprehensive analysis of the floristic composition of the communities, understanding its sinusoid structure, developmental rhythm, and some other characteristics.

Thus, for example, the appearance of Impatiens noli-tangere L. in the herbage of beech forests, growing on the tops of ridges and mounds, is evidence of the presence of leakage water at a depth of 1-2 m; a detailed analysis of the structure of meadow-bog communities allows the differentiation of the most inundated sectors: they are, for example, demarcated by growths of Menyanthes trifoliata, Phragmites communis, and Calla palustris L., while a substantial proportion of legumes (clover and alfalfa) in the herbage of mixed-grass meadows is evidence of loamy soils; the presence in a pine forest of a well-developed undergrowth (of Corylus avellana, Tilia cordata, etc.) and of broad-leaved species (oak, hornbeam) in the tree stand indicates a hardening of the mechanical structure of the soils, and in some cases also the presence of loam or limestone at a depth of up to 2-3 m below the sands.

It is interesting to note that in a number of investigations the indicators used were not plant communities, but elementary landscape units (of the "physiographic unit" type suggested by Nichols [1917]), i.e., the combination of definite communities with recognized elements of the relief. This broad concept of an indicator can be especially effective in the complex physicogeographical deciphering of aerial photographs, when the investigator is of necessity faced with vegetation in combination with the relief.

The actual procedure of indicator research under the conditions of the forest zone has a number of special features. The first of these is the possibility of the preliminary construction of indicator schemes. Prior to departure for field studies, the concordance of the most widely distributed plant communities, with particular en-

A Portion of a Generalized Preliminary Geologic Indicator Table, Compiled on the Basis of Data in the Literature

| Plant community | Conditions indicated | | Author |
|---|---|---|---|
| | mechanical composition of deposits | depth of bedding of groundwaters, m | |
| Beech forests | Loams, sometimes underlain with limestones; in the Predkarpat'e and Zakarpat'e, on thinnish debris loams, underlain with skeletal rocks | 5 | Vorob'ev, 1953; Gavrusevich, 1958; Grin', 1950; Kozhevnikov, 1931; Kosets, 1947; and others |
| Hornbeam forests | Loams of varying thickness, sometimes underlain with limestones | Usually 5, sometimes 2 (moist and humid types) | Vorob'ev, 1953; Grin', 1950; Kosets, 1953; and others |
| Hornbeam-oak, oak-hornbeam forests | Loams, sometimes sandy loams, with a close under-layer of clays | 2 | Bykov et al., 1936; Zhukov, 1949; Kozhevnikov, 1931, 1939; Kosets, 1953; and others |
| Oak woods | Loams, sometimes sandy loam—loam deposits in floodplain areas | At different depths from 1 to 5 | Vorob'ev, 1953; Kozhevnikov, 1931, 1939; Konovalov, 1948 |
| Oak and hornbeam forests with the inclusion of pine | Sandy loams and sands, underlain at a depth of up to 2 m with loams or with sublayers of loams and clays | The same | Kozhevnikov, 1931, 1939; Lavrinenko, 1954; Shmidt, 1948; and others |
| Pine forests with a substantial proportion of broad-leaved species (oak, hornbeam) | Sandy loams or sands, underlain at a depth of up to 3 m with loams or sublayers of clays and loams | The same | Lavrinenko, 1954; Pogrebnyak, 1944; Shmidt, 1948; and others |

vironmental conditions is elucidated on the basis of a study of the literature, and possible indicators are noted. The material obtained is compiled into a table, which shows plant communities and the conditions to which they correspond.

A preliminary geologic indicator table of this type was compiled on the basis of an analysis of literature on forest types (Vorob'ev, 1953; Grin', 1950; Zhukov, 1949; Kozhevnikov, 1931, 1939; Kosets, 1947, 1953; Pogrebnyak, 1944; and many other papers). We present above a portion of a preliminary indicational scheme, compiled on the basis of data in the literature (see table).

During the process of field investigations, we combined ecological profiling, a description of key standard plots, and indicator mapping. Of leading importance in this context was the complex ecological profiling, consisting of a detailed geobotanical description of the vegetation according to ecological profiles, and taking into account geomorphological, hydrogeological, and geological features of the site.

The length of profiles varied from 250 m to 1 km and more. In the majority of the profiles, a series of hand-drilled bore-holes or exploration pits were sited, by means of which it was possible to obtain material for the characterization of subsoils, soils, and groundwaters. Very detailed geobotanical descriptions were made at the drill-hole points. This was necessary because of the significant variegation of the relief and the great lithological variability of the maternal soil-forming rocks.

In using the method of ecological profiles, we based ourselves on the hypothesis that different soils and plant communities develop on a particular type of rock under different conditions of the relief (Viktorov, 1955,

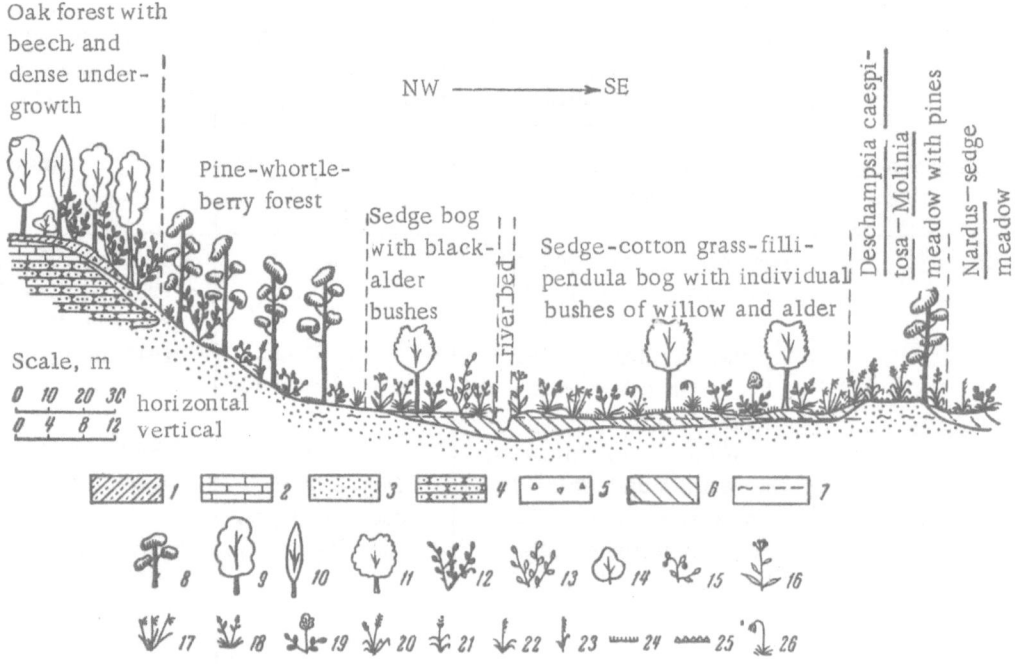

Ecological profile through the floodplain of a river: 1) sandy loam; 2) limestone; 3) sands; 4) sandstone; 5) rock debris; 6) peat; 7) level of groundwaters; 8) pine; 9) oak; 10) beech; 11) black alder; 12) Corylus avellana; 13) willow; 14) Cornus sanguinea; 15) Vaccinium myrtillus; 16) forest bulrush; 17) rush; 18) sedges; 19) oak-leaved filipendula; 20) Deschampsia caespitosa; 21) Molinia coerulea; 22) Nardus; 23) reed; 24) green mosses; 25) Polytrichum; 26) cotton grass.

1960), i.e., a system of biocenoses forms within the field limits of a single rock. These biocenoses taken in conjunction form an ecological-genetic series (Dokhman, 1936), the individual components of which correspond to different stages in the weathering of the particular rock. It is thus possible to establish ecological-genetic series on the basis of the type of soil formation, the depth of groundwaters, etc. Systems of biocenoses arising on different rocks are themselves different.

The task of ecological profiling is thus to establish the pattern of changes in plant communities within the range of a definite rock area (Viktorov, 1955), characterized by recognized soil, lithological, and hydrogeological conditions, and also systems of biocenoses within the areas of different rocks.

A picture of the profiles obtained is given in the figure, which clearly indicates the relationship between vegetation and the mechanical composition of the rocks. On the slope of the ridge, broad-leaved forest (oak-hornbeam-beech), developed on thin sands close to underlying limestones, is quite clearly replaced by pine-whortleberry woodlands, this being associated with outcrops of thick sands. In the floodplain of the river, the close proximity of the water table is responsible for the development of meadow-bog communities. Pines inhabit scattered sandy islands.

Complex profiling revealed the main patterns in the association of vegetation with physical-geographical, particularly geological-hydrogeological conditions.

The vegetation in both profiles and individual key plots is detailed by a series of standard descriptions, representing a detailed description of the vegetation, soils, and subsoils in small plots—sample plots (such descriptions are usually compiled on a separate form). In describing these plots, a possibly more complete study is made of ecological conditions, and in addition significant attention is paid to the detailing of the plant community, the completeness of the floristic list, the study of seasonal aspects, the vigor of individual plants, and so on. Standard descriptions, carried out in large numbers, provided the basic factual material which facilitated a quantitative evaluation of the connection both of communities and individual species with specific environmental conditions.

Geobotanical indicator mapping has as its purpose the elucidation of the degree of coincidence of the limits of individual communities and their complexes with particular lithological types of deposits and specific combinations of hydrogeological conditions. It provides a practical verification of designated patterns. The maps are compiled on the basis of the wide utilization of aerial photography data and are subsequently interpreted for some purpose or other.

Standard descriptions are classified in the office initially on the basis of their phytocenotic character, i.e., all similar communities are grouped together and are analyzed. The indicator validity—the degree of coincidence with the objects of indication—is elucidated for each association or group of associations. On the basis of an analysis of the descriptions, a composite table of communities is constructed with a brief descriptive account and reference to the conditions in which it is encountered. Later, indicator tables, reflecting the indicator role of the vegetation, are compiled. These mention 1) the indicator features (the species composition of the plant community, its outstanding features, the dominant plants, sometimes the indicator plants and the relief), and 2) the conditions indicated by these communities (mechanical composition of the deposits and depth of bedding of groundwaters). Such tables show the indicator role of the vegetation and serve as a recognized key for determining the mechanical composition and the depth of the water table. For verification and increased accuracy of indicators, a comparison was made of indicator tables, compiled on the basis of field data, and data in the literature. In the field, the indicator significance of the communities differentiated was tested in control standard plots: a diagnosis of mechanical composition and the depth of the water table was made on the basis of the vegetation, and this was then tested by a pit or a borehole.

It thus seems feasible to mention now the following special features of geobotanical indicator research in forest regions:

1) substantially narrower vertical limits of hydrologic indicator research (no deeper than 5 m);
2) limitation of the possibility of lithologic indicator prognoses to the top 10-m layer;
3) prominent significance of a complex profiling in research methods
4) necessity for a broad phytocenological approach to the study of indicator communities.

All these features render indicator investigations in forest areas much more complex than in the arid zone, and special procedures need to be elaborated.

## Summary

Indicator research with the mechanical composition of soils and subsoils in the forest zone is possible only to depths no greater than 10 m. Where solid rocks underlie friable alluvia at a depth less than 10 m, the use of indicators to determine mechanical composition is possible only down to the roof of the solid rock. The maximal limits of hydrologic indicator prognoses are even narrower; they do not exceed 5 m.

Indicator investigations in the forest zone should be based on a very careful analysis of floristic lists, because even slight variations in the abundance of a given species may be of great importance in the evaluation of ecological conditions. In many cases, elementary landscape units (of the Nichols "physiographic unit" type) have to be used as indicators, rather than plant communities.

## Literature Cited

Abramova, T. G. 1954. "On the relationship between the vegetation cover of bogs and the structure of the upper layers of peat deposits." Uchenye zap. LGU, seriya biol., No. 34.

Biské, G. S. 1949. "Experience in applying aerovisual observations in a survey of tertiary deposits in Karelia." Izv. Karelo-finsk. fil. Akad. Nauk SSSR, No. 4.

Bykov, P., Dryuchenko, M., Kozhevnikov, P., and Pyatnitskii, S. 1936. Forest Crops in the Forest-Steppe Part of the Ukrainian SSR. Kiev.

Viktorov, S. V. 1955. The Use of the Geobotanical Method in Geological and Hydrogeological Research. Moscow, Izd. Akad Nauk SSSR.

Viktorov, S. V. 1960. "Vegetation as an indicator of lithological and soil-geochemical conditions in deserts." In collection: Problems of Indicator Geobotany. Moscow, Izd. MOIP.

Viktorov, S. V., Vostokova, E. A., and Vyshivkin, D. D. 1959. A Short Guide to Geobotanical Surveying. Izd. MGU.

Vorob'ev, D. V. 1953. Forest Types in the European Part of the Ukrainian SSR. Kiev, Izd. Akad. Nauk UkrSSR.

Voronkova, L. F. 1955. "Experience in using the geobotanical method in compiling a lithological map of old alluvial deposits.", Tr. VAGT, seriya geobot., No. 1. Moscow, Gosgeoltekhizdat.

Vostokova, E. A. 1955. "The application of the geobotanical method in hydrogeological investigations in deserts and semideserts." Tr. VAGT, seriya geobot., No. 1. Moscow, Gosgeoltekhizdat.

Vysotskii, N. K. 1904. "Some geobotanical observations in the Northern Urals." Pochvovedenie, Vol. 6, No. 2.

Gavrusevich, A. N. 1958. "Types of beech forests in the Northern Opol'e and Rastoch'e." Nauchn. tr. Zakarp. lesn. opytn. stantsii, Vol. 1. Uzhgorod.

Grin, F. O. 1950. "Pattern of forest vegetation in Ternopol' province." Bot. zhur. Akad. Nauk UkrSSR, Vol. 7, No. 1.

Dokhman, G. I. 1936. "Experience in the ecological-genetic classification of vegetation in the Ishimsk forest steppe." Byull. MOIP, Nov. seriya, otd. biol., Vol. 45, No. 3.

Zhukov, A. B. 1949. "Oakwoods of the Ukrainian SSR and means of their regeneration." In book: Oakwoods of the USSR, Vol. 1. Moscow.

Zauér, L. M. 1959. "On the problem of using indicator plants in geology." Vestn. LGU, No. 24.

Karpinskii, A. M. 1840. "Can living plants be indicators of the mountain rocks and formations on which they occur, and does the plant habitat justify special attention in geognosy?" Zh. sadovodstva, No. 3, 4.

Kozhevnikov, P. P. 1931. "Forest types and forest associations in Podolia." In book: P. S. Pogrebnyak. Forest Growing Conditions in Podolia [in Ukrainian]. "Vid. n.-docl. in-tu lisiv gosp. ta agrolisomelior.," ser. nauk. vidan; No. 10, Kharkiv.

Kozhevnikov, P. P. 1939. "Oak forests in the forest steppe of the European part of the USSR." Tr. VNIILKh, Vol. 1.

Konovalov, N. A. 1948. "A short guide to the types of oak forest in the central forest steppe." Uchenye zap. Ural'sk. gos. univ., No. 4. Sverdlovsk.

Kosets', M. I. 1948. Beech forests in Western Podolia [in Ukrainian]. Bot. zhur. Akad. Nauk UkrSSR, Vol. IV, No. 3-4.

Kosets', M. I 1953. "A survey of the forest vegetation of L'vov Province, Ukrainian SSR." [in Ukrainian]. Bot. zhur. Akad. Nauk UkrSSR, Vol. X, No. 4.

Mazing, V. V. 1953. "On methods for the study and use of vegetation as an indicator of the effect of drying and other changes in the environment." In book: Jubilee Collection of the Society of Naturalists [in Estonian; summary in Russian]. Tallin.

Mazing, V. V. 1955. "Experience in determining the degree of drying in bog forests on the basis of the type of vegetation." Tr. Inst. lesa Akad. Nauk SSSR, Vol. 31. Moscow—Leningrad, Izd. Akad. Nauk SSSR.

Matyuk, I. S. 1953. "The effect of soil conditions on the development of the root systems of tree and shrub species." Pochvovedenie, No. 5.

Ososkov, P. A. 1899. "The distribution of low-chalk, iron-containing rocks in the region of the Zasurskii forests." Materials Toward an Understanding of the Geological Structure of the Russian Empire, No. 1. Izd. MOIP.

Ososkov, P. A. 1909-1912. "Relationship of forest vegetation to the geological composition of indigenous rocks." Lesnoi zhur., part 39, Nos. 2-5, 8-9 (1909); part 41, Nos. 3-4 (1911); part 42, Nos. 4-5 (1912).

Pankratov, Yu. A. 1958. "Some patterns in the location of forest types in the kame landscape of the Karelian isthmus." Lesnoi zhur., No. 6.

Pogrebnyak, P. S. 1944. Principles of Forest Typology. Kiev, Izd. Akad. Nauk UkrSSR.

Pogrebnyak, P. S. 1951. "Investigations of soils and root systems in the forests of the Poles'e in the Ukrainian SSR" [in Ukrainian]. Pratsi inst. lisivnitsva, Vol. 2. Kiev.

Soldatov, A. G. 1955. Root Systems of Tree Species. Kiev, Gossel'khozizdat UkrSSR.

Shmidt, V. É. 1948. Forest Crops in the Main Types of Forest. Moscow—Leningrad, Goslesbumizdat. Photo-tipografiya Goslestekhizdata.

108

Apalia, D. 1957. Rytu Lietuvos banguoto fliuvioglaciolinio keiminio reljefo plotu angalijos indicatorinis savybes [in Lithuanian]. Tr. Akad. Nauk Lithuan. SSR, B3 (11).

Krause, W. 1957. "Pflanzengesellschaften als Anzeiger der Standortbedingungen." Umschau, No. 3.

Krüdener, A. 1950. "Forstlicher Standortanzeiger. Auslese zum Gebrauch im Walde." B. Neumann.

Mraz, K. 1958 (1959). Vegetaini typy joko pomůcka při mapovani kvarternich pokryvnych ůtvarů a zvětrali-onových plaśtu. Anthropozoicum, Vol. 8.

Nichols, G. 1917. "The interpretation of certain terms and concepts in the ecological classification of plant communities." Plant World.

Sykora, Z. 1959. Rostliny v geologickem výzkumu. Praha.

# EXPERIENCE IN THE UTILIZATION OF FOREST VEGETATION
## AS AN INDICATOR OF DEPOSITS
## OF THE GLACIAL COMPLEX

### N. G. Moskalenko, L. N. Tagunova, and V. I. Turmanina

In geological and engineering-geological mapping in regions where glacial deposits are located, there is considerable interest in obtaining accurate information on the distribution limits of morainic and fluvio-glacial layers. However, the delimitation of different types of glacial deposits is frequently difficult, especially in afforested districts. In some cases, the difficulties of mapping can be overcome by using geobotanical indicational methods.

The present authors investigated the northeastern part of Moscow province and the south of Yaroslavl' province for a period of two years with the aim of studying the relationship between the vegetation cover and the lithological types of deposits in the glacial complex. The geobotanical indicational investigations were the first carried out in these areas. Some data are however presented in the general botanical literature for these areas on the association of plant communities with soil-forming rocks of differing mechanical composition. Thus, many investigators mention the association of pine forests with sandy deposits, and of spruce and oak forests with a loamy surface or morainic deposits (Alekhin, 1925, 1947; Konovalov, 1929; Kravchenko, 1953; Lyubimova, 1957; Smirnov, 1940; Flerov, 1898, 1902; Shakhanin, 1945, and others).

We used the normal procedure for this type of work (Viktorov, 1955a, b; Vostokova, 1953, 1955) in our investigations. The procedure adopted includes detailed descriptions of the vegetation, accompanied by the sinking of pits or drill holes, ecological profiling, and the compilation of geobotanical maps. In view of the fact that geological surveys were being carried out on the same territory as the geobotanical indicator investigations, the authors had the opportunity of comparing geobotanical and geological data.

The greater part of the regions where the work was done comprises a humpy outwash plain (absolute elevation 135-155 m), included within the Meshchersk and Nerl'sk-Klyaz'minsk lowlands. The height of individual hillocks does not exceed 10 m. They alternate with numerous bog hollows, and in localities where these develop further, wide bog massifs are developed. Tertiary deposits are represented by inequigranular sands of fluvioglacial origin. In the regions studied in Moscow province, the thickness of the fluvioglacial deposit layer reaches 20-30 m. Sublayers and lenses of clays and loams of lacustrine-bog origin are found here among the sands.

A hillocky relief is typical for a substantial section of the areas studied in Yaroslavl' province. The relative elevation of the watersheds is 20-40 m. The highest hillocks reach 50-70 m. This is a region where morainic deposits have developed, represented primarily by loams; of course, the morainic ridges are composed of loams, alternating with sands and gravel-pebble material. Only small outcrops of moraines are found within the territory studied in Moscow province. Morainic deposits only emerge on to the surface at the village of Bol'shie Zherebtsy.

In the regions studied, there is contact between the sloping hillocky morainic plain and the slightly hummocky outwash-plain areas; the relief shows significant leveling and therefore the extremities of deposits of different types, taking into account only geological-geomorphological conditions, are difficult to establish.

Under these circumstances, the extremities of different deposits can be clearly traced from the vegetation cover. The nature of the vegetation can also be used to evaluate changes in the thickness of fluvioglacial deposits, which lie under the washed-out moraine.

Investigations showed that different types of pine forests, and localities with a substantial admixture of birch, are associated with fluvioglacial deposits. In districts of moraine development, pine forests are found only on bog sectors. The distribution of different associations of pine forests is associated with specific geomorphological and hydrogeological conditions.

The most elevated parts of the outwash plain (the tops of dune hummocks and osars, and the upper sections of their slopes), where groundwaters occur at a depth of 5-10 m, are occupied by pine forests with Antennaria dioica (L.) Gaertn., Hieracium pilosella (L.), Calamagrostis epigeios (L.) Roth., and Calluna vulgaris (L.) Hill. Patches of lichens (Cladonia rangiferina (L.) Web., C. silvatica (L.) Hoffm., C. alpestris (L.) Rabh., and Cetraria islandica (L.) Ach.) occur on the soil cover.

The central parts of the slopes are covered with pine forests, with mountain cranberry and green mosses (Pleurozium schreberi (Willd.) Mitt., Hylocomium splendens Hedw. and Dicranum undulatum Ehrh.). Groundwaters lie here at a depth of 3-5 (up to 7) m. The mountain cranberry is dominant in the grassy-shrub layer. Green mosses frequently form a solid cover.

Pine forests with bilberry and green mosses are associated with the lower parts of the slopes and the flat sections of the outwash plain, where loam sublayers occur in the sandy deposits or the sands are underlain by loams. The forests grow under conditions where the water table is rather close to the surface (1.5-2, sometimes 3 m). Spruces frequently occur within the tree stands of these forests. The presence of spruce is evidence of the close bedding of rocks of heavy mechanical composition, either underlying sands at a depth of 1-3 m, or forming sublayers in sand deposits.

Pine forests with Vaccinium myrtillus L., V. uliginosum L., and Polytrichum commune L. develop on depressed sectors of the outwash plain, where the groundwaters occur at a depth of 0.5 to 1 (2) m.

Pine forests with bog-type shrubs and sphagnum grow on the margins of bog basins. The grass-shrub layer is dominated by Ledum palustre L., Chamaedaphne calyculata (L.) Moench., Vaccinium uliginosum L., and Eriophorum vaginatum L. Various species of sphagnum moss are dominant in the soil cover.

The distribution of pine forests in relation to changes in geological and geomorphological conditions and humidity conditions are illustrated by a profile dissecting the watershed sector and valley of the Kupavenka River (Fig. 1, profile 1).

Spruce + narrow-leaved forests and narrow-leaved + spruce forests, with an admixture of broad-leaved species and elements of oakwood grasses in the grassy cover, are associated with moraine deposits in the regions studied in Yaroslavl' province. These forests have a secondary character and develop in the place of completed fellings of spruce + broad-leaved and broad-leaved + spruce forests. Spruce-birch forests and spruce-aspen forests, with an admixture of pine and individual oaks, predominate in regions where terminal moraines develop, and have a markedly disintegrated ridge-hummock relief and a variegated lithological structure (alternation of boulder loams, sands, and sandy-gravel material). Groundwaters occur here at great depths (10-15 m). An undergrowth of hazel (Corylus avellana L.) shows good growth in these forests. The grass cover is formed by Calamagrostis arundinacea Roth., Rubus saxatilis L., and oakwood grasses, such as Aegopodium podagraria L., Asarum europaeum L., Galeobdolon luteum Huds., and Stellaria holostea L.

Spruce + narrow-leaved forests and narrow-leaved + spruce forests, excluding pine, develop in the lowest morainic deposits, consisting of loams (covered in places with thin sandy loams).

The well-drained slopes of morainic hillocks are covered mainly with birch-spruce and spruce-birch forests. The undergrowth usually includes hazel, Lonicera xylosteum L., and Sorbus aucuparia L. The grassy cover consists of various species of oakwood grasses, the dominant species being Asarum europaeum, Galeobdolon luteum, and Aegopodium podagraria, with Carex pilosa Scop. being dominant in the driest habitats. Spruce-aspen forests, with an admixture of Alnus incana (L.) Moench., are located on the solid tops and lower parts of

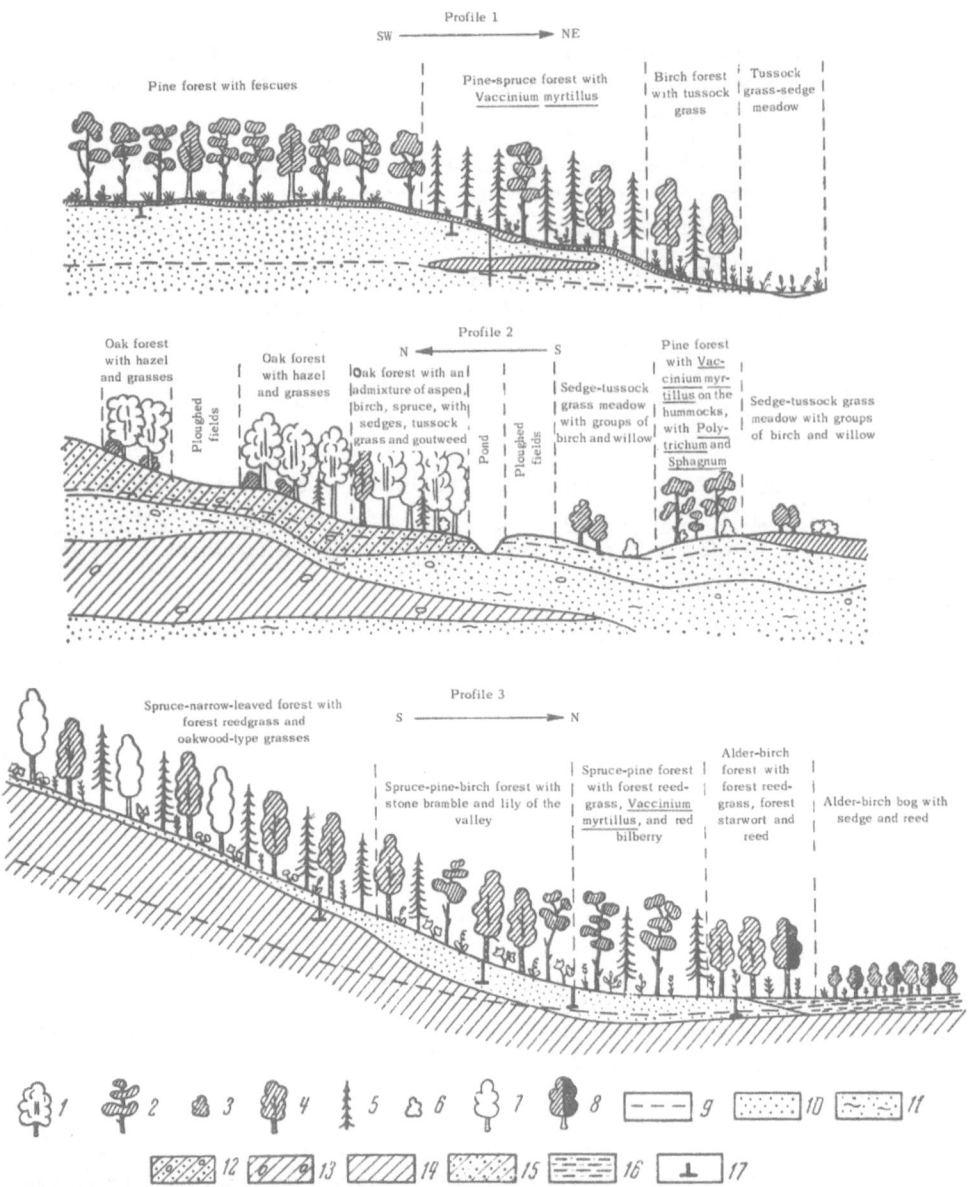

Profile 1

SW ⟶ NE

Pine forest with fescues    Pine-spruce forest with *Vaccinium myrtillus*    Birch forest with tussock grass    Tussock grass-sedge meadow

Profile 2

N ⟵ S

Oak forest with hazel and grasses    Ploughed fields    Oak forest with hazel and grasses    Oak forest with an admixture of aspen, birch, spruce, with sedges, tussock grass and goutweed    Pond    Ploughed fields    Sedge-tussock grass meadow with groups of birch and willow    Pine forest with *Vaccinium myrtillus* on the hummocks, with *Polytrichum* and *Sphagnum*    Sedge-tussock grass meadow with groups of birch and willow

Profile 3

S ⟶ N

Spruce-narrow-leaved forest with forest reedgrass and oakwood-type grasses    Spruce-pine-birch forest with stone bramble and lily of the valley    Spruce-pine forest with forest reedgrass, *Vaccinium myrtillus*, and red bilberry    Alder-birch forest with forest reedgrass, forest starwort and reed    Alder-birch bog with sedge and reed

Fig. 1. Distribution of the vegetation cover in relation to the mechanical composition of soil-forming rocks and the level of subsurface waters: 1) oak; 2) pine; 3) hazel; 4) birch; 5) spruce; 6) goat willow; 7) aspen; 8) black alder; 9) level of groundwaters; 10) sands; 11) sands with a seam of clays; 12) loams, interspersed with boulders; 13) boulder loams; 14) loams; 15) sandy loam; 16) peat; 17) boreholes.

the slopes of hillocks, where drainage is difficult and there is a perched water table. There develops an undergrowth of varied species composition: hazel, *Daphne mezereum* L., mountain ash, *Euonymus verrucosa* Scop., *Padus racemosa* (Lan.) Gilib., and other species. *Aegopodium podagraria*, *Aconitum excelsum*, and *Equisetum silvaticum* L. are dominant in the grassy cover.

Finally, oak forests are associated with outcrops of morainic loams at the village of Bol'shie Zherebtsy (Fig. 1, profile 2). An undergrowth of hazel shows good development in these forests. The oaks have normal regeneration, evidence for this being the viable regrowth. Species of oakwood grasses (*Carex pilosa*, *Galeobdolon luteum*, *Asarum europaeum*, and other species) are dominant in the grass cover.

In the course of our investigations, we succeeded in tracing the changes in plant communities in relation to changes in the thickness of the mantle of fluvioglacial deposits overlapping the moraine. A. G. Trutnev (1937) terms this stratification of tertiary deposits of differing genesis bipartite drift. He mentions the wide distribution of bipartite drifts in the northwestern portion of the European part of the USSR.

In regions of morainic development overlapped by fluvioglacial deposits more than 1 m in thickness, spruce-pine forests with Vaccinium myrtillus, forest reedgrass, and green mosses are located. Where the thickness of the fluvioglacial deposits does not exceed 1 m, the spruce-pine forests are replaced by spruce + small-leaved forests, with the inclusion of elements of oakwood grasses in the grass cover.

We shall examine profile 3 (see Fig. 1) for an illustration of the patterns in vegetation distribution in relation to the thickness of sands of fluvioglacial origin, underlain by morainic loams.

Spruce + small-leaved forest with oakwood grasses develops on the slope of morainic hummocks, where, according to data from boreholes, sands are underlain with morainic sands at a depth of 0.7 m. Lower down the slope, the thickness of fluvioglacial deposits exceeds 1 m; spruce-pine-birch forest with Rubus saxatilis and Convallaria majalis grows here, while under conditions of a closer water table this forest is replaced by spruce-pine forest with forest reedgrass, Vaccinium myrtillus, and mountain cranberry. Where there is a further reduction in the depth at which groundwaters are bedded and in the thickness of sands, the spruce-pine forest is replaced by aspen-birch forests with reedgrass, reed (Phragmites communis Trin.), Stellaria nemorum L., and Aegopodium podagraria. Under conditions where groundwaters are bedded closest to the surface (0-0.5 m), an aspen-birch bog develops, with sedges and reed.

The regular pattern in the distribution of plant communities in relation to the thickness of sand deposits, underlain by loams, is clearly demonstrated by the data in the table.

Association of Plant Communities to Specific Mechanical Types of Glacial Deposits (as % of the total number of descriptions of the particular community)

| Plant communities | Sands of substantial thickness | Sands underlain with loams and clays at a depth of >1 m, or sands with seams of loams and clays | Sandy loams more than 1 m in thickness, underlain with sands | Thinnish sands, underlain with loams | Loams |
|---|---|---|---|---|---|
| Pine forest with Antennaria, Festuca ovina and Hieracium pilosella . . . . . . . | 87 | 13 | — | — | — |
| Pine and birch-pine forest with mountain cranberry . . | 80 | 20 | — | — | — |
| Spruce-pine and birch-pine forest with inclusion of spruce, Vaccinium myrtillus and reedgrass . . . . . . | 17 | 70 | 9 | 4 | — |
| Spruce + small-leaved forest with infrequent pine, reedgrass and mixed grasses. . . | — | 87 | — | 13 | — |
| Spruce + small-leaved forest with reedgrass and oakwood grasses. . . . . . . . . . . . | 8 | 33 | — | 59 | — |
| Spruce + small-leaved forest with oak and oakwood grasses. . . . . . . . . . . . | — | 10 | — | 10 | 80 |

Fig. 2. Geobotanical indicator map (A): $I_1$) pine and birch-pine forest with mountain cranberry and reedgrass; sands 5 m in thickness, groundwaters at 3 m; $I_2$) spruce-pine forest with Vaccinium myrtillus and green mosses; sands 1-3 (5) m in thickness, underlain with loams or with seams of loams, and groundwaters at 1-3 m; $I_3$) boggy birch-pine forests with cotton grass, bog shrubs, and sphagnum, peat 0.5 m in thickness, underlain with sands; groundwaters at 0.5-1.0 (1.5) m; $II_1$) spruce-pine forests with individual oaks and mixed grasses; sands 1-2 (3) m in thickness, underlain with loams; groundwaters at 5 m; $II_2$) pine-spruce-small-leaved forests with an admixture of oak, hazel, and oakwood grasses; loams, overlain with thinnish sands (up to 1 m); groundwaters at 5 m; $III_{1,2}$) alder woods; peat up to 1 m in thickness or muddy and gleying sands; groundwaters at 0-1 (1.5) m; $IV_{1,2,4}$) pine, alder-birch, and birch bogs; peat from 0.5 to 5-7 m in thickness; water level at 0-0.5 (1) m; stagnant moisture; $IV_3$) sedge-reed bogs; peat from 0.5 to 5-7 m in thickness; groundwaters at 0.5 m; flowing moisture stream; $V_{2,3}$) tussock grass-mixed gass and mixed grass-sedge meadows; sands, frequently muddy and gleying, sometimes in the process of peat formation.

Map of tertiary deposits (B): $1 + hQ_{IV}$) lacustrine-bog deposits; peat, peat-forming sands and sandy loams; $alQ_{IV}$) alluvial floodplain deposits; sands inequigranular, in places muddy and gleying, peat; $lQ_{IV}$) lacustrine deposits of a present-day lake terrace; $eoQ_{IV}$) eolian deposits: sands small- and medium-grained; $alQ_{III}$) alluvial deposits of territory above the floodplain: sands inequigranular, in places with the sparse inclusion of pebbles and gravel,

or peat; $lQ_{III}$) lacustrine deposits of an ancient lake terrace: sands fine-grained, uniform; $kfrmlg\ Q_{III}$) lacustrine-glacial deposits, forming kames; sands inequigranular, gravel, pebbles; $fgl\ Q_{III}$) fluvioglacial deposits of outwash-plain fields: sands inequigranular, in places with gravel and pebbles; $gl\ Q_{II}$) glacial deposits of the bottom moraine: loams with gravel, pebbles, and boulders.

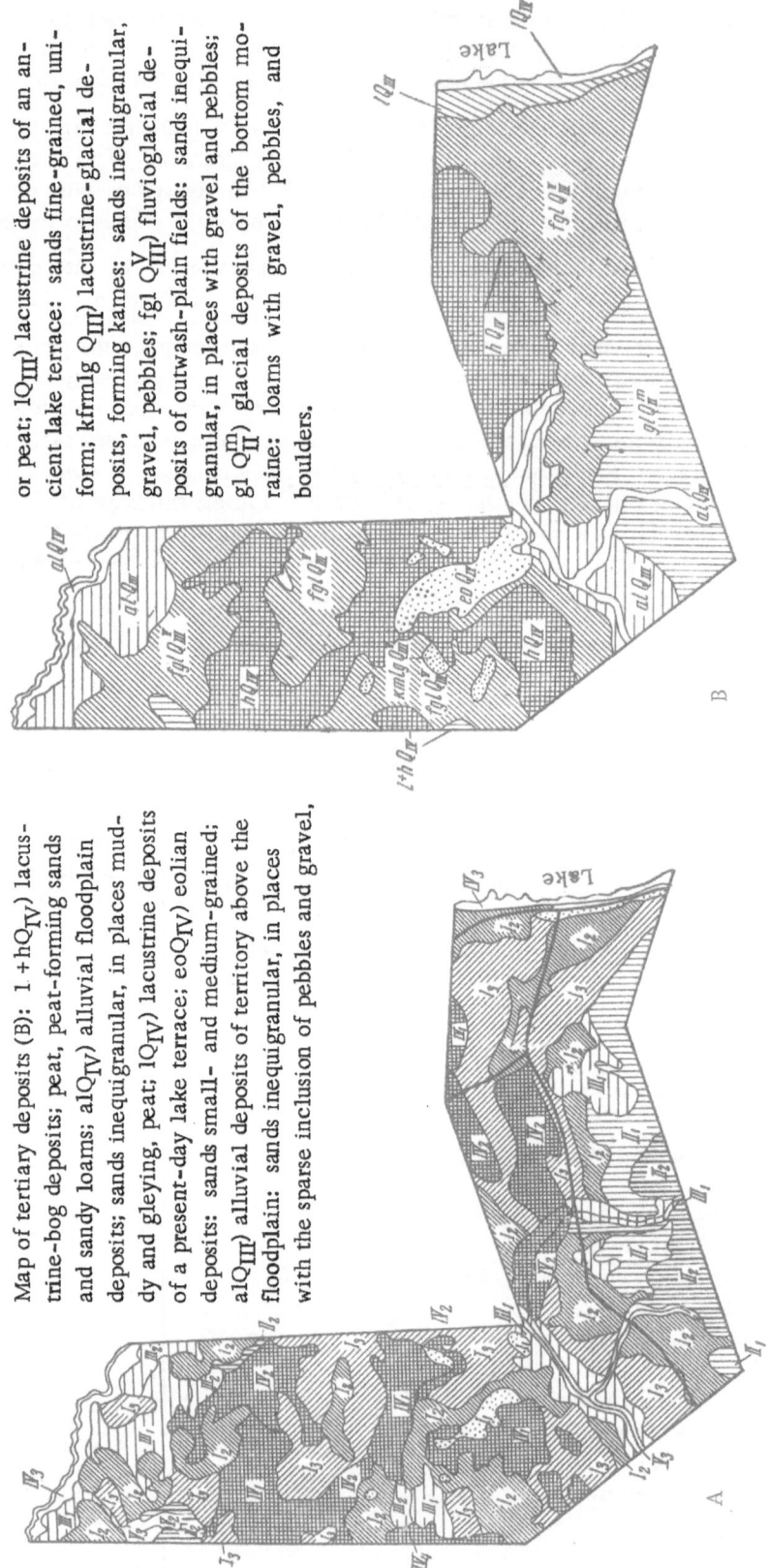

114

The table, which was compiled on the basis of an analysis of numerous detailed geobotanical descriptions, shows that pine forests with Antennaria, Festuca ovina L., and Hieracium pilosella, and pine forests and birch-pine forests with mountain cranberry are most frequently distributed on sands of fluvioglacial origin of substantial thickness. On sands with seams of loams or underlain by loams at a depth of more than 1 m, the most frequent communities are spruce-pine, birch-pine and pine forests with spruce, with Vaccinium myrtillus and forest reedgrass in the grass cover, and spruce + small-leaved forests with pine, and a grassy cover made up of forest reedgrass, mixed grasses, and V. myrtillus. Spruce + small-leaved forests with a few oaks, and a grassy cover made up of forest reedgrass, mixed grasses, and elements of oakwood grasses, show an association with loams, covered with sands up to 1 m in thickness; in exceptional cases, spruce + small-leaved forests with oak, a grassy cover made up of oakwood grasses, occur on loams.

The possibility of using the geobotanical indicator method in the mapping of genetically different types of deposits of the glacial complex was confirmed by an analysis of the geobotanical indicator map which we compiled and maps of tertiary deposits (Fig. 2). A comparison of these maps shows that it is possible, on the basis of geobotanical data, to find the distribution areas of various types of glacial deposits. Thus, for example, different associations of pine forests are typical for fluvioglacial deposits; spruce-pine forests with Vaccinium myrtillus and mixed grasses are located on thinnish (1-3 m) fluvioglacial deposits, underlain with moraines; a thin mantle of fluvioglacial deposits is evidenced in the landscape only through the nature of the vegetation cover; spruce + small-leaved forests with oak and platyphyllous grasses are confined to morainic deposits, overlaid with fluvioglacial deposits up to 1 m in thickness.

Contours on geobotanical and geological maps do not coincide completely; thus, for example, on the basis of geobotanical survey data, the distribution of lacustrine-bog deposits and peat ground could be traced more precisely (forest and grassy bogs—$III_{1,2}$ and $IV_3$, boggy forests—$I_2$).

It thus seems possible to utilize the vegetation cover as an indicator of genetically different types of deposits of the glacial complex, and this assists the identification of glacial deposits of differing mechanical composition in geological mapping.

Summary

While studying the forests of Moscow and Yaroslavl' provinces, the authors concluded that areas with different types of glacial deposits can be identified on the basis of the type of the vegetation cover. Various pine associations are typical on fluvioglacial deposits. Spruce and pine forests with bilberry and mixed grasses occur on shallow (1-3 m) fluvioglacial deposits underlain by moraines. Such areas can be recognized in a locality from the character of their vegetation covers, without any drilling or mining operations. Finally, spruce + small-leaved forests with an admixture of oak and platyphyllous grasses are confined to morainic debris (occasionally overlaid by fluvioglacial deposits no less than 1 m deep). These observations have been used in geological mapping.

Literature Cited

Alekhin, V. V. 1925. "The flora and vegetation of the Moscow area." In collection: The Moscow Territory. Moscow.

Alekhin, V. V. 1947. The Vegetation and Geobotanical Districts of Moscow and Neighboring Provinces. Izd. MOIP.

Viktorov, S. V. 1955a. The Use of the Geobotanical Method in Geological and Hydrogeological Research. Moscow—Leningrad, Izd. Akad. Nauk SSSR.

Viktorov, S. V. 1955b. "A brief survey of the history of the development and present-day position of the geobotanical method in geology." Tr. VAGT, No. 1. Moscow, Gosgeoltekhizdat.

Vostokova, E. A. 1953. Vegetation as an Indicator of Geological and Hydrogeological Conditions in the Semi-Desert and Desert in Connection with their Reclamation. Candidate Dissertation, Moscow.

Vostokova, E. A. 1955. "The application of the geobotanical method in hydrogeological investigations in deserts and semi-deserts." Tr. VAGT, No. 1. Moscow, Gosgeoltekhizdat.

Konovalov, N. A.1929. "Types of forest in experimental forests of the Podmoskov'e." Tr. po lesnomy opytnomu lesu (TsLOS), No. 5.

Kravchenko, B. A. 1953. The Forests of Moscow Province. Moscow—Leningrad.

Lyubimova, E. L. 1957. "A survey of the vegetation of the natural regions of Moscow province." Tr. Inst. geografiya Akad. Nauk SSSR, Vol. 21, No. 2.

Smirnov, P. A. 1940. "The flora and vegetation of the central industrial region." Materials Toward an Understanding of the Fauna and Flora of the USSR. Izd. MOIP, nov. seriya, otd. botaniki, No. 1 (9), Moscow.

Trutnev, A. G. 1937. "On bipartite drifts." Pochvovedenie, No. 4.

Flerov, A. F. 1898. "A survey of the vegetation of Pereyaslavsk district, Vladimir province." Protokoly Mosk. obshch. ispyt. prirody, Nos. II—III.

Flerov, A. F. 1902. The Flora of Vladimir Province. Moscow.

Shakhanin, N. I. 1945. Botanical and geographical characteristics of Yaroslavl' province. Uchenye zap. Yaroslavsk. gos. ped. inst., No. 6.

# PROSPECTS FOR GEOLOGIC INDICATOR RESEARCH
# IN THE INSULAR FOREST STEPPE
# OF CENTRAL SIBERIA

## E. L. Lyubimova and N. A. Khotinskii

Hardly any geologic indicator investigations have as yet been carried out on the vast territory of Siberia. There are thus great prospects in the establishment of investigations on these lines. Of particular interest in this respect are the forest-steppe and steppe basins of the southern part of Central Siberia. In this area, under conditions of a sharply continental climate, a complex geological structure and substantial site variability, the vegetation cover clearly reflects particular environmental conditions.

The basins represent depressions of the indigenous relief, filled up with a thick layer of loose sediments of differing composition and age, and bearing varying vegetation.

Preliminary data obtained by us permit the differentiation, with a sufficient level of accuracy, of Jerassic and Devonian rocks[*] from the specific features of the vegetation cover. Within the Kansk and Krasnoyarsk forest steppes, these rocks are the most characteristic, and frequently emerge at the surface, showing marked difference in lithology.

Jurassic deposits are widely distributed in these regions. Outcrops of continental carboniferous Jurassic rocks, represented by light grey, white, and yellowish sandstones and sands, sand-clays and sand- gravel layers, carry steppe vegetation. True dry, shallow-turfed steppes are developed. The grass cover of the steppes is low (the average height of the grass layer is 20-30 cm) and sparse (projective cover, 40-60%). The steppe vegetation is dominated by Festuca pseudovina Hack., Koeleria gracilis Pers., Stipa decipiens P. Smirn., Artemisia frigida Willd., Potentilla acaulis L., Veronica incana L., and small steppe sedges (Carex duriuscula C.A.M. and C. korschinskyi Kom.). The majority of the species belong to the true steppe xerophytes species with Mongolian-Siberian and Mongolian-Daur areas of distribution playing a considerable role in the composition of cenoses, together with European species, as is clearly shown in the table.

Of particular interest for geologic indicator investigations are the characteristic mountain and-desert-type steppes, which occur on surface outcrops of strongly carboniferous marlaceous rocks, comprising a Devonian seam which is red or variegated in color. The best-known Devonian outcrops occur in the basins of the Kacha Rivers (west of Krasnoyarsk) and Kan Rivers (in the region of the town of Kansk).

As regards appearance, species composition, ecology, and origin, the steppes on Devonian outcrops differ sharply from the steppe vegetation associated with outcrops of Jurassic strata. The vegetation cover of the stony, desert steppes is very sparse, the plants forming only isolated cushions, while the projective covering is 20-30% on the average. All the plants are low-growing with thick roots, and are frequently in a pillowlike form. Even the normal steppe plants acquire clearly expressed xeromorphism. For example, Thalictrum foetidum L. is only 5-8 cm in height, and has small, very hairy, bluish-violet leaflets, whereas normally this plant reaches a height of 15-30 cm, and has slightly hairy, dull green leaves.

---

[*]Outcrops of Permo-Carboniferous and Cretaceous rocks were encountered rather rarely and their identification from geologic indicator data requires additional data.

## Distribution of Plants in Relation to Steppe Types

| Steppe types | Cover, % | Average height, cm | Number of steppe and dry-steppe species, % | Number of mountain and mountain-steppe species, % | Number of desert species, % | Number of Siberian-Mongolian and Daur-Mongolian species, % | Number of European and Eurasian species, % |
|---|---|---|---|---|---|---|---|
| Stony desert steppes on outcrops of red and variegated Devonian strata | 20—30 | 8—15 | 47 | 46 | 7 | 81 | 19 |
| True shallow-turfed steppes on outcrops of continental Jurassic rocks | 40—60 | 15—30 | 84 | 16 | — | 70 | 30 |

The following can be singled out among the plants of these steppes: several mountain and mountain-steppe species, for example, Stellaria cherleriae (Fisch.) Williams, Gypsophila patrinii Ser., Leontopodium campestre (Ldb.) Hand.-Maz., Patrinia sibirica Juss., and many others. These plants form the main group of species, in both number and abundance (see table). In addition, we encounter plants typical of saline soils, as well as desert species, for example, Ephedra monosperma C.A.M. and Eurotia ceratoides (L.) C.A.M., which impart an uncharacteristic appearance to these cenoses.

The third group of plants includes such true steppe and dry-steppe species as Agropyron turczaninowii Drob. and Artemisia frigida Willd., their actual number being quite considerable, though the majority of the plants are found very infrequently and are sparsely distributed.

The role of Mongolian-Daur and Mongolian-Siberian elements (more than 80% of the total number of species) shows a still greater increase in the species composition of the stony steppes. The number of European and Eurasian species shows a clear decrease as compared with the true shallow-turfed steppes.

The peculiar stony desert steppes within the Krasnoyarsk and Kansk forest steppes are also closely linked with surface outcrops of red or variegated rocks of the Devonian group.

On watersheds or slanting slopes, where the indigenous rocks are covered with strata of quaternary loess-type loams, the influence of the indigenous rocks on the vegetation cover decreases in proportion to an increase in the thickness of the covering deposits. This relationship is clearly traced on the scheme which we have compiled (see figure). Stony desert steppes are developed in areas where Devonian deposits are located on rock outcrops. Wormwood-feather grass steppes, made up of Stipa decipiens P. Smirn. and Artemisia glauca Pall., occur on slanting slopes, covered with, loess-like loams up to 1.5 m in thickness, while watersheds, where the average thickness of tertiary deposits is 2-4 m, are populated with meadow steppe.

It is possible to differentiate shallow-turfed steppes on rock outcrops, wormwood-feather grass steppes on slanting slopes, and meadow steppes in watersheds, all on Jurassic deposits under similar conditions.

We thus see that loess-type and covering loams, even of rather small depth, moderate the influence of indigenous deposits on the vegetation cover.

Plant cenoses associated with surface outcrops of rocks therefore have the greatest geologic indicator significance.

In addition to Jurassic deposits, it is sometimes also possible to discover areas where saliferous Cambrian deposits are bedded, by means of the vegetation cover. Thus, sites are encountered with solonchak vegetation in the Kansk depression, north of the Kan River, in valleys and lacustrine basins (the southern shore of Lake

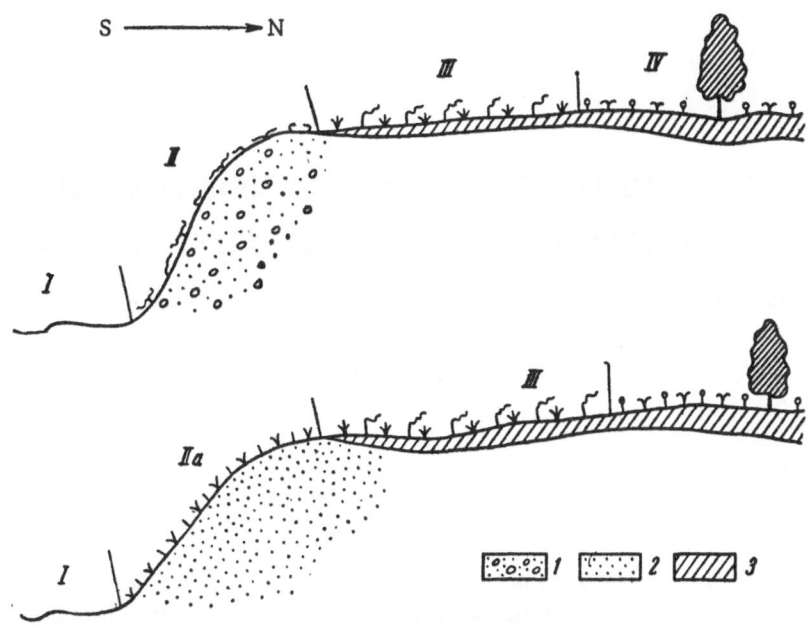

Scheme of the distribution of the vegetation cover on outcrops of Devonian
and Jurassic rocks:  I) river valley; II) stony desert steppe; IIa) true shallow-
turfed steppe; III) wormwood-feather grass steppe; IV) plowed fields on the
site of meadow steppes in a complex with birch groves.  1) Variegated Devonian stratum; 2) pale-yellow and pale-grey sands and friable Jurassic sandstones; 3) loams.

Ulyu-Kol', and the valleys of the Taina, Antsyr', and Kurysh Rivers).  According to the data of the salt division
of the Krasnoyarsk joint expedition of the SOPS, these places have salt sources or boracic waters with a high
content of salts (up to 3-4%).  Sodium chloride or calcium chloride predominate within the salt complex of
the majority of these waters.

The outlets of saline waters are confined to zones of dislocation of stratum fissures and are associated
with saliferous horizons of Cambrian deposits.  In spite of their depth of bedding, these rocks are clearly marked
by the vegetation.  Thick solonchaks appear, with growths of succulent saltworts, Salicornia herbacea L., and
various species of annual saltworts, and in autumn these form clearly marked crimson or red patches amongst
the greenish-brown steppe vegetation.  Substantial areas are occupied by groupings of wormwood scrubs, made
up of Artemisia nitrosa Web.; in some places we find huge growths of Lasiagrostis splendens (Trin.) Kunth.,
almost attaining the height of a man.  The vegetation has a desertlike character.

We thus see that the occurrence of solonchak vegetation in the Kansk forest steppe is caused by the salinization of the particular sectors with highly mineralized sodium chloride and calcium chloride waters, whose
genesis is associated with deeply bedded Cambrian deposits.

The information obtained from reconnaissance work in the forest steppes of the Soviet Union demonstrated
the great possibilities from applying the indicator method for studying the characteristics of the physical-
geographical environment of these regions and for solving a range of practical problems.

Summary

The authors studied forest-steppe plots in the Kansk and Krasnoyarsk regions.  It was found that stony
desert steppes are confined to Devonian red and variegated beds, while true shallow-turfed steppes are
associated with continental carboniferous Jurassic deposits.  Saliferous Cambrian rocks in the Kansk depression
are indicated by halophyte communities (Salicornia herbacea, Suaeda spp.) and Artemisia nitrosa communities.
Halophytes occur in combination with Lasiagrostis splendens near salt-water crops on lines of tectonic disturbances.

119

# THE RELATIONSHIP BETWEEN VEGETATION
# AND ROCK LITHOLOGY IN THE DALDYNSK REGION
# OF THE YAKUTIAN ASSR FROM AERIAL PHOTOGRAPHS

## N. L. Zagrebina

The application of aerial-survey data in studying various elements of the natural complex is opening up great opportunities for using the vegetation as an indicator of geological structure and, in particular, rock lithology.

This particularly affects the forest zone of our country, where the vegetation has for long been considered an obstacle in using aerial photographs in geological research. It is for precisely this reason that, up to 1938-1939, there was no experience in applying aerial photography in geological research (Miroshnichenko, 1946).

Unfortunately, up to the present time, aerial photographs in the forest zone have been used by geologists only to provide a topographical basis.

It is true that in some contributions, especially the handbooks on the conduct of aerogeological research, the need to use the vegetation as a demarcating feature has been mentioned (Krasnov, 1954; Petrusevich, 1954; and others). But in order to utilize the vegetation as a demarcating feature in a successful way, the geologist should set out to clarify the relationships of vegetation with geological structure.

In view of the complexity of the factors governing the distribution of vegetation, ecologists and geobotanists should be attracted to geological work, since they may assist in identifying the relationship between vegetation and rocks within this complex chain (Viktorov, 1947).

Joint work of this type was carried out by the Yakutian expedition of the Laboratory of Aerial Methods of the USSR Academy of Sciences under the direction of V. M. Barygin in 1955-1956. The work was methodical in character and was linked with prospecting for indigenous diamond deposits.

The region where the expedition operated is located in the northeastern part of the Siberian platform and occupies the basin of the central part of the course of the Daldyn River.

Its geological and geomorphological structure has been studied in detail, and appears in papers by L. N. Zveder (1958), E. N. Kornutova (1959), I. I. Krasnov and V. L. Masaitis (1955), and others.

There is information on the vegetation cover in an article by V. B. Sochava (1957). Some aspects of the association between vegetation and relief and rocks have been discussed in papers by N. L. Zagrebina (1960, 1960a) and A. N. Lukicheva (1960).

The rocks which make up the region can be divided into three groups: normal sedimentary, pyroclastic, and igneous (Saarsadskikh and Popugaeva, 1955).

The most widely distributed among these are rocks of the sedimentary complex, the dominant types being limestones, marls, and dolomites, with almost horizontal bedding of the strata. The pyroclastic rocks include, somewhat provisionally, the kimberlites, which are represented by serpentinic eruptive breccia. The igneous rocks in the region covered are predominantly olivine varieties of dolerites.

These rocks differ markedly in their physical and chemical properties, and this is clearly reflected in their patterns of weathering. At high latitudes, traps are most stable against weathering, predominantly due to frost, and this can be explained by the fact that they always develop definite relief forms, while their eluvial and eluvial-deluvial deposits are most frequently present as clumpy placers, among which silt and gruss sectors are found. Kimberlites are the least stable toward weathering processes, this being due to their breccial texture. Rocks of the carbonate stratum occupy an intermediate position in this respect, and it is also possible to differentiate within this stratum bands of more or less stable rocks, leading to a steplike pattern in the horizontal bedding conditions. In spite of the marked differences between all the rocks which make up the region, there are very clear differences between rocks of the trap formation and rocks of the sedimentary complex in the types of vegetation. These differences appear in the following aspects: varied vigor of a species on different rocks under similar relief conditions; association of individual plant species with specific rocks; quantitative inclusion of a species within the structure of a particular plant community; and in the structure of the vegetative cover.

Some of these differences can be established in aerial surveys and allow the utilization of the vegetation as a deciphering feature for particular rocks.

We shall now deal in somewhat greater detail with each of the above-mentioned differences in the vegetation cover on traps and on carbonate rocks.

The differences in vigor of species apply in the first place to the principal forest species of this region, Larix dahurica Turcz. As a rule L. dahurica on carbonate rocks, irrespective of its position on the relief, is represented by markedly depressed forms, 8-12 m in height, and with weakly developed crowns and dry, warped tops. Larch forms of this type form a sparse forest type of vegetation (Norin, 1961) on carbonate rocks, with a crown canopy of 0.1-0.4 and a distance of 20-40 m between trees.

Larches on traps, markedly broken down by denudation processes which are usually confined to the central parts of the slopes of ridges, attain a height of 14-16 m, with a crown canopy of 0.8-0.9 and a distance of 4-10 m between trees. These differences in the structure of the tree layer are clearly reflected in aerial photographs.

As regards certain plant species being confined to definite rocks, it was noted that Dryopteris iragrans (L.) Schott., Linnaea borealis L., and Potentilla inquinans Turcz. occur only on traps. On the other hand, Cystopteris fragilis (L.) Berh., Dryopteris robertiani (Hoffm.) C.Christens, and Woodsia glabella R.Br. seem to be associated with rocks of the sedimentary stratum.

Although such differences in the vegetation cover are not directly reflected in aerial photographs, they can give real assistance in work on the ground.

It is possible to cite quite a number of examples of different quantitative proportions of individual plant species included in the structure of communities on carbonate and igneous rocks. Thus, Ledum palustre L. is dominant in the shrub layer of larch forests and thin forests on traps, while Vaccinium uliginosum L. predominates in thin larch forests on carbonate rocks with close bedding of the indigenous rocks. In the soil layer of larch forests on traps, a green-moss layer made up of Rhytidium rugosum (Hedw.) Kindb., Ptilium cristacastrensis (L.) Ne Not., and Pleurozium schreberii (Willd.) Mitt. develops. The shrubby lichens Cladonia silvatica (L.) Hoffm., C. alpestris (L.) Raben., and C. rangiferina (L.) Web. dominate in the soil layer in thin larch forests on carbonate rocks with close bedding of the indigenous rocks. In thin larch forests on rather thick drift beds of carbonate rocks, complexity in the soil layer is observed, with the lichens being predominant on the elevated parts of the microrelief and green mosses (Tomenthypnum nitens (Schreb.) Loeske, Aulacomnium turgidum (Wahlend.) Schwaegr., and A. acumitatum (Lind. et Arn.) Par.) developing on the depressed parts of the microrelief.

The predominance of lichens in the soil cover of thin larch forests on carbonate rocks imparts a bright tone to photographic images, and this is a good feature for deciphering sedimentary strata (Fig. 1).

The differences in the structure of the vegetation cover are evident because on the carbonate rocks, as opposed to the traps, the vegetation cover is represented by regularly alternating fragments of associations. This is caused by the regular alternation of soil and ground conditions, associated with the specific relief forms characteristic for the carbonate stratum, namely the structural denudation scarps and run-off hollows (dells).

The differences in habitat conditions on traps and sedimentary rocks lead in the final analysis to the fact that certain associations are located on igneous rocks and other associations on sedimentary rocks. Thus, for

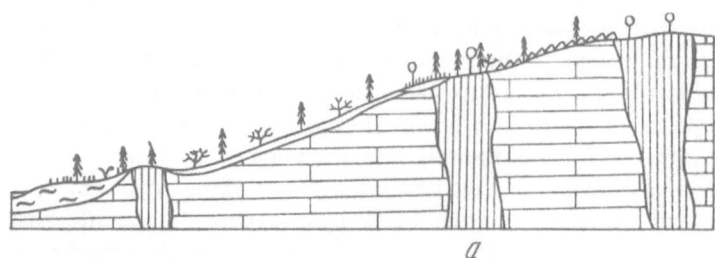

Fig. 1. Relationships of vegetation with traps. a) Trap intrusions.

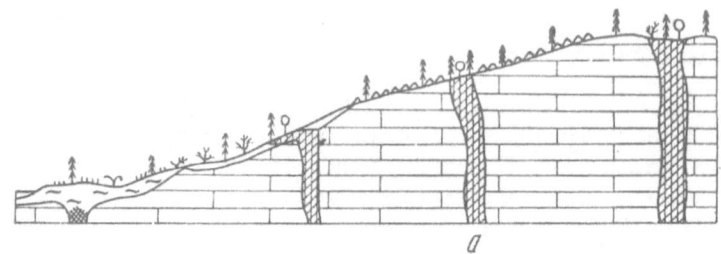

Fig. 2. The association of vegetation with kimberlites. a) Kimberlite bodies of rock.

example, on trap intrusions confined to the watershed surface of a ridge, shrub growths develop, made up of Betula middendorfii Trautv. et Mey, Alnaster fruticosus Ldb., and Salix kolymensis O. v. Semm., with the inclusion of Larix dahurica in the form of bushes. On carbonate rocks encircling such intrusions, thin forests of larch, Vaccinium uliginosum, and lichens develop, with driads and small sedges. Depressed larch—Ledum palustre thin forest, with inclusion of spruce, is developed on the clumpy placers of traps in the lower parts of ridge slopes. A thin moss-lichen larch forest with the shrubs Betula exilis Sukacz., Arctous alpina Niedenz, Vaccinium uliginosum L., as well as other species, develops on sites surrounding intrusions within the carbonate-rock area.

There thus appear to be clear associations of vegetation with two types of rock, sedimentary and igneous. Using these relationships, traps can be quite easily differentiated from carbonate rocks (Fig. 1) on the basis of the type of vegetation on aerial photographs, the tone of the photographic image, and its configuration, irrespective of their position on the relief.

The problem of the relationship of vegetation with the third type of rock, the kimberlites, has already received adequate attention in the literature (Zagrebina, 1961; Kobets, 1960; Lukicheva, 1960), and we shall therefore merely mention that the clearest association of vegetation with kimberlites occurs under circumstances where the kimberlite body of rock emerges to the surface with its eluvium, as appears possible when it occurs in the central part of the slope of a ridge with southern or southwestern exposure (Fig. 2).

We should mention in conclusion that it is possible to assess the horizontal bedding of rocks of the sedimentary stratum from the character of the distribution of the vegetation in this region. As already mentioned above, different bands of carbonate rocks react differently to weathering. Bands of rocks which are less stable toward weathering are broken down more rapidly, while the more stable rocks form small escarpments.

On the rims of these escarpments, there is a bushy larch undergrowth with a crown canopy up to 0.6, which differs sharply form the surrounding thin larch forests with a crown canopy of 0.2-0.3.

The close association of the bushy larch growths with the rims of escarpments produces a characteristic stripiness on aerial photographs, the strips being parallel to one another. It is feasible to decide, on the basis of this parallelism, on the horizontality of the rocks forming the carbonate stratum (Fig. 2a).

The possibility is not excluded that a more detailed study of the region will show up other relationships, too, reflecting some or other features of the geological structure.

Summary

The author has studied a number of cases where various rocks are demarcated by the vegetation in the basin of the Daldyn River. It was, for example, found that thickets of Betula middendorfii, Alnaster fruticosus, Salix colimensis, and Ledum palustre are confined to trap intrusions, while thin forests of Larix dahurica are associated with carbonate rocks surrounding these intrusions.

Literature Cited

Viktorov, S. V. 1947. "Biological indicators in geology." Usp. sovr. biologii, Vol. 23, No. 2.

Zagrebina, N. L. 1960. "Deciphering vegetation on aerial photographs in the diamond-bearing regions of north-western Yakutia." In collection: The Application of Aerial Methods in Prospecting for Indigenous Diamond Deposits. Moscow, Izd. Akad. Nauk SSSR.

Zagrebina, N. L. 1960a. "On the relationship of vegetation to geomorphological and geological structure in the basin of the central part of the course of the Daldyn River." Tr. Labor. aerometodov Akad. Nauk SSSR, IX. Moscow–Leningrad, Izd. Akad. Nauk SSSR.

Zagrebina, N. L. 1961. "The landscape method of discovering kimberlite bodies (using the northern taiga landscape of Western Yakutia as an example)." In collection: The Use of Aerial Methods in the Investigation of Natural Resources. Moscow–Leningrad, Izd. Akad. Nauk SSSR.

Zveder, L. N. 1958. "Some features of kimberlite formations in the north of the Siberian platform." Izv. Sib. otdel. Akad. Nauk SSSR, No. 4.

Kornutova, E. N. 1959. "The application of aerial methods in geological-geomorphological research in the basin of the Vilyui River." In book: Materials on the Geology and Geomorphology of the Siberian Platform.

Kobets, N. V. 1960. The Application of Aerial Methods in Prospecting for Kimberlite Bodies in the Daldynsk Diamond-Bearing Region of the Yakutian Region. Moscow–Leningrad, Izd. Akad. Nauk SSSR.

Krasnov, I. I. 1954. "The application of aerial methods in geological surveying." In collection: Handbook of Procedures in Geological Surveys and Exploration (VSEGEI). Moscow, Gosgeoltekhizdat.

Krasnov, I. I., and Masaitis, V. L. 1955. "Tectonics of the Olenek-Vilyui watershed in connection with the structure of the marginal zones of the tungus syneclise." In book: Materials on the Geology of the Siberian Platform. Moscow, Gosgeoltekhizdat.

Lukicheva, A. N. 1960. "The vegetation cover as an indicator of kimberlite cores." Geologiya i geofizika, No. 1.

Miroshnichenko, V. P. 1946. Aerial Surveying. Moscow–Leningrad, Gosgeolizdat.

Norin, B. N. 1961. "What is Forest Tundra?" Bot. zhur., Vol. 46, No. 1.

Petrusevich, M. N. 1954. Geological Survey and Exploration Work on the Basis of Aerial Methods. Moscow, Gosgeoltekhizdat.

Saarsadskikh, N. N., and Popugaeva, L. A. 1955. "New data on the occurrence of ultrabasic magmatism in the Siberian platform." Razvedka i okhrana nedr, No. 5.

Sochava, V. B. 1957. "Taiga in the northeastern part of the Central Siberian Highlands." Bot. zhur., Vol. 42, No. 9.

# THE USE OF THE GEOBOTANICAL METHOD
# FOR THE DIFFERENTIATION OF ELEMENTS OF RIVER VALLEYS
# IN THE NORTHERN PART OF YAKUTIA

## L. S. Demidova

In the summer of 1961, the joint division of the All-Union Aerogeological Trust (VAGT) investigated a region comprising the basins of a number of rivers in the northern part of Yakutia.

The geobotanists of the expedition were faced with the task of establishing indices of different river-valley levels on the basis of plant associations.

In this context, we shall present some preliminary data on the vegetation as an indicator of elements of valleys of differing age.

Sectors of the low-lying floodplain of a river valley were studied, and could be divided into the following four zones according to the predominant vegetation:

Zone I is formed by the sector of the low floodplain near the riverbed.  It is characterized by the presence of thinnish groupings of Chamaenerium latifolium (L.) Th. Fr. et Lange and Roegneria jacutensis (Drob.) Nevski.  Perennially frozen ground is present here at a substantial depth, the substrate consisting of coarse gravel.

Zone II occupies the part of the low floodplain which is drenched at high water.  It is characterized by the development of associations of Salix jacutica + Vicia cracca + Equisetum palustre, Salix jacutica + Equisetum palustre, or a brome association (of Bromus sibiricus Drob.).  Perennially frozen ground is developed at a depth of one meter in this zone.  The soils are most frequently sandy.

Zone III is formed by the central part of the low floodplain and is characterized by the development of herbaceous mixed-grass willow (Salix jacutica) groves.  Perennially frozen ground occurs at a depth of 50 cm in this part of the floodplain.  The soils are sandy loams or loams.

Zone IV, comprising the terrace part of the low floodplain, is clearly differentiated by the presence of mixed-grass willow groves.  Perennially frozen ground in this section of the low floodplain occurs at a depth of 50-60 cm, as in the third zone.  The soils are usually floodplain, lamellar meadow soils.

An analysis of the above indicates that in different parts of the floodplain there develop associations where the edificators are different species of willow.  Thus, Salix jacutica Nas. is dominant in the part of the floodplain near the riverbed, S. hastata L. in the central part, and S. pyrifolia Ldb. in the terrace part.

In addition to the indicator role of the above-mentioned willow groves in which they act as indicators of different levels of a low-lying floodplain, they can also be used as indicators of the thickness of friable alluvial deposits and the underlying coarse gravel, and this can have great significance in prospecting for diamond escarpments.  Thus, the Salix jacutica —Vicia cracca—Equisetum palustre association is an indicator of 2-m thickness of the friable deposits and an almost 4-m thick coarse-gravel bed.  The herbaceous mixed-grass willow forest is located on 6-7-m thick deposits, underlain by coarse gravel 1.2 m in thickness.  The thickness of friable deposits under the mixed-grass willow forest fluctuates from 4 to 5 m, while the thickness of coarse gravel is equal to 2 m.

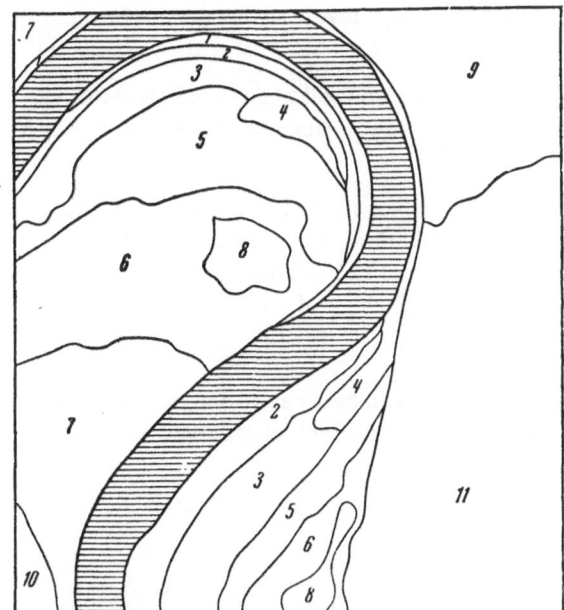

Scheme for a geobotanical map of a sector of the valley of the Syungyuyudé River, compiled on the basis of aerial photograph deciphering. Low floodplain: 1) thin grouping of Roegneria and Chamaenerium latifolium; 2) Vicia cracca—Equisetum; 3) herbage—mixed grass willow forest; 4) mixed grass willow forest; High floodplain: 5) green-moss larch forest; 6) complex of associations of green-moss larch forest, thin green-moss larch forest, and thin larch forest with Ledum palustre and lichens; 7) alder-mountain cranberry larch forest; 8) sedge bog; 9) complex Erica—Ledum palustre and cotton grass—sedge associations; 10) complex of thin larch forest, with lichen—Ledum palustre, sphagnum—sedge, and cotton—grass associations. Terrace I.

This variability in the thickness of friable deposits and coarse gravel in different communities of a low-lying floodplain is explained by the differing periods of high water.

The vegetation permits the differentiation not only of different levels of a low floodplain but also allows the clear distinction of a low floodplain from a high one, which can have particular significance if there are no contrasts in the relief. Thus, a high floodplain differs markedly from a low one by the occurrence on the former of Larix dahurica. Where there are loam or clay soils, and perennially frozen ground occurs at at depth of 20-40 cm, there develops a thin forest of larch, Ledum palustre, and lichens. Where sandy-loam or sandy soils occur on a high floodplain (in which case perennially frozen ground is encountered at a depth of 50-70 cm), a Larix dahurica—Equisetum scirpoides + Arctous erythrocarpa association can be found.

Other frequently found associations are Eriophorum angustifolium—Carex caespitosa and larch with green mosses (Ptilidium ciliare (L.)Hampe., Aulacomnium palustre (L.) Schwgr., Camptothecium nitens Schor., Aulacomnium trugidum (Wahlend.) Schwaegr., and Hylocomium splendens Br. eur.).

It is possible to differentiate those areas which are relatively young and those which are relatively old by means of the vegetation on a high floodplain, as on a low floodplain. Thus, all associations of non-boggy larch woods (Larix—horsetail, larch—Arctous, larch—green moss, etc.) are indicators of a comparatively young section of a high floodplain. But these sections of a high floodplain may in turn be younger than the part of the valley where the following complex of associations is located: a) cotton grass—sedge and b) thin larch forest with shrubs; bog areas are widely distributed in such areas. Finally, the oldest section of a high floodplain will be the part of the territory where cotton grass—sedge or willow—sedge bogs are developed exclusively.

Communities in a high floodplain can be deciphered just as well in aerial photographs as vegetation zones of varying age in a low floodplain (see figure).

The association of the vegetation with the age of territories can also be clearly traced on terraces. Thus, the following associations were noted: Ledum palustre—mountain cranberry larch forest with the inclusion of mosses and lichens, Vaccinium uliginosum—green moss larch forest, thin sphagnum—sedge larch forest.

The second terrace is characterized by the development only of a normal or thin larch forest, while the second layer is always dominated by shrubs (Vaccinium uliginosum, mountain cranberry, Ledum palustre, Empetrum nigrum).

We thus see that shrubs and sphagnum mosses join together as high floodplain edificator plants in associations found on the first terrace. The vegetation of the second terrace is characterized by the complete dominance of shrubs.

All the above-mentioned associations are thus indicators of different age levels in the valleys of the Yakutian rivers studied.

It should be emphasized that, as in other geobotanical indicator work, we used plant associations rather than individual plant species as indicators of the age of elements of river valleys, since vegetation units taken aggregately give a more complete reflection of habitat conditions.

It must again be mentioned that all the associations described are shown up well in aerial photographs. This assists geologists and geomorphologists to distinguish different floodplain levels and terraces of differing age, which are difficult to differentiate in nature. And this in turn constitutes helpful material in prospecting for diamond escarpments, whose location is closely linked with the geomorphology of river valleys.

## Summary

In the river valleys of Central Yakutia, a marked relationship is observed between different plant communities and different levels of floodplains and terraces. This facilitates geomorphological observations.

# THE ASSOCIATION OF VEGETATION WITH
# THE THICKNESS OF ELUVIAL AND DELUVIAL DEPOSITS
# ON THE USTYURT PLATEAU AND THE MANGYSHLAK PLAIN

## D. D. Vyshivkin

In describing the plant cover of districts with outcrops of stony rocks, many researchers have noted the association of individual plant groupings or communities with these outcrops or with the thin, slightly developed soils on them.

However, in all such work, hardly any concrete data on the thickness of friable deposits under particular communities are given, except for cases where there is close bedding of solid rocks or the emergence of rocks directly to the surface.

Because of this, we have made an attempt to study to what extent vegetation is associated with eluvial and deluvial deposits of differing thickness. In order to exclude the effect of other factors as much as possible, the work was carried out in areas which are quite similar in physical-geological conditions—the Ustyurt plateau and the lowland part of the Mangyshlak peninsula, the surfaces of which are covered with Neogenic limestone shell rocks.

In 1926, the soil and plant cover of the territory under investigation was studied by B. A. Borneman and M. A. Spiridonov (1929), G. I. Dolenko (1930), and F. P. Rusanov (1930). Their published reports include many concrete description of the vegetation and drill holes, and we were able to use some of this data in the present study. We present below the principal results of our observations.

Sheer slopes, composed of limestone shell rocks, oölitic limestones, and Neogenic marls, are almost deprived of vegetation. Only in a few crevices can one find individual specimens of Ephedra and other pioneer chasmophytes.

Vegetation does develop on limestone outcrops with a moderate slope, but the vegetation is very sparse, the ground cover being less than 5%. The plants in this case occur both in fissures and in the thin silt which in places covers the limestone outcrops. Zygophyllum macropterum C.A.M., Lasiagrostis caragana Trin. et Rupr., Teucrium polium L., Zygophyllum turcomanicum Fisch. et Mey., Nanophyton erinaceum (Pall.) Bge., and Anabasis truncata (Schrenk) C.A.M. may be encountered on limestone outcrops. Shrubs such as Atraphaxis spinosa L. and Convolvulus fruticosus Pall. are also quite well represented, while Salsola arbuscula Pall. and Eurotia ceratoides (L.) C.A.M. are rarer. The development of Atraphaxis spinosa and Convolvulus fruticosus on limestone outcrops in border scarps of Ustyurt is also reported by S. V. Viktorov (1951, 1955).

In certain cases, trees may also grow on such habitats, including Rhamnus sintenisii Rech. and Morus sp., although they are never found where there is a more or less thick horizon of friable deposits.

On the more saline sectors of limestone outcrops, groupings of Salsola chiwensis M. Pop. may develop, together with Anabasis truncata and isolated bushes of A. salsa. A close vegetation cover almost never develops where the eluvial deposit is thin (less than 50 cm).

With greater thickness of eluvial and deluvial deposits, the plant cover usually closes up more or less and forms clearly indentifiable plant communities. Two successional series may be noted: one for the more salinated habitats, and the other for the more demineralized habitats. An Artemisia formation develops on the more demineralized habitats, and an Anabasis salsa formation on the saline habitats.

Relationship of Plant Communities with the Thickness of Eluvial and Deluvial Deposits on a Plateau
Composed of Neogenic Limestone

| Plant communities | Thickness of eluvium, cm | | | | Indicated thickness, cm |
| | number of descriptions, % | | | | |
| | 0-50 | 50-125 | 125-200 | more than 200 | |
|---|---|---|---|---|---|
| Pioneer groupings of chasmo-phytes and shrubs. . . . . . . . . | 10*⁄(100) | — | — | — | 0-50 |
| Anabasis salsa formation (as a whole). . . . . . . . . . . . . . . | 3⁄(8.7) | 18⁄(51.5) | 9⁄(25.8) | 5⁄(14) | 50-200 |
| Wormwood formation (as a whole). . . . . . . . . . . . . . . | — | 20⁄(32.8) | 20⁄(32.8) | 21⁄(34.4) | More than 125 |
| A. salsa without ephemerals. . . . . . . . . . . . | 2⁄(15.4) | 8⁄(61.5) | 3⁄(23.1) | — | 50-125 |
| A. salsa with the inclusion of Eremopyrum orientale . . . . . . | — | 5⁄(31.2) | 10⁄(62.5) | 1⁄(6.3) | 125-200 |
| A. salsa with inclusion of ittsegek. . . . . . . . . . . . . . . | — | — | — | 4⁄(100) | More than 200 |
| A. salsa with A. truncata. . . . . . . . . . . . . . | 4⁄(28.6) | 8⁄(57.1) | 2⁄(14.3) | — | 0-125 |
| Shrub-wormwood community. . . . . . . . . . . . . | — | 4⁄(100) | — | — | 50-125 |
| Salsola rigida—wormwood community. . . . . . . . . . . . . | — | 8⁄(100) | — | — | 50-125 |
| Agropyron sibiricum—wormwood community. . . . . . . . . . . . . | — | 2⁄(28.6) | 3⁄(42.8) | 2⁄(28.6) | 50-200 |
| Wormwood association . . . . . . . | — | 1⁄(6) | 8⁄(47) | 8⁄(47) | More than 125 |
| Anabasis aphylla—wormwood community. . . . . . . . . . . . . | — | — | 3⁄(50) | 3⁄(50) | More than 125 |
| Kochia prostrata—wormwood community. . . . . . . . . . . . . | — | 1⁄(12.5) | 1⁄(12.5) | 6⁄(75) | More than 200 |
| Salsola arbuscula community. . . . . . . . . . . . | — | 2⁄(28.6) | 4⁄(57.1) | 1⁄(14.3) | 50-200 |
| Nanophyton erinaceum community. . . . . . . . . . . . | 1⁄(11.1) | 4⁄(44.5) | 3⁄(33.3) | 1⁄(11.1) | 50-200 |

*Numerator—number of descriptions; denominator—frequency of particular community.

If we examine these formations as a whole, no clear relationship between them and specific thicknesses of eluvial and deluvial deposits is evident, but all the same as one notes that the thickness of eluvial and deluvial deposits under A. salsa populations is less than under wormwood. Thus, the thickness of eluvia under A. salsa in some cases does not exceed 50 cm, and thicknesses in excess of 200 cm are observed only in 14% of the cases. Wormwood however was never found on plots where the eluvial thickness was less than 50 cm, and it was observed in 34.4% of the cases where the thickness of the eluvial and deluvial stratum exceeded 200 cm (see table).

If we examine the associations included within the structure of these basic formations, their association with a specific thickness of eluvial and deluvial deposits becomes completely clear.

On sectors of the plateau which are covered with a thin eluvial and deluvial stratum, pure A. salsa populations develop, a characteristic feature being the very weak development also of Eremopyrum orientale (L.) Jaub. et Spach in the community. Two-thirds of the plots occupied by pure A. salsa growths studied were characterized by an eluvial thickness of 50-125 cm, although the maximum depth of bedding of limestone was noted at 170 cm. Strong evidence of gypsum formation is usually observed on these plots even at shallow depths (about 50 cm), and in places interlayers or even horizons of gypsum are found.

Substantial development of ephemerals, particularly Eremopyrum orientale, is noted in A. salsa growths on eluvia of greater thickness. The limestone platform in such cases is usually bedded no closer than 90 cm down, and in a few cases it is not uncovered even by a 2-m-deep drill hole. The thickness of friable deposits most frequently (62.5% of cases) fluctuates from 125 to 200 cm.

Areas populated by A. salsa with the inclusion of Anabasis aphylla or other plants with long roots are found comparatively rarely. The thickness of eluvia in such areas exceeds 2 m.

Communities and groupings of Anabasis truncata are also located in saline habitats (apart from pure A. salsa growths), where there are thinnish eluvia.

Very frequently, groupings of A. truncata with a substantial inclusion of a corky grey lichen (genus Acarospora), with a cover reaching 80-90%, occupy strongly gypsiferous soils, the so-called bozyngens (term proposed by G. I. Dolenko). The abundance of A. truncata in such areas is particularly noted, other higher plants being virtually absent on such habitats. Bozyngens are distributed as small patches within an Anabasis salsa formation. There are few lichens among the A. truncata groupings along the slopes. Isolated shrubs and some herbaceous plants are encountered. The thickness of eluvia under all A. truncata communities and groupings does not usually exceed 125 cm.

On less saline habitats, an increase in the thickness of friable deposits is indicated by a substantial increase in the abundance of grey wormwood (Artemisia terrae-albae Krasch.) and grasses. Thus, according to the data of G. I. Dolenko (1930) and B. A. Borneman and M. D. Spiridonov (1929), feather-grass groupings with the inclusion of Lasiagrostis caragana and wormwoods, develop on slopes where the thickness of silt does not exceed 1 m, and is frequently less than 50 cm. Stipa barata, Agropyron sibiricum (Willd.) P. B., and A. fragile are dominant in these groupings. The prevalence of wormwoods (Artemisia fragans Willd. and A. terrae-albae Krasch.) usually does not exceed a few specimens.

Where the thickness of eluvial-deluvial deposits is somewhat greater (from 50 to 125 cm), a shrub-wormwood community occurs. The shrubs include Convolvulus fruticosus, Atraphaxis spinosa, and more rarely, Eurotia ceratoides. Agropyron sibiricum and Stipa sareptana grow occasionally. Artemisia has the abundance level of sor[1].* A shrub—Artemisia community is frequently located on sinkholes with leached soils.

An association of Artemisia with the inclusion of Salsola rigida Pall. is associated with sites with close bedding of limestones (from 50 to 125 cm), possibly connected with the somewhat greater salinization of these sites.

With an increase in thickness of friable deposits, shrubs completely disappear, apart from Salsola arbuscula, which in conjunction with wormwood forms an independent group of communities.

Where the eluvial and deluvial deposits are of moderate thickness (50-200 cm) the wormwood populations show a considerable abundance of grasses, especially Agropyron sibiricum and, more rarely, Stipa capillata L., although in some sites, for example, around Fort Shevchenko, the prevalence of S. capillata attains the level of sor[3]. The grasses are usually distributed irregularly, patches with a comparatively dense grass cover and other patches with a rather thin cover being found. Wormwood, however, is always present at the sor[1]-sor[2] level of abundance. An increase in the grass component of the community is observed in wormwood populations in small, weakly developed depressions.

The closest bedding of limestone under grass—Artemisia communities was observed at a depth of 75 cm. The thickness of friable deposits fluctuated from 125 to 200 cm in 43% of sites described.

The substantial thickness of friable deposits is indicated by the presence of pure wormwood populations, and of wormwood with Anabasis aphylla, which is encountered with a frequency of sol.-sp. The thickness of

---

*Sor—Salt marsh.—Tr.

the friable deposits under these communities exceeds 125 cm in all cases, and in 50% of the cases observed exceeded 200 cm, while in some boreholes limestone was not found even at a depth of 7 m.

The clearest evidence of the great depth of eluvia under Artemisia communities is the presence in them of Kochia prostrata (L.) Schrad., a plant with a long root system and, for the particular conditions, comparatively glycophilic (as compared with Salsola rigidi, Anabasis aphylla, and Salsola chiwensis). Three-quarters of the plots studied, in which Artemisia terrae-albae was found together with K. prostrata, had friable deposits more than 2 m in thickness, and in some cases the thickness was greater than 7 m.

The absence of any association of groupings and communities of Nanophyton erinaceum with closely bedded limestone was rather unexpected, as such an association might well have been anticipated in the light of the view that N. erinaceum is a plant of rocky habitats. From our observations, the thickness of friable deposits under N. erinaceum communities fluctuated very widely and in some cases was very substantial. The plant was most often encountered where the thickness of friable deposits was 50-200 cm. In all cases, the soil under N. erinaceum communities was very stony.

It should be noted in conclusion that the detailed study of the vegetation cover in areas where Neogenic limestone deposits are located on the Ustyurt plateau and the Mangyshlak plain can give quite a precise picture of the thickness of eluvial and deluvial deposits. Large vegetational units—formations or groups of formations—provide a very good approximation of the thickness of friable deposits. Much more exact conclusions can be reached if smaller taxonomic units—associations or groups of associations—are used as indicators of the thickness of eluvial and deluvial deposits.

Summary

The distribution of communities on limestone plateaus is closely related to the thickness of the eluvial mantle covering the limestone. Zygophyllum macropterum, Teucrium polium, Anabasis truncata, Atraphaxis spinosa, Convolvulus fruticosus, and other species are plant indicators of early stages of the erosion of carbonate rocks. The thickness of the eluvial layer under pure thickets of Anabasis salsa varied from 50 to 125 cm. Where A. salsa occurred in combination with Eremopyrum orientale, the eluvium was no less than 90 cm thick, reaching in some cases 200 cm. The thickest eluvial mantle was observed under communities of Artemisia terrae-albae, with an admixture of Kochia prostrata, Anabasis aphylla, and species of the genus Stipa.

Literature Cited

Borneman, B. A., and Spiridonov, M. D. 1929, "A survey of the soils and vegetation of the Mangyshlak and Buzacha peninsulas." Materialy KÉI, No. V. Leningrad, Izd. Akad. Nauk SSSR.

Viktorov, S. V. 1951. "Vegetation as an indicator of lithological conditions in the Northern Ust'-Urt and in the steppes of Western Kazakhstan." Byull. Mosk.obshch. ispyt. prirody, novaya seriya, otd. biol., Vol 56, No. 1.

Viktorov, S. V. 1955. The Use of the Geobotanical Method in Geological and Hydrogeological Research. Moscow—Leningrad, Izd. Akad. Nauk SSSR.

Dolenko, G. I. 1930. "A brief description of the landscape regions of the western Ust'-Urt and the Mangyshlak plain." Materialy KÉI, No. IV, Chap. 2. Leningrad, Izd. Akad. Nauk SSSR.

Rusanov, F. P. 1930. "A survey of the vegetation of the western Ust'-Urt and Mangyshalk plain." Materialy KÉI, No. IV, Chap. 2. Leningrad, Izd. Akad. Nauk SSSR.

# SOME DATA ON THE RELATIONSHIP
# OF THE VEGETATION COVER WITH ELUVIAL
# AND DELUVIAL FORMATIONS
# ON THE GREAT BALKHAN RIDGE

## G. M. Proskuryakova

The watershed line of the Great Balkhan ridge runs at an average height of 1500-1700 m above sea level, while individual points and peaks reach 1886 m. To the north of the watershed, the ridge is broken by steep, perpendicular (up to 500 m in height in places) scarps, while the southern slope descends smoothly, forming a peculiar hummocky "plateau," dissected with deep gullies and rocky ravines running southwards.

The characteristic type of vegetation in the upper zone of the ridge is steppe, developed on montane-chestnut soils (Bobrov, 1931) or brown soils (Korovin and Rozanov, 1938), at an elevation above 1300 m on the northern slope and above 1500 m on the southern slope.

The soils of the Great Balkhan ridge, particularly its upper steppe section, are quite uniform in type, being in the main characterized by the thickness of the silt layer overlying limestone, and the level of humification of the top soil layer.

To describe the soil profile, we may cite the following profile which was made in the middle part of a smooth, slanting hillock slope 2 km east of the Arlan mountain:

0-2 cm—layer of shallow rubble chippings;
2-10 cm—very loose dry loam, dark, dusty-grey with occasional rubble chippings;
10-75 cm—solid loam, dark, brownish-grey, fairly moist, structure clearly of small lumps, with occasional small fragments of limestone, permeated with abundant grass roots;
75-100 cm—clay of still greater solidity, paler in color, dusty grey, dry, structure not clearly defined or structureless, roots less frequent;
Below 100 cm—solid, continuous limestone platform.

The silt layer in this profile thus reaches a thickness of almost 100 cm.

The thickness of the silt layer in general varies and may sometimes reach 400 cm.

The thicker deluvial-colluvial (Bilibin, 1938) soils with a substantial layer of silt are usually found because of the prevalence of surface wash-down of silt particles from the higher parts of the slope. These soils are rich in nutrient substances and have good moisture and thus there is profuse development of grasses, the group best adapted to the climatic conditions of the steppe zone of the Great Balkhan and thus strongest from the competitive point of view.

With a decrease in the thickness of the silt layer and the consequent deterioration in the water and nutrient regime of the grasses, there are signs of depression, and a decrease in abundance, projective covering, vigor rating, etc. On thin soils, grasses lose their dominant position and are transferred from the role of edificators to the group of associated or occasional species in wormwood scrubs, which occupy these poor, thin eluvial soils.

As an example of a thin wormwood soil, we may cite the description of a soil profile, taken at the same place as the previous one, but in a steeper, stonier section higher up on the slope.

0-2 cm—very loose, dry, grey-pale yellow loam, many roots;

10-25 cm—solid loam, light, grey-brown, little moisture, less roots;

25 cm and below—large, flat lumps of limestone, lying very tightly. Wormwood roots squeezing between the lumps.

Besides the difference in thickness of the silt layer (100 and 25 cm), the soil profiles cited differ also in the degree of humification of the upper horizons. The thin eluvial soils developed under Artemisia are usually less humified.

Wormwood has a strong root system and can provide itself with moisture even on stony and rocky soils, and thus all rocky sites with a gravel substrate and a thin silt layer are inhabited by wormwood, which does not undergo the strong depressive effect experienced by grasses.

The character of the vegetation is thus a very clear indicator of the properties of the substrate. A gravelly substrate silt layer, 20-40 cm in thickness, develops under wormwood, and the thinner the herbage and the lower the vigor rating of Artemisia, the less is the thickness of this layer.

Under grass—mixed herbage communities there are rich deluvial and colluvial structural soils with a thick (70 cm or more) silt layer, and the more luxuriant the grass stand and the greater the projective covering, the thicker is the silt layer.

Steppe grass—mixed herbage communities occupy hollows, the smooth slopes of hillocks and valleys. Even a small accumulation of silt along scarcely noticeable erosion lines, descending along slopes and formed as a result of temporary water currents, is sufficient for the development of a community of Stipa and Festuca sulcata with mixed herbage. As a result, the vegetation cover gives a detailed representation of the erosion network, especially in summer and toward autumn, when the yellowing feather grass and fescue can be clearly differentiated on the bluish-grey background of Artemisia.

These communities are formed by several species of Stipa, F. sulcata, and some regular herbage species (Euphorbia seguieriana Neck, Centaurea squarrosa Willd., etc.). The edificators of the community, Stipa lessingiana Trin. et Rupr. and S. szowitsiana Trin., frequently act as dominants also, being inferior in this respect only to Festuca sulcata Hack., which is an indispensable and constant component both of grass—mixed herbage communities and of wormwood scrubs.

Steppe areas with predominance of Stipa szowitsiana are typical of the lower limits of the steppe zone. Higher up, especially on the plateau itself, S. lessingiana is dominant in the grass—mixed herbage communities, permanently replacing the previous species under these conditions.

S. szowitsiana has a greater elevational range than S. lessingiana and is distributed right up to the highest points in the Greater Balkhan, being encountered throughout the entire steppe zone, though not as an edificator but merely as an associated species.

The structure of the herbage in grass—mixed herbage sectors is characterized by the presence of three layers:

Layer I (40-70 cm) is formed by the reproductive shoots of grasses and the stems of some herbage species.
Layer II (10-40 cm) is formed by the fescue turf and the vegetative mass of the other grasses and herbage.
Layer III (up to 10 cm) is formed by ephemerals.

The value of the average total projective cover of the grass—mixed herbage community reaches 57.5 ± 1.4%, and in a moist year can go up to 67 ± 1.5%.

The total projective covering in steppe sites varies markedly: the limits seem to be 30 to 80%.

Grasses are especially responsive to the change in the character of the substrate, and they are responsible for an increase in the total projective covering. The grasses in the community give a 30-80% cover, 40-50% being the most frequent range.

A high-percentage total projective covering (70-80%) is shown by communities on very rich soils, with a humified silt layer more than 75 cm in thickness. This is caused by the vigorous development of feather-grass

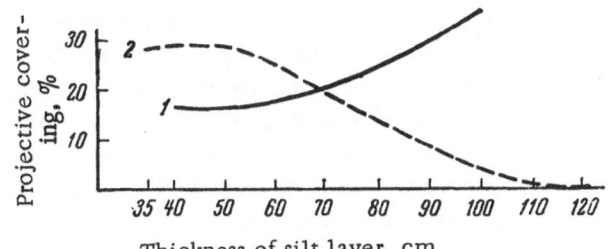

Thickness of silt layer, cm

Relationship of projective covering of grasses and wormwood to the thickness of the silt layer. 1) Grasses; 2)
wormwood.

and Festuca sulcata, the latter being a species with high sensitivity to the substrate; on thick deluvial soils F.
sulcata increases the projective covering to 40-50%, sometimes even to 70%, compared with the usual 10-20%.
Grasses show much weaker development on thinner soils with a silt layer of less than 50 cm.

The lowest percentage cover is shown by grasses on stony soils, where the silt layer reaches 50-60 cm
in only a few cases, and is as a rule 20-30 cm. Wormwood usually develops on such soils, and the role of grasses
is reduced as compared with the grass — mixed herbage community.

The edificator of wormwood growths is Artemisia maritima (s.l.), and it is accompanied by all those
species of Stipa, herbage, and Festuca, which were mentioned for the grass—mixed herbage community.

Artemisia also frequently develops in grass—mixed herbage sites (it was recorded in 41% of the plots
studied), but its role in the herbage is negligible. The Artemisia has clear depression symptoms and in most
cases shows less than 10% cover.

At the same time, Artemisia in wormwood scrubs, with an average projective covering of 42.5 ± 1.1%,
occupies an average of 30%, i.e., more than half of the total.

The projective covering of grasses in wormwood scrubs is usually around 10%, although in a few cases it
can reach 40%.. The insignificant role of grasses—the most responsive group to humidity—in the herbage of
wormwood scrubs is reflected if we compare the values for total projective covering in moist and dry years.
Whereas grass—mixed herbage communities show a difference of 10% (1958 compared with 1959), wormwood
scrubs increase their average total projective covering in a moist year only to 43 ± 1%, i.e., by less than 1%.

Artemisia has an average projective covering of around 30%, but in individual cases this can be increased
even as high as 80%. This is associated with especially favorable conditions for its development and primarily
with substantial thickness of the silt layer of the soil. However, tracing the increase in Artemisia cover in relation to increasing thickness of the silt layer of the soils cannot be done in most cases, since the increasing
influence of the grasses depresses Artemisia and counteracts the effect of the favorable soil conditions.

An illustration of the relationship of the projective covering of grasses and wormwood with the thickness of the silt layer is provided by theoretical curves, obtained by the "sliding averages" method (see figure).
The curves are constructed on the basis of empirical values for the projective covering of grasses and wormwood (for 200 plots), determined for each class interval (5 cm) as the arithmetic average of the values occurring in each interval.

The external appearance of the grass turfs in the two communities differed. Whereas the vigor of grasses
in the grass—mixed herbage community was rated at 3, 4, or 5 points (on a five-point scale), with a predominance of 4 or 5 points, the dominant rating of grass vigor in the wormwood population was 3. There was a
tendency to an increase in vigor with an increase in the thickness of the silt layer; the same applies also to
wormwood, especially on disturbed sites from which the grass cover has been removed artificially. Only two
grasses develop well in wormwood scrubs: Oryzopsis holciformis M.B. and Stipa caucasica Schmalh. The latter is the most petrophilic species among the Balkhan feather-grasses, and has an exceptionally wide ecological
range, as a result of which it occurs almost from the level of the plain (300 m above sea level) to the peak itself (1886 m above sea level).

In spite of the good condition of Stipa caucasica in wormwood populations, it nevertheless is still experiencing substantial depression, a good illustration of this being provided by old burnt sites. In the place of the burnt wormwood growths, in which Stipa caucasica is present also, pure groupings of this feather-grass are developed, and show the entire area of the burnt plots, with a total projective covering up to 60%, and sometimes even 70%. It would be a mistake, however, to expect to find thick silty soils under these Stipa growths, though such soils are characteristic for other communities of feather-grass. Under the conditions of the region studied, S. caucasica never develops independent pure groupings, and therefore sites where it shows mass development constitute an accurate indicator of the prior presence of wormwood on thin soils (20-55-cm silt layer).

Wormwood populations, in the same way as grass—mixed herbage communities, are virtually two-layered (layer I, 70 cm—reproductive shoots of grasses; layer II, 45-50 cm—Artemisia; layer III, up to 10 cm—ephemerals and ephemeroids), since by the time the first layer has formed, the third has almost completely ceased to exist. In the second half of May and the beginning of June, the uniform grey background of wormwood is brightened by Ranunculus linearilobus Bge., which in conjunction with Bongardia scrysogonum forms variegated yellow meadows, blossoms of Iris drepanophylla Aitch. et Baker, and sites with exceptionally thick (sp. cop' rating) beds of Astragalus testiculatus Pall. and A. velatus Trautv., with pink-white flowers comprising dense inflorescences.

In the extreme case, wormwood populations are very thin (10-30% total projective covering), and the species composition of the community is completed by a few markedly petrophilic species, such as Acantholimon korolkovii Rg. et Schmalh. and Lagochilus balchanicus.

Wormwood growths usually occupy the clearly defined elements of the relief—the tops of hummocks and ridges—and form belts in the upper parts of slopes, although they can also be located along slopes if the substrate is gravelly or is underlain at a shallow depth by indigenous rock.

Wormwood growths alternate with the grass-mixed herbage community, and this produces a peculiar complex, constituting an accurate reflection of the character of the soil substrate and the proximity of indigenous rocks.

The fact that there is a close association between two communities, which differ widely in physiognomy, and eluvial-deluvial formations of varying thickness may be of considerable significance for the economic development of the Great Balkhan territory, especially in identifying sectors which are suitable for agriculture.

Summary

In the Great Balkhan mountains, shallow stony soils are indicated by communities with dominance of wormwood (Artemisia maritima s.l.), whereas more developed soils are indicated by feather-grass communities (Stipa lessingiana and S. szowitsiana). Wormwood growths destroyed by fire are replaced by communities of S. caucasica. By using these indicators, areas can be determined where the soils are relatively deep, rich, and suitable for agricultural purposes.

Literature Cited

Bilibin, Yu. A. 1938. The Principles of the Geology of Placers. GONTI NKTP SSSR. Moscow—Leningrad.
Bobrov, E. G. 1931. "The vegetation of the Great Balkhan mountains." Tr. Bot. sada Akad. Nauk SSSR, Vol. 44.
Korovin, E., and Rozanov, A. 1938. "The soils and vegetation of Central Asia as a natural productive resource." Tr. SAGU, Vol. 17. Tashkent.

# THE INDICATOR SIGNIFICANCE
# OF THE VEGETATION COVER FOR MAPPING
# RIVER TERRACES IN THE DZHEZKAZGAN-ULUTAU DISTRICT
# OF CENTRAL KAZAKHSTAN

## V. N. Ostrovskii and V. P. Olekseenko

The Dzhezkazgan-Ulutau district occupies the western extremity of Central Kazakhstan (Sary-Arka). From the geomorphological aspect, the district is an isolated section of the Kazakh hilly-plain area, and is bounded on the west, the north, and the south-southwest by the Turgai trough, the Tengiz depression, and the Sarysu depression, respectively. The district has a hummocky relief. The climate is markedly continental and dry.

The modern hydronetwork of the Sarysu and Ishim basins and Lake Tengiz is largely confined to ancient valleys and dissects, at a depth of up to 25-30 m, the rocks of the friable lacustrine-alluvial cover which has developed on the bottoms of the ancient valleys. The surface of this loose covering layer, formed in pre-quaternary times, shows comparatively slight differentiation (the latter increases with proximity to present-day rivers) and has a clear gradient toward the axis of the valley. In the work of earlier investigators, such surfaces were frequently regarded as sites of third (and, sometimes, fourth) above-floodplain terraces of quaternary age. The river valleys of the western slope Turgai River basin) have been regarded as hilly plain and sculptured lowland (in the upper part of the river basin) and as Neogenic stratal lowlands (middle and lower parts of the river basin), and frequently have a deeply serrated canyonlike shape.

The morphological structure of the bases of the present-day valleys are similar in all the basins. Two terraces above the floodplain develop within them, as well as high (meadow) and low (riverbed) floodplains. Traces of inundation are not always preserved on the high floodplain, and as a result, it is sometimes regarded as the first terrace above the floodplain.

In addition to the morphological and lithological characteristics of the formations mentioned, a specific vegetation cover is typical. The use of this fact in field research substantially facilitates the study of terraces; in certain cases the vegetation cover is the sole criterion for identifying elements of the valley. Geobotanical observations therefore acquire considerable significance within the complex of geomorphological research.

We present below lithological and hydrogeological information on the elements of present-day river valleys, with a description of the characteristic plant associations.

Low (riverbed) floodplains are up to 200 m in width. Their surface is uneven, and has rain channels and hummocky deposits of loose material. The composition of the vegetation of the floodplains adjoining riverbeds is determined by two factors: the lithology of the floodplain deposits and the depth of bedding of groundwaters. Three types of floodplain are differentiated:

1. Floodplains which are mainly built up of sand and pebble deposits, covered with clays up to 5 m in thickness (Zhaksa-Kon and Zhilanda Rivers). Groundwaters are located at a depth of up to 4 m, with a pressure head of up to 1.5-2 m. The presence of compressed clay deposits leads to good stability of the banks of water stretches, and the latter are therefore very extensive.

2. Floodplains built up of pebble-sand-clay deposits (Karakingir, Sarykingir, and other rivers). The level of groundwaters is at a depth no greater than 1 m, the banks of water stretches are quite stable, and the water stretches are of substantial size.

3. Floodplains built up of sandy alluvium (Akmai and Kumol Rivers).

The vegetation cover of the low floodplains is characterized by the greatest variety and range in species composition. Associations incorporating meadow grasses are dominant, together with shrub groupings, developed on meadow-chestnut, alluvial-meadow, meadow-bog, and other soils.

The vegetation in floodplains of the first type primarily consists of communities of Agropyron repens (L.) P.B. and Bromus inermis Leyss, with rich mixed herbage of the meadow type. The projective covering of the grass vegetation reaches 100%. Populations of willow (Salix serrulatifolia E. Wolf., S. wilhelmsiana M.B., and other species) are observed on the banks of water stretches. Floodplains of this type can be called meadow types.

Floodplains of the second type are characterized by the predominant development of brush, including the above-mentioned willow, Lonicera tatarica L., Rosa canina L., and Elaeagnus angustifolia L., forming compact tugai vegetation. The composition of the lower grass layer of the vegetation includes various sedges (Carex sp.), reed (Phragmites communis Trin.), Typha angustifolia L., Agropyron repens, and other plants.

In comparison with the vegetation of the first two types of floodplain, that of the third type is relatively poor, because of the occurrence of constant soil erosion and redeposition of sand. We find here low-growing reeds with their characteristic creeping rhizomes, and adaptation against sand drift, and also weed and incidental species. The projective covering of the vegetation does not exceed 10%.

The high floodplain (in the floodplain terrace) reaches 300-400 m in width. It is usually around 1.5 m higher than the surface of the low floodplain, while the height of the rim above the normal water level in the water stretches is 2-2.5 m. High floodplain sites are usually flat, with rain channels and crescent lakes. The high floodplain is built up of sands, sandy loams, and loams. The depth of bedding of groundwaters fluctuates from 1.5 to 3-3.5 m. The vegetation of high floodplains is characterized by the presence of communities of xerophilic grasses, such as Elymus angustus Trin., Lasiagrostis splendens (Trin.) Kunth., and Agropyron pectiniforme Roem. et Schult., and also associations of A. repens and wormwood, incorporating xerophilic meadow mixed herbage. The soils belong to the meadow type. Two ecological groups may be distinguished (according to L. N. Beideman) in the vegetation of high floodplains: 1) phreatophytes and trichohydrophytes, plants which are associated with groundwaters or with their capillary fringe; 2) ombrophytes, which are plants not associated with groundwaters.

The first group includes such species as Elymus angustus, Agropyron repens, Lasiagrostis splendens, while the second group includes Agropyron pectiniforme, Echinops ritro L., Eryngium planum L., and other plants. When the second half of the summer commences, the water deficiency in the soil begins to exert its effect, ombrophytes cease or slow up their development, while phreatophytes and trichohydrophytes continue to grow throughout the season.

On aerial photographs, meadow and riverbed floodplains can most frequently be identified by dark to black strips, in which it is sometimes possible to identify sectors with different tone intensities: the darker sectors represent the riverbed floodplains. Phreatophytes have the greatest degree of significance in the development of these aerial photographic images.

The first above-floodplain terrace develops almost everywhere in the middle and lower river courses. Its width goes up to 1.5-2 km, while the height of the rim above the summer level of the stretches is 3.3-3.5 m. The surface of the terrace area is flat and there is a clearly evident rim at the back. The terrace in the lower part of the profile is made up of obliquely laminated sands with coarse gravel, while the upper section is composed of brown loams with gypsum inclusions. The thickness of the loam mantle is equal to 2.2-2.5 m. The level of groundwaters in the first above-floodplain terrace occurs at a depth of 3-3.5 m.

In the vegetation cover of the first above-floodplain terrace, representatives of the family Chenopodiaceae are dominant, and these grow on solonchak-type solonetzes with a takyrlike surface, the terrace being therefore designated as the takyr type.

Associations of Atriplex cana C.A.M., Anabasis salsa (C.A.M.) Benth., with black wormwoods (artemisia terrae-albae Krasch. and A. pauciflora Web.), and Suaeda physophora Pall., develop on a large scale on the terrace. Lasiagrostis splendens, Rudbeckia hirta, and Agropyron pectiniforme are also found. Complexity of the

The Distribution of the Vegetation Cover on Floodplains

| Relief | Morphology and morphometry of the relief | Lithology of the alluvium | Depth of bedding and type of ground-waters | Soil-vegetation cover |
|---|---|---|---|---|
| Low floodplain | Width of up to 200 m, numerous crescent lakes and dry watercourses observed on the surface | Sand-pebble deposits with argillaceous material, clays, loams, sandy loams | From 0 to 3 m, fresh | Communities of willow, Lonicera tatarica, Rosa canina, Agropyron repens, Bromus inermis, reed with meadow mixed herbage, developed on alluvial-meadow, meadow-bog, and other soils |
| High floodplain | Width up to 400 m, 1.5 m below the surface of the low floodplain; surface flat, with occasional rain channels and crescent lakes | Sands, sandy loams, loams, with coarser material confined to the lower section of the profile | From 1.5 to 3.5 m, fresh | Communities of Lasiagrostis splendens, Elymus angustus, Agropyron pectiniforme, Artemisia schrenkii, developed on more or less saline meadow types of soils |
| First above-floodplain terrace | Width up to 1.5-2 km; height of the rim above the summer level of the water stretches, 3.3-3.5 m. Surface of the area is flat | Obliquely laminated sands, overlaid with loams 2.2-2.5 m in thickness | From 3 to 5 m, fresh, rarely salinated | Communities of Atriplex cana, Artemisia terrae-albae and A. pauciflora, Anabasis salsa, Suaeda physophora, developed on solonchak-type solonetzes with a takyrlike surface |
| Second above-floodplain terrace | Width 59-200 m, 6-6.5 m scarp, surface sloping markedly toward the valley | Sand and gravel deposits, overlaid with loams | More than 5 m | Communities of Stipa capillata, S. lessingiana, Festuca sulcata, Artemisia sublessingiana, A. incana, and Spiraea hypericifolia |

vegetation cover is characteristic: Anabasis salsa and Suaeda physophora are located in small plots, some tens of square meters in area, among Atriplex cana communities. The projective covering of the vegetation of the first above-floodplain terraces is 30-40%. First-terrace areas may be identified in aerial photographs in the form of pale-grey strips, while examination with a magnifying glass indicates the patchy character of the picture.

The second above-floodplain terrace is noted only in downstream valleys, where it develops in 50-200-m wide strips in a number of sites. The surface of the terrace has a marked slope in the direction of the thalweg of the valley and is dotted with small, well-rounded pebbles. The terrace is composed of sandy-coarse gravel alluvium, covered with a thin layer of loams. The depth of bedding of groundwaters exceeds the thickness of the alluvial mantle.

The vegetation of the second above-floodplain terrace is made up of steppe species, which develop on the pale-chestnut soils and are not associated with the groundwaters. The most widely distributed are Stipa capillata—Artemisia sublessingiana, Festuca sulcata—Stipa capillata with Spiraea hypericifolia, and other associations, the dominant species of which are wormwoods (Artemisia sublessingiana (Kell.) Krasch. and A. incana Kell.), feather-grasses (Stipa capillata L. and S. lessingiana Trin. et Rupr.), fescue (Festuca sulcata Hack.), and dropwort (Spiraea hypericifolia L.). It should be noted that the composition of the vegetation of floodplains and part of the first above-floodplain terrace is subject to rather sharp changes depending on fluctuations in the mineralization of the groundwaters; we will not, however, dwell on this, since the alluvial waters of the majority of rivers are fresh. The above-mentioned geobotanical, morphological, lithological, and other features of various elements of the relief of river valleys are shown in the table.

## Summary

The river terraces of the region investigated can be identified by the character of their vegetation cover, even when they show marked leveling or are represented by small fragments. Thus on the second above-flood-plain terrace, steppes incorporating various species of _Stipa_ and _Artemisia_ are dominant; associations of halophytes of the family Chenopodiaceae dominate the first above-floodplain terrace, while the floodplains themselves are characterized by meadows (sand-pebble floodplains), brush thickets (sand-pebble-clay floodplains), and reeds (sandy floodplains).

# THE USE OF VEGETATION AS AN INDICATOR
# IN DECIPHERING SOILS FROM AERIAL PHOTOGRAPHS

## E. V. Leont'eva

Vegetation has been used as an indicator of soils for a very long period, probably from the very inception of agriculture. In contrast to other objects of indications, communities as a whole are most often used in identifying soils, rather than individual species or aspects of their morphology, vigor, and phonological development. This trend was noted as early as 1904 by Vysotskii and was incorporated in I. V. Larin's procedural instructions on the mapping of agricultural lands (1924).

The present communication is based on data collected during the work of the Kazakh expedition of the Laboratory of Aerial Methods of the USSR Academy of Sciences on the territory of Northern Kazakhstan, (North-Kazakhstan, Kokchetavskaya, and Akmolinsk provinces) in 1954-1956. A section of the data and studies, carried out under the direction of A. S. Preobrazhenskii in this area, has already been published (Preobrazhenskii, 1957; Tolchel'nikov, 1957; Semenova, 1958).

A. S. Preobrazhenskii attached great significance to the study of the association of vegetation with soils. In one of his papers on Northern Kazakhstan (1959), he wrote that deciphering soils from the vegetation can be very effective and reliable (as reliable as the interpretation of the vegetation), and suggested that soil identification should be based on a comprehensive study of the interrelationships between soil and plant and other components of the landscape. Thus, in deciphering vegetation, everything depends on using those features of the vegetation cover which can be differentiated in aerial photographs as indicator characteristics (terminology of Viktorov, Vostokova, Vyshivkin, 1961).

In the district where the work was carried out, it is possible to decipher with almost complete accuracy, on the basis of direct characteristics, species of trees and shrubs (except in the undergrowth), and some groups of herbaceous vegetation (sedge and reed growths), the photographic image (in black-and-white aerial photographs) reflects differences in the illuminated and shaded herbage mosaic, determined by its structure. However, only comparatively large units of the vegetation cover can be easily identified from the tone in a small photographic image, i.e., groups of associations united by the dominance of plants of various ecological and morphological groups within their composition. More detailed deciphering is effected on the basis of the relationship of the vegetation with other components of the landscape (Vinogradov and Leont'eva, 1957; Leont'eva, 1961).

By special deciphering, we can artificially distinguish a single landscape component of particular interest in aerial photographs, while consciously or unconsciously using other associated components and elements as direct influence on the photographic image. This is confirmed if only by the fact that normally sites where herbage has been mowed are deciphered with much greater difficulty than sites with undisturbed herbage.

Changes in the vegetation cover reflect changes in the soil cover and other conditions of the habitat which are also significant factors in soil formation, such as the soil-forming rocks and the microrelief and the associated changes in moisture regime, temperature of the soils and the ground air layer, etc.

The character of the habitat environment is known to control not only the species composition but also the structure of phytocenoses.

The structure of communities can in fact also be used as an indicator characteristic, since it depends on the morphological structure, vigor, and phenological state of the individual components which are dominant in projective abundance. The morphological structure will in this context be a principal factor in reflecting the plants' belonging to a definite ecomorphological group of species.

The difference in the photographic images of ecomorphological groups of communities is magnified by the fact that under communities including a high proportion of hydrophytes and mesophytes there will be soils with a surface richer in humus and moisture, covered with litter, of course, but through the normal annual mowing or grazing, rather than cutting. Correspondingly, under more xerophytic plants, the soils have less humus, are usually drier in summer, frequently with a takyr-type surface disintegration, and are almost free of litter.

Of no less importance is the lag in the phenological development of plants in the more humid habitats, this being quite evident in summer thanks to the overall appearance: all depressions, even small ones, are differentiated by their green color on the faded steppe. According to the data of A. P. Fedoseev (1956) and A. P. Fedoseev and G. G. Beloborodova (1958), this lag comprises 5-15 days in Festuca sulcata and Stipa, judging by the flowering dates. The relationship within the photographic image changes in other seasons (Vinogradov and Leont'eva, 1957; Vinogradov, 1960).

The examples given merely illustrate the general pattern, although the structure of communities can reflect precisely even slight changes in the soil cover. Sharp differences lead to changes of communities, even communities pertaining to different ecological groups, while small differences result in the modification of microcenoses, or induce mosaicism in the vegetation cover, as understood by F. Ya. Levina (1958) and N. D. Yaroshenko (1951). Such differences are usually visible in large photographs. The details of the pattern can sometimes be measured in millimeters or fractions of a millimeter. Even slight solonetz formation in the rich mixed herbage-feather grass steppe zone is reflected in a thinning of the herbage, and F. sulcata rather than Stipa becomes the dominant component, while the more xeromesophilic herbage is replaced by Artemisia spp.

With successful survey conditions, this type of mosaicism is very clearly seen on aerial photographs.

It has sometimes proved possible to observe in aerial survey work, or even in land work, still finer differences in the structure of phytocenoses, appearing in the form of changes in the vigor of all the individual plants in small plots, there being the same relative proportions of species in these plots, as well as in the whole phytocenosis. Such mosaic patches, evaluated in photographs, are frequently associated with rodent damage.

The cultivated vegetation, especially in the first years after plowing virgin lands or old pastures, is frequently a better indicator than the natural vegetation for deciphering soils in aerial photographs. This can be explained by the fact that cultivated plants remain in a weak condition on unfavorable soils, while under natural conditions there occurs a replacement of some species by others which develop well under conditions which are unfavorable for cultivated plants. It is partly plowing itself, the turning of the furrow, which accentuates soil differences in the upper, most root-saturated layer. Plowing soils for many years levels the microrelief and the mosaicism is reduced.

Plantations of forest species, or nurseries and forest belts, are an even better indicator of soils. Their use is, of course, restricted, since they occur only in the forest steppes, and even there only in rather small areas.

In some cases where soils are directly shown on aerial photographs (as in plowed fields prior to the appearance of plant seedlings, or in bare fallow), the soils can be deciphered better than soils with a vegetation cover. Solods (degraded solonetzes) can usually be especially well differentiated when the solothized horizon is exposed. Not infrequently, however, exposure of soil by plowing merely eliminates those differences in the soil cover which were evident from the vegetation.

It should be mentioned in conclusion that for the indication of soils, as for any other landscape component, we do not use the structure of some individual phytocenosis in isolation, but rather the structural relationships. The identification of the structure of the vegetation is itself provisional and is carried out so as to facilitate the determination of characters for deciphering individual units of the vegetation cover and the corresponding soils. The relationships of the indicator characteristic identified allows it to be used in different zones (which cannot be said about individual species or even indicator phytocenoses) and at the same time confirms the need for a comprehensive geographical approach to the deciphering of soils and vegetation from aerial photographs.

Summary

      The structure of communities is an indicator characteristic which can be used in the deciphering of soils. The dominance of species belonging to particular ecological groups is of special importance in this context. Thus, for example, soils are always richer in humus and more humid under hydrophyte and mesophyte communities of xerophytes. Cultivated vegetation and vegetational litter in the first years after plowing are also valuable characters for the deciphering of soils, since differences in the vegetative vigor of cultivated plants (as well as weeds) are clearly evident on aerial photographs.

Literature Cited

Viktorov, S. V., Vostokova, E. A., and Vyshivkin, D. D. 1961. "Some problems in the theory of indicator research." In book: Problems of Indicator Geobotany. (Summaries of Papers at the Conference on the Problems of Indicator Geobotany.) Izd. VSEGINGEO, VAGT, and MOIP.

Vinogradov, B. V. 1960. "Changes in the appearance of the vegetation of Northern Kazakhstan on aerial photographs depending on the date and time of aerial photography." Vestn. LGU, No. 6.

Vinogradov, B. V., and Leont'eva, E. V. 1957. "The use of aerial methods for studying the vegetation of Northern Kazakhstan." In book: Materials Toward the Use of Aerial Methods in Studying the Soils and Vegetation of Northern Kazakhstan. Moscow—Leningrad.

Vysotskii, G. N. 1904. "On a map of types of habitat." In book: Contemporary Problems of Russian Agriculture. SPb.

Larin, L. V. 1924. "A contribution to the procedure of agricultural surveys in the Kirgiz SSR." Sovetskaya Kirgiziya, No. 10.

Levina, F. Ya. 1958. "Complexity and mosaicism of vegetation." Bot. zhur., Vol. 43, No. 12.

Leont'eva, E. V. 1961. "Studying the vegetation for the exploitation of the forest steppes and steppes." [Citation incomplete in Russian text.]

Preobrazhenskii, A. S. 1957. "Experience in studying the geography and topography of the soils of Northern Kazakhstan with the use of aerial photographic survey data." In book: Materials Toward the Use of Aerial Methods in Studying the Soils and Vegetation of Northern Kazakhstan.

Preobrazhenskii, A. S. 1959. "On the use of aerial methods in soil research." Tr. Labor. aérometodov Akad. Nauk SSSR, No. VII.

Semenova, N. N. 1958. "Identification of some soil complexes of Northern Kazakhstan in aerial photographs." Pochvovedenie, No. 8.

Tolchel'nikov, Yu. S. 1957. "On the problem of the soils of some steppe sinkholes in Northern Kazakhstan." Pochvovedenie, No. 5.

Fedoseev, A. P. 1956. "The mean annual dates of mass flowering of pasture grasses in the Kazakhstan plain." Tr. Kazakhstanskogo n.-i. gidrometeorol. inst., No. 7. Leningrad.

Fedoseev, A. P., and Beloborodova, G. G. 1958. Agrometeorological Conditions of Plant Growth and Development. The Microclimate of the Northern Part of the Kazakh Hummocky Plain. Leningrad.

Yaroshenko, P. D. 1961. Geobotany. Moscow—Leningrad, Izd. "Vysshaya shkola."

# DETERMINATION OF SOIL AND TERRAIN CONDITIONS FROM THE VEGETATION COVER (USING THE VOLGA-AKHTUBINSK FLOODPLAIN AS AN EXAMPLE)

## L. S. Rodman

The question of the possible use of vegetation as an indicator of soil conditions is not new. A quite detailed systematic treatment of papers devoted to the problem of the association of plants with specific soils was provided by V. V. Petrov (1959), and we shall therefore describe only the basic approaches.

The first approach is the finding of plants which are indicators of the borders of soil zones and subzones. The contributions of F. I. Ruprecht (1866), V. N. Drobov (1914), and V. M. Bogdanov (1934) should be mentioned here.

The second approach might be termed the geobotanical indication of specific soil properties, such as mechanical structure, humidity, pH value, etc. We may mention here the work of V. P. Nogtev (1932) and M. F. Korotkii (1912), carried out in humid regions, and equally could cite other, more numerous investigations carried out in arid zones, and devoted mainly to studying the possible use of the vegetation for determining the salinization of soils. Among the numerous researchers who have worked on this problem, we should first of all mention S. K. Chayanov (1909), B. A. Keller (1911), and B. V. Fedorov (1930), who provided both a theoretical basis to these relationships and concrete classification schemes. Later studies of a similar nature were carried out by U. Malina (1952), for the Mugan' depression, and by N. I. Akzhigitova (1958), for the Andizhan district.

The third approach consists of the indication of specific taxonomic soil units from the vegetation. There has been very little work of this type (Korotkii, 1912; Larin, 1953; Tsatsenkin, 1953), although practical soil workers quite frequently use the vegetation cover as a character of assistance in soil mapping especially in arid regions. Work of this category also includes the investigations which we carried out in the Volga-Akhtubinsk floodplain, although there were substantial differences in procedure from the previously noted investigations, and in particular the work of S. K. Chayanov, B. A. Keller, and B. V. Fedorov. These workers seek to establish through the vegetation a special classification of the salinization of soils—a classification of soils on the basis of the vegetation cover, i.e., to express soil properties through the medium of vegetation rather than the characteristics of the soil itself. The evaluation of soil on the basis of vegetation permits a more complete expression of the properties of its vegetation cover, which essentially is also the objective of the majority of soil investigations.

Our work differed from the above principally by the fact that since this aspect was of secondary importance in researches on soil their objective was to discover the relationship of specific units of the vegetation cover to taxonomic soil categories, identified from the soil characteristics themselves. This treatment of the problem is justified in view of the fact that identical factors govern the genesis of soil and vegetation, and consequently the present-day genetic classification of soils should in itself reflect the properties of the vegetation layer.

The relationship of soil with plant communities in the Volga-Akhtubinsk floodplain was studied during two-year (1958 and 1959) geobotanical investigations by the All-Union Aerogeological Trust. The normal procedures of indicator research were employed. During the period, 800 standard descriptions were made on actual plots of associations, with information from soil boreholes. In addition, to correlate the data more precisely, a detailed large-scale soil and botanical survey was carried out on 11 plots, each 25-30 hectares in area. The soil part of the research was carried out by the soils group of the "Yuzhgiprovodkhoz" Institute of which the director of studies is A. A. Popov. His earlier work on the systematics of the soils (Popov, 1960) was accepted by us.

Our studies were based on the investigations of V. P. Nogtev (1932) and L. G. Ramenskii (1938), devoted to the analysis of the relationships of vegetation with the soils of floodplains, and on the data of S. A. Vladychenskii (1953), R. R.Krinitskaya (1957), M. V. Markov (1938, 1951), and other research workers on the association of soils to the ecology of floodplain plants.

The data we obtained provided confirmation that the forest-meadow vegetation of the Volga-Akhtubinsk floodplain is characterized by clear relationships with soils and can be used for determining the limits of soil types. The most simple aspect is the differentiation of soil types and subtypes by the vegetation. Soils generally correspond to large units of the vegetation cover (formations and classes of associations), which are quite clearly identified by their floristic composition. Thus, for example, the floodplain alluvial soils subtype, which is differentiated within the floodplain meadow soils type, and is associated with the steppe-meadow class of associations, while the forest-meadow soils subtype is linked with the floodplain oakwood class of associations.

Moving to the moist-meadow floodplain soils type, mixed herbage polydominant meadows acquire absolute dominance, though they are replaced in very moist sites by a group of formations of grass communities on bog soils. Shallow types of soils are linked with specific groups of associations. Thus, the development of a group of reed associations is typical for gley floodplain moist meadow soils, in contrast to the coarse herbage polydominant meadow associations associated with gleyish soils of this type.

The largest areas of the Volga-Akhtubinsk floodplain are occupied by the floodplain-meadow dark soils subtype, which includes many varieties of soil. This subtype is on the whole associated with meadows of moderate elevation and may be quite clearly identified from other soils by the vegetation cover. The simple identification of this subtype, in view of its wide distribution, is, however, inadequate for practical purposes.

Using this subtype as an example, let us examine the feasibility of the indication of smaller taxonomic soil categories by the vegetation. It should be mentioned immediately that such categories are associated with smaller units of the plant cover than mentioned already, i.e., with individual associations. Thus, an association of Galium verum L., Carex melanostachya M. B., and Calamagrostis epigeios (L.) Roth. develops on dark, loamy meadow soils of moderate thickness; Bromus inermis Leyss.—Agropyron repens (L.) P.B. develops on heavy loams of moderate thickness; sedge—brome, Euphorbia uralensis Fisch. et Link.—sedge—brome, and various other associations develop on thick heavy loams; and on thick clay soils there are associations of brome, sedge, Senecio jacobaea L., and Lythrum virgatum L.

With the appearance of gleying symptoms, Hierochloë odorata (L.) Wahlb.—Lythrum virgatum—Senecio jacobaea—Bromus inermis and other associations are noted on the thick, dark, meadow clay soils with weakly gleying surface, the associations being characterized by an increase in the role of the mixed herbage and the replacement of the gramineous basis (A. repens and B. inermis being dislodged by H. odorata and Digraphis arundinacea (L.) Trin.). Even a simple enumeration of soil types and the associations linked with them shows the possibility of indicator research by means of the vegetation, when a sufficiently detailed study is made of the highly subdivided taxonomic units of the soil cover.

The problem of the degree of reliability and stability of the relationship of communities with specific soil types is of considerable current interest. Analysis of data shows that in those cases where soil types are differentiated on the basis of the morphological characteristics of the soil profile and are associated with differences in mechanical composition, groundwater level, and the occurrence of stagnation, leading to gleying of the soils, their identification on the basis of the features of the plant cover is very reliable. However, when we proceed to differentiate saline and nonsaline variants of soil types, the identification of indicator associations is frequently very difficult and the accuracy of the method shows a marked decrease.

This phenomenon is associated with the characteristics of the vegetation of the region. The specific, very rigid conditions affecting the formation of the flora of Volga-Akhtuba, with its late, prolonged flooding and its dry summer, are responsible for the extreme floristic poverty and the resultant almost complete absence of the species which normally indicate salinization of soils.

A. P. Shennikov (1930) mentions the "unusual similarity in the species composition of the herbage (Volga meadows) at different levels." This unusual similarity is the result of the fact that the flora of the Volga meadows is dominated by species with a rather wide ecological range. According to Whittaker (1957), these species "are formed by a large number of genetically varied populations and their significance as indicators is slight, in contrast to species characterized by great genetic uniformity of their populations, which develop a clear association with specific ecological factors, are consistent components of a small group of communities, and are the most stable indicators of habitat conditions." Where it is impossible to use species as indicators of salinity, the indicator role of communities becomes of primary importance, since they in turn are characterized by great floristic similarity. Differences between associations are expressed, as noted by A. P. Shennikov (1930), in different quantitative proportions between particular species, and subsequently in variations in the onset of developmental phases and in the level of development. The indication of salinization by means of communities is consequently also very complex.

The qualitative nature of the salinity of floodplain soils, characterized by the predominance of $SO_4$ ions, still further complicated indication according to the vegetation cover, since the toxicity of sulfates for plants is relatively slight, in comparison with Cl' ions. It is also necessary to take into account the fact that, according to the work of Ya. F. Dubovik (1951), an increase in the content of sulfate ions in the soil inhibits the uptake of chlorine ions into the plant. Until very recently, soil classifications of salinity did not take into consideration the specific effect of the type of salinity on the plant organism (Kovda, Egorov et al., 1960), which has still further complicated the indication of saline soil types by the vegetation.

All the above facts, together with the relatively low percentage content of salts in the floodplain soils, are responsible for the absence of associations developing exclusively on saline or nonsaline varieties of particular soil subtypes.

Let us examine a few concrete examples. One of the most widely distributed communities within the A. repens meadow group is that of A. repens and sedge, which usually occupies the summits of ridges in the sloping-ridge part of the inner floodplain. An analysis of 15 standard plots showed that this community is associated with floodplain-meadow, dark-colored (15 cases), thick soils (13 cases), and heavy loams (12 cases). In nine cases out of 15, the soils were salinated. This substantial proportion of saline soils can be explained by the characteristics of the habitat of this association. The ridges of the sloping-ridge part of the inner floodplain, with which the community is associated, are the first to be freed from high floods which are salinated as a result of the intensive evaporation, while during the period when they are already free of water, and the surrounding areas are inundated, they become wicks of evaporation and consequently draw in salts. At the same time, they are less drenched by flood and rain water than are flat areas or inter-ridge depressions (Letunov, 1942; Plyusnin, 1938).

A comparison of the floristic lists of a community on saline and demineralized soils demonstrates the somewhat greater profusion of species in the saline habitats. This enrichment depends mainly on mixed-herbage species.

One of the most widely distributed associations of the A. repens meadow group is that of A. repens and Eleocharis palustris (L.) R.Br., which is confined to the inner sections of the floodplain. In the northern areas of the floodplain, this association develops in three principal types of habitat. In the first case, on the inner flat floodplain, it is characterized by a very sparse floristic composition and low frequency of species other than the dominant. The soils are thick, moderate to heavy loams, and are salinated, dark-colored, and meadow in type. On the inner, microhummocky floodplain (second case), the community develops on the microelevations. The floristic list is somewhat richer. The soils are thick clays, moderately solonchak in type. A comparison of the soils associated with Agropyron repens—Eleocharis palustris associations on the flat and microhummock inner floodplain indicates the hardening of the mechanical composition and an increase in the degree of salinization in the soils under the conditions of the microhummock relief.

When the A. repens—E. palustris association develops on the sloping-ridge inner floodplain, the reverse is observed: the thickness of the soils decreases, the mechanical structure is slightly loosened, and the salinization disappears. These changes are also reflected in the plant cover, and although the dominants of the community

144

remain as before, the species composition of the community changes. Species such as <u>Potentilla</u> <u>bifurca</u> L. and <u>Carex</u> <u>schreberi</u> Schrank appear, although numerically they are few. We thus see that a description of the soil relationships of <u>A.</u> <u>repens</u> and <u>E.</u> <u>palustris</u> association indicates its broad range, connected with the existence of some associations or even smaller systematic units, linked with special soil varieties and habitat conditions.

In the southern regions of the floodplain (districts around the village of Kharabali-Petropavlovka), the community is associated with dark-colored, floodplain-meadow soils (24 cases), thick or moderately thick nonsaline soils (20 cases), heavy loams (11 cases), or clays (6 cases). In this case also, the typical habitat is the inner floodplain. But whereas in the north, <u>E.</u> <u>Palustris</u>—<u>A.</u> <u>repens</u> meadows occupy the innermost sites of the floodplain (ecologically extreme conditions of existence), in the south they cede this position to mixed herbage—<u>E.</u> <u>palustris</u>—<u>A.</u> <u>repens</u> associations, and move to the more peripheral areas, this being associated with the predominance of nonsaline soils.

One of the associations replacing it on the saline soils of the inner areas of the floodplain is made up of <u>A.</u> <u>repens</u>, <u>E.</u> <u>palustris</u>, and <u>Asparagus</u> <u>officinalis</u> L., which is rather closely associated with thick or moderately thick, heavy loam, saline soils (31 cases out of 41).

The examples given clearly show the whole complexity of the problems involved in indicator research into the soil cover on the basis of vegetation in a floodplain, with determination of salinization. A plant community ceases to be an indicator unit when we are dealing with the separation of saline and nonsaline varieties of a particular soil. A substitute can obviously be provided by some sufficiently subdivided units of the vegetation cover, reflecting very narrow differences in the floristic composition, on the one hand, and the topography of the particular habitat governing the water salinity regime, on the other. Nichols' "physiographic units" may be used, although their interpretation is highly precise. This approach can also be used to deal with the abovementioned saline and nonsaline variants of sedge-grass and <u>E.</u> <u>palustris</u>—grass meadow associations on the sloping-ridge and microhummock inner floodplain. In the evaluation of salinity by vegetation, considerable attention should also be paid to the geobotanical zonation of a territory (Viktorov, 1947), involving the identification of indicator regions, i.e., territories within which units of the vegetation cover, examined as indicators of soil conditions, have a sufficiently narrow indicator significance. In conclusion, we shall deal briefly with the practical significance which discovering the value of plant communities as indicators of soil conditions can have for accelerating and facilitating soil-improvement investigations.

The study of the soil associations of plant communities permits the determination of their concrete indicator significance in the role of indicators of soil varieties. The indicator schemes thus obtained can be used by soil scientists in the initial stages of their work for providing a general orientation of the distribution of different soils in the site studied and the rational location of soil boreholes, and also in later stages for tracing the borders of contours or for identifying specific soil types with signs of gleying.

Particularly great significance attaches to the use of indicator relationships when aerial methods are applied in soil surveys; these methods require a comprehensive analysis of the aerial photographic picture of the locality, in the formation of which the vegetation plays a leading role.

Summary

Under the conditions of the Volga-Akhtubinsk floodplain, the simplest procedure for the assessment of soils by their vegetation consists of the differentiation of soil types and subtypes according to their geobotanic characteristics. Different groups and classes of associations are linked with different soils; thus, for example, the alluvial plain soil subtype is related to meadow-steppes, while the forest-meadow soil subtype is associated with the class of floodplain oakwood associations. A division of soil types and subtypes into individual units can be effected, but in this case individual associations or even smaller units of the vegetation cover are used as indicators, rather than groups of associations. Finally, in the indication of salinization, landscape features as well as geobotanical features have to be used; small structural landscape elements, the so-called "physiographic units," are utilized.

## Literature Cited

Akzhigitova, N. I. 1958. "Plant indicators of salinization." Uzb. biol. zhur., No. 9.

Bogdanov, V. M. 1934. "A contribution to the problem of plant indicators of the steppe belt of the Northern Caucasus." Sov. botanika, No. 2.

Viktorov, S. V. 1947. "Geobotanical zonation as one of the methods of geological research." Byull. Mosk. obshch. ispyt. prirody, otd. biol., Vol. 52, No. 2.

Vladychenskii, S. A. 1953. "Characterization of salinization in the soils of the Volga-Akhtubinsk floodplain and delta." Pochvovedenie, No. 3.

Drobov, V. P. 1914. "The steppes of Akmolinsk province." In book: Preliminary Report on Botanical Research in Siberia and Turkestan in 1913. Pg.

Dubovik, Ya. F. 1951. "The salinity resistance of plants in saline soils." Bot. zhur., Vol. 36, No. 1.

Keller, B. A. 1911. "Botanical and geographical investigations in the Zaisansk district of Semipalatinsk province." Tr. pochv.-bot. éksped. issled. kolon. raionov Aziat. Rossii, No. 10.

Kovda, V. A., Egorov, V. V., Muratova, V. S., and Strogonov, B. P. 1960. "Classification of soils according to degree and type of salinization in connection with the salinity resistance of plants." Bot. zhur., Vol. 45, No. 8.

Korotkii, M. F. 1912. "A contribution to the problem of the distribution of meadow and forest vegetation in relation to soil (on the basis of research in Toropetsk district in 1908)." In book: Materials from a Study of the Vegetation of Pskov province, Carried Out under the Direction of V. N. Sukachev, Vol. 1, Pskov.

Krinitskaya, R. R. 1957. "Characteristics of the recovery of couch grass on liman meadows of the Western Caspian area in connection with problems in their improvement." Byull. MOIP, otd. biol., Vol. 62, No. 4.

Larin, I. V. 1953. Determination of Soils and Agricultural Lands by the Vegetation Cover in the Steppe and Semi-Desert Volga and Ural Interfluve. Moscow—Leningrad, Sel'khozgiz.

Letunov, P. A. 1942. Soils of the Volga-Akhtubinsk valley and the Volga delta. Collection in Memory of Academician V. R. Williams. Moscow—Leningrad.

Malina, U. 1952. "Determination of the salinization of soils by the vegetation." Sots. sel'sk. khoz. Azerbaidzhana, No. 6.

Markov, M. V. 1938. "Natural conditions of vegetation development in a floodplain," Pt. I. Tr. BIN Akad. Nauk SSSR, Vol.3, No. 4.

Markov, M. V. 1951. "Natural conditions of vegetation development in a floodplain." Pt. 2. Uchenye zap. Kazansk. gos. univ., No. 1.

Nogtev, V. P. 1932. Plants as Approximate Indicators of Soil Lime Supplies. Nizhni Novgorod.

Petrov, V. V. 1959. "A survey of data on the use of the natural vegetation as an indicator of soil conditions." Izv. TSKhA, No. 4.

Plyusnin, I. I. 1938. The Soils of the Volga-Akhtubinsk Floodplain. Stalingrad, obl. kn. izd.

Popov, A. A. 1960. "Systematics of the floodplain soils of the Volga-Akhtubinsk floodplain and the delta of the Volga River." Pochvovedenie, No. 5.

Ramenskii, L. G. 1938. Introduction to the Complex Soil and Geobotanical Research of Terrain. Moscow, Sel'khozgiz.

Ruprecht, F. I. 1866. Geobotanical Investigation on Chernozems. SPb.

Whittaker. 1957. "Plant populations and the principles of botanical indicator research." RZh Biologiya, No. 12 (abstract No. 50411).

Fedorov, B. V. 1930. "Determination of the degree of soil salinization by the vegetation cover." Golodnostepskaya Agricultural Station, Solonchak section, No. 10. Tashkent.

Tsatsenkin, I. A. 1953. "Experience in the complex geobotanical and soil mapping of meadows and hayfields in districts of the Caspian area with the use of aerial photographs." Bot. zhur., Vol. 37, No. 3.

Chayanov, S. K. 1909. "A brief report on the soils and vegetation of the Temirskii experimental field in connection with the problem of evaluating the soils of virgin steppes by the vegetation covering them." Izv. MSKhI, Book 4.

Shennikov, A. P. 1930. Volga Meadows of Central Volga Province. Materials of Geobotanical Research in 1914-1921 in the Former Simbirsk Province. Leningrad, Izd. Ul'yanovsk. okrzemupravleniya.

# ON THE RELATIONSHIP OF MEADOW VEGETATION
# WITH SOILS IN THE STUPIN DISTRICT
# OF MOSCOW PROVINCE

## L. V. Motorina

The discovery of concrete forms of soil-vegetation relationships has great theoretical and practical significance.

In 1959-1960, a joint geobotanical and soil* study was carried out on the meadows of Stupin district, Moscow province. Both the watershed areas and the valley of the Oka River were covered by the investigations.

The Stupin district is located in the south of Moscow province and from the geomorphological aspect consists of a slightly hummocky plain with a well-developed erosion pattern. The watershed areas are represented by small plateaus or hilly elevations 5-30 m above the level of the river floodplains. A substantial area occurs as slightly sloping ground with an incline of 1-2 to 3-4°.

The soil cover of the district is dominated by sod-podzolic soils of differing degrees of podzolization and gleying, alternating in the watersheds with pale-grey, strongly podozolized soils.

Of the meadow vegetation on the watershed plateaus and slopes, Anthoxanthum odoratum and Pimpinella saxifraga formations, including a series of associations, are most widely distributed.

The most widely distributed on sloping elevations and slopes, and more rarely located on flat areas of the plateau, are populations of Agrostis vulgaris, particularly Agrostis vulgaris + Anthoxanthum odoratum + Pimpinella saxifraga, although this gives way in places with intensive pasturing to an association of Agrostis vulgaris + Anthoxanthum odoratum + mixed herbage and an association of Agrostis vulgaris + mixed herbage. All these are usually associated with medium sod, slightly podzolized soils, and more rarely with medium sod, moderately podzolized soils.

The results of chemical analyses show a rather acid reaction (especially in the moderately podzolized soils) and an overall depletion of potassium and phosphorus (Table 1).

Anthoxanthum odoratum formations are most characteristic for flat areas, the predominant association being A. odoratum + Melampyrum nemorosum + Centaurea jacea, which is most frequently found along the margins of forests or in forest glades in young birch groves of the park type.

The soils are medium sod, slightly podzolized types, with signs of gleying. They are similar to the previously described soils in agrochemical characters (see Table 1).

Festuca pratensis + Vicia cracca + Trifolium pratense + T. repens and F. pratensis + Dactylis glomerata + V. cracca associations occur on level sites on the edges of gullies. They are associated with medium sod, slightly podzolized soils characterized by a more pronounced sod horizon and a substantially higher content of phosphorus and potassium (see Table 1).

---

* The soil studies were carried out by the department of soil science of the Timiryazev Agricultural Academy under the direction of Prof. M. N. Pershina and N. N. Poddubnaya, Candidate of Agricultural Sciences.

TABLE 1. Chemical Indices of the Soils of Inland Meadows

| Site | Associations | Soil | pH | $K_2O$ | $P_2O_5$ |
|------|-------------|------|-----|--------|----------|
| | | | | (mg per 100 g soil) | |
| Elevations on the plateau, slopes | Agrostis vulgaris + Anthoxanthum odoratum + Pimpinella saxifraga | Medium sod, slightly podzolized | 4.8-5.3 | 5.5-11.0 | Traces |
| Plateau (without forest) | Anthoxanthum odoratum + Melampyrum nemorosa + Centaurea jacea | Same with signs of gleying | 4.8 | 10.0 | Traces |
| Rims of gullies | Festuca pratensis + Vicia cracca + Trifolium pratense | Medium sod, slightly podzolized | 5.7 | 13.7 | 3.65 |
| Same | Festuca pratensis + Dactylis glomerata + Vicia cracca | Same | — | 46.0 | 22.0 |
| Troughs, small gullies (up to 2.5 m in depth) | Festuca pratensis + Festuca rubra + Agrostis vulgaris + Centaurea jacea | Silt with submerged horizon | 5.2 | 4.0 | 2.5 |
| Deep gullies (more than 3 m deep) | Deschampsia caespitosa + Alapecurus pratensis + associations | Silt | 4.8 | >20.0 | 2.5 |
| | Deschapsia caespitosa + Agrostis vulgaris + mixed herbage | Medium sod, slightly podzolized | 4.6-4.9 | 6.7-9.2 | 1.25-3.65 |

An association of Festuca pratensis + F. rubra + Agrostis vulgaris + Centaurea jacea is usually associated with the upper reaches of gullies and their small branches.

The soils, predominantly silt with a submerged humus horizon, are low in acidity and low in phosphorus and potassium.

Deschampsia caespitosa + Alopecurus protensis + mixed herbage and D. caespitosa + Agrostis vulgaris + mixed herbage associations are the most widely distributed in deep gullies (2.5-4 m in depth).

In the first case, the soils are medium sod, weakly podzolized and gleying; in the second (pasture), the soils are of two types: a medium sod, slightly podzolized type, and a silt type with a submerged horizon.

The slightly and moderately podzolized soils of gullies differ substantially from the weakly and moderately podzolized soils of the watershed plateau by their well-expressed humus horizon and the masking of the podzolized horizon. They differ also by their heavier mechanical structure and their slightly greater acidity.

Parallel soil and botanical investigations were carried out in the floodplain of the Oka River, the floodplain having clearly differentiated riverbed, control, and terrace sections.

The riverbed section is dominated by plant communities formed primarily by rhizomatous and stoloniferous plants. Mixed herbage—quackgrass—timothy and brome—Agrostis alba—fescue associations have the widest distribution.

In the first case (in which the dominant components are Phleum pratense L., Agropyron repens (L.) P.B., Berteroa incana(L.)D.C., and Achillea millefolium L.), the herbage is slightly thin (projective covering, 75-85%) and comparatively low in height (40-60 cm on the average).

TABLE 2. Chemical Indices of the Soils of the Floodplain of the Oka River

| Site | Associations | Soil | pH | $K_2O$ | $P_2O_5$ |
|------|-------------|------|-----|--------|----------|
| | | | | (mg per 100 g soil) | |
| Riverbed flood-plain | Phleum pratense + Agropyron repens + mixed herbage | Thin, meadow soil of the stratified floodplain | 6.7 | 3.6 | 9.0 |
| Central floodplain, ridged | Festuca rubra + Trifolium pratense + mixed herbage | | 6.7–6.9 | 3.6 | 10.0–17.5 |
| Elevated-flatland | Libanotis intermedia—Heracleum sibiricum + Chaerophyllum prescottii | Thick meadow-sod soil of granular floodplain on a submerged stratified horizon | 5.0–6.9 | 3.6–6.2 | 10.0–15.0 |
| Trough-flatland | Festuca pratensis—Poa pratensis + Festuca rubra + mixed herbage | | 6.5–6.8 | 3.6 | 7.5–12.5 |
| Deep troughs | Bromus inermis + Alopecurus pratensis | Thick, meadow-sod gleying | 6.7 | 4.5 | 15.0 |
| Sinkholes of the flat floodplain | Festuca pratensis—Alopecurus pratensis + Trifolium pratense | Thick, meadow-sod, gley | 7.0 | 3.6 | 5.0–7.5 |
| Terrace floodplain | Deschampsia caespitosa—Agrostis canina + mixed herbage | Deep gley sod | 6.5–7.0 | 3.6–4.5 | 5.0–10.0 |

The soil is the thin sandy meadow-sod type of a stratified floodplain. The thickness of the sod does not exceed 5 cm, and there is a significant amount of sand through the entire profile.

The role of mixed herbage and legumes increases in the brome—Agrostis alba—fescue association (made up of Festuca pratensis L., F. rubra L., Bromus inermis Leyss., Agrostis alba L., Phleum pratense L., Galium verum L., Libanotis intermedia Rupr., Trifolium pratense L., etc.). The projective covering is 90-95%, and the height of the herbage is 70-100 cm.

The soil cover shows some changes: sod formation is more clearly expressed, the thickness of the sod being about 9 cm. The soil is the sandy loam, moderately thick meadow-sod type of a stratified floodplain.

Both soil varieties have a satisfactory acidity level, but are depleted in nutrient substances in a form available to plants (Table 2). The central floodplain is characterized by a trough-ridge relief, which changes into a flat plain. Seven plant associations and three soil varieties are the most typical. On ridges, in shallow (1-1.5 m) troughs and in the flat areas, the soils are light-loam, thick, meadow-sod types, characteristic of granular floodplain on submerged stratified horizons; in the deep troughs there are thick, meadow-sod gleying soils; and thick, meadow-sod gley soils occur in sinkholes of the flatland area.

Five associations are linked with the first soil variety.

Festuca rubra + Trifolium pratense + mixed herbage and Bunias orientalis + Festuca rubra associations are distributed on the ridges, the second having arisen as a derivative of the first as a consequence of incorrect agricultural exploitation.

A submerged stratified part of the soil occurs here in most cases, beginning at a depth of 65-70 cm. The soil reaction is close to neutral. The content of phosphorus is above average, and the potassium level is very low (Table 2).

On the level ridge complex (elevated-flatland section), we find a tall mixed-herbage + grass population up to 1.4-1.7 m in height (made up of Libanotis intermedia Rupr., Heracleum sibiricum L., Chaerophyllum prescottii D.C., etc).

The soil variety is the same as above apart from a few morphological differences: the stratified part of the soil profile occurs no lower than 35-45 cm, indicating the somewhat greater sod formation of these soils.

In shallow trough, on slopes, and on levelled sites in the central floodplain, we find an association of Festuca pratensis + F. rubra + Poa pratensis + mixed herbage, legumes being an important constituent of the herbage.

The soils are light-loam, thick meadow-sod soils of granular floodplain on a submerged stratified horizon. The latter is usually observed at a depth of 35-50 cm. Some changes are noted in the structure of the soil under the mixed herbage—fescue—meadow grass associations. The structure of the submerged stratified section and the low-lying horizons (in some cases, part of the sod horizon also) is lumpy and cloddy, in contrast to the granular structure of the entire profile under the associations examined above. The soil reaction is close to neutral. The phosphorus level is lower than on ridges. The potassium content is particularly low (see Table 2).

Under these conditions we also find an association of Rumex confertus + Poa pratensis + Festuca rubra; in our view, this constitutes a modification of the mixed herbage—fescue—meadow grass association which had arisen under the influence of many years' cutting for hay and the lack of adequate management of the meadows.

In deep (2-4 m) troughs in the central floodplain, we find a Bromus inermis + Alopecurus pratensis association, characterized by a tall (90-100 cm), dense grassy population, with a very small proportion of mixed herbage and an almost complete absence of legumes.

The soils blanket is submitted by meadow-sod, gley soils in which slight signs of gleying occur, approximately from a depth of 60 cm.

Shallow sinkholes in the flatland section of the central floodplain show an association of Festuca pratensis + Alopecurus pratensis + Trifolium pratense, incorporating such hydrophilic plants as Filipendula ulmaria (L.) Maxim., Galium uliginosum L., and others.

The soil is a light-loam, thick, meadow-sod, gley type. Signs of gleying in the form of bluish and ochre spots occur from 30-35 cm, and there is a blue gley horizon beginning from 60-90 cm down.

In the terrace section of the floodplain, two main associations are encountered: Festuca pratensis + Agrostis vulgaris + mixed herbage, incorporating Deschampsia caespitosa, and D. caespitosa + Agrostis canina + mixed herbage, together with sedges. Typical for both are light-loam, deep-sod, gley soils of the terrace floodplain, characterized by their degree of gleying. Whereas in the first case, signs of gleying occur from a depth of 15-30 cm, in the mixed herbage—Agrostis canina—Deschampsia caespitosa association the soils typically show gleying from the surface downwards. In other respects, they are approximately the same (see Table 2).

It is apparent from the above that specific plant communities are associated with definite soil types. For example, Agrostis vulgaris meadows of blind crecks are, in the main, associated with medium sod soils, depleted in nutrient elements available to plants. Typical Anthoxanthum odoratum meadows are most frequently encountered on medium sod, slightly podzolized, gleying soils, also depleted in nutrient elements. Festuca pratensis meadows of watersheds are also associated with medium sod, slightly podzolized soils, but the latter have a more pronounced sod horizon and a better supply of nutrient elements. Specific soil types in the Oka River floodplain, typical for the riverbed, central, and terrace sections of the floodplain, are also linked with specific plant communities, which occur only in particular conditions. For example, Alopecurus pratensis + Bromus inermis associations are confined to the thick meadow-sod gleying soils of troughs with deep groundwaters. A mixed herbage + Alopecurus pratensis + Festuca association is located on thick meadow-sod soil in boggy ground.

Gleying is observed from a depth of 15-30 cm under a mixed herbage + Agrostis vulgaris + Festuca pratensis association incorporating Deschampsia caespitosa, which occurs on gley soils of the terrace floodplain, while gleying starts from the surface under the mixed herbage + Agrostis canina + D. caespitosa association, and so on.

It would hardly be correct to speak of the above-mentioned plant communities as indicators of soil types in the broad sense.  However, under the specific conditions of the Stupin district and neighboring areas,  the phytocenoses surveyed do have definite indicator significance.

Summary

In the district investigated, communities of Agrostis vulgaris and Anthoxanthum odoratum are indicative of medium sod soils, very poor in nutrient elements.  Associations in which Festuca pratensis is dominant develop on soils of the same type which are, however, richer and more fertile.  In the floodplain of the Oka River, plant communities were found to be closely related to the soil varieties typical of the riverbed, central, and terrace parts of the floodplain.  As well as the correlation between the vegetation and taxonomic soil units, there is a well-expressed relationship between the character of the vegetation cover and the individual morphological characters of the soils.

# THE ASSOCIATION OF CERTAIN PLANT COMMUNITIES
# WITH SPECIFIC SOIL TYPES IN THE DESERT STEPPES
# OF CENTRAL KAZAKHSTAN

## N. V. Pavlova

The investigations discussed in this paper comprise part of the problem, "Biological complexes of new development regions, and their rational utilization and improvement," with which the joint biological expedition of the USSR Academy of Sciences in Central Kazakhstan was concerned. The work was carried out on the territory of the station located in the Zhara-Arkinsk district of Karaganda province, comprising the foothills of the Koksengir Mountains; this represents the section of the Central Kazakhstan hilly area in the northern part of the western Prisarysuisk depressed plain. Using the methods of indicator geobotany, we attempted to establish some of the regular patterns in the association of plant communities with specific soil types in the district investigated. From the vegetation point of view, the territory of the station belongs to the Eurasian steppe zone, and the subzone of wormwood-sod-grass desert steppes (Lavrenko, 1947, 1956), a characteristic feature of which is the combination of zonal (wormwood—feather grass and Artemisia lercheana—feather grass) communities, associated with the contours of hills and on different forms of light-chestnut soils (Popov, 1940), with Artemisia pauciflora, Atriplex cana, Anabasis salsa, and Nanophyton erinaceum on saline soils of the solonetz and solonchak types. As a result of the investigations carried out, it was established that phytocenoses, the edificators of which are various species of feather grass (Stipa lessingiana Trin. et Rupr., S. sareptana Becker, and S. rubens P. Smirn.) and certain species of wormwood (Artemisia sublessingiana (Kell.) Krasch. and A. marschalliana Bess), as well as white wormwood (A. lercheana Web.) in combination with Stipa lessingiana, S. sareptana, and Festuca sulcata Hack., are associated with different forms of light-chestnut soils, having a series of specific characteristics. Thus, the thickness of the humus horizon in soils under feather-grass communities, the edificators of which are the above-mentioned species of Stipa, is greater ($A_1$—10-26 cm) than in soils under A. lercheana communities, where the thickness of this horizon does not as a rule exceed 20 cm (6-10-12 cm). The degree of solonetzization is also different: solonetzization of the B horizon in soils under feather-grass communities (Festuca sulcata—A. lercheana—Stipa, or F. sulcata—A. lercheana—S. capillata) is either entirely absent or is very weakly expressed; in soils under A. lercheana communities (F. sulcata—Stipa—A. lercheana, F. sulcata—S. capillata—A. lercheana, etc.), the degree of solonetzization of this horizon is increased (especially under pure A. lercheana phytocenoses, where solonetzes are normally developed); with the aim of obtaining more precise information on this relationship, we prepared a three-meter-deep trench in a site with a F. sulcata—A. lercheana—Stipa community, which was interspersed with patches of S. lessingiana and A. lercheana. We found under S. lessingiana a typical light-chestnut soil without signs of solonetzization: $A_1$—0-20 cm, effervescence with 10% HCl begins at 20 cm and continues to the bottom of the profile (100 cm); a gypsum horizon occurs at a depth of 70 cm. There is a medium solonetz under A. lercheana: $A_1$—0-9 cm; the solonetz horizon B is from 9-26 cm; soil effervescence with 10% HCl begins at 16 cm. Thus, the introduction of A. lercheana into a Stipa community indicates an increase in solonetzization in light-chestnut soils.

Different combinations of the above-mentioned species do not permit any judgment to be made of the degree of alkalinity of the soils, which depends on the depth of bedding of the carbonate and gypsum horizons. The latter may be located at different depths in these soils, depending on the mechanical structure (the lighter the mechanical structure, the deeper these horizons are bedded). There does, however, exist a group of species which may be indicators of an increase in the degree of soil alkalinity. Thus, if Artemisia sublessingiana, A.

austriaca Jacq., Ferula songorica Pall., Caragana balchaschensis (Kom.) Pojark., and Stipa capillata L. enter S. capillata communities (F. sulcata—Artemisia lercheana—S. capillata, A. lercheana—F. sulcata—S. capillata, and others), this always constitutes evidence of an increase in alkalinity. The carbonate horizon in these cases descends from the normal level for light-chestnut soils (25-35 cm) to 43-58-60-69 cm. The soils under such phytocenoses thus seem to be substantially more alkaline.

It is interesting to note that the introduction of petrophytic herbage (Ephedra and other genera) in combination with S. lessingiana and Kochia prostrata (L.) Schrad. into the above-mentioned communities, the edificator of which is Artemisia lercheana, is always associated with a reduction in the solonetz properties of the light-chestnut soils: in these cases horizon A is not directly followed by a $B_2$ solonetz horizon, but by a transitional AB horizon, under which a nonsolonetz $B_1$ horizon occurs.

A distinctive characteristic of light-chestnut soils under Stipa rubens—Artemisia marschalliana communities (the edificators of which are S. rubens and A. marschalliana) is their meadow character, as evidenced by the greater humification of the upper horizons, the very strong alkalinity (no soil effervescence with 10% HCl within the profile range, 150 cm), and the moistness of the lower horizons.

Some features of the salt composition of the soils under different phytocenoses are of interest. We shall take four communities as examples: A. lercheana—Stipa, A. lercheana—S. capillata, S. capillata—A. lercheana, and a community representative of solonchak vegetation: Atriplex verrucifera M.B., Limonium gmelinii Ktze., L. suffruticosum (L.) Ktze., and Suaeda physophora Pall.

Light-chestnut soil under the A. lercheana—Stipa community was the least salinated. This soil is a weak solonchak type, since the horizon containing more than 0.3% readily soluble salts occurs at a depth of 80-150 cm. Bicarbonates, mainly of calcium, predominate in the upper horizons down to 16-20 cm, with gypsum occurring below 20 cm. The light-chestnut soils under the A. lercheana—S. capillata and S. capillata—A. lercheana communities, as regards degree of salinization, belong to the solonchak, light-chestnut group, since the salinated horizon occurred at a depth of 30-80 cm, at the first sampling date.

At the end of August, their degree of salinization is increased; they acquire the features of solonchak, light-chestnut soils, since the salinated horizon shifts toward the surface, to a depth of 5-30 cm. We thus see that the degree of salinization in soils under A. lercheana—S. capillata and S. capillata—A. lercheana communities is very dynamic and changes during the course of the growth period. Furthermore, about 6-8 times more salts are accumulated at the end of August in the upper horizons of the soil profile under the A. lercheana—S. capillata community than in May, whereas the dry residue in the upper horizons under the S. capillata—A. lercheana community is only 2-3 times greater in August than at the end of May. The degree of salinization under the A. lercheana—Stipa community does not change during this period, in spite of some movement of salts through the profile. Salinization is in all cases sulfatic. Under the saltwort community, the edificators in which are Atriplex verrucifera and Limonium gmelinii, the soil is a solonchak, from the level of salinization. The latter is chloridic, and is increased substantially at the second sampling date.

Salinization of soils thus increases sharply when we move to communities of the solonchak vegetation type. On the territory studied, these communities include the following: Artemisia nitrosa—Atriplex and Camphorosma—Atriplex phytocenoses in combination with Limonium gmelinii, Halosnemum strobilaceum, and Suaeda communities, and also annual saltwort communities, associated with different forms of solonchak soils, or dry meadow soils, depending on the reasons responsible for the salinization of the soil. The presence of saltworts in any community indicates that a solonchakization process is in progress in the soil.

In the district where our studies were carried out, the soils with the most clearly expressed solonchak properties are associated with Atriplex cana communities (edificator—Atriplex cana C.A.M.): these are typically shallow (the solonetz horizon begins 5-6 cm from the soil surface), or, more rarely, medium (solonetz horizon 12 cm from the soil surface) solonetzes, which agrees with data in the literature (Popov, 1940). Under Artemisia pauciflora communities, in which Atriplex cana, Anabasis salsa (C.A.M.) Benth., and Psathyrostachys juncea (Fisch.) Nevski are present together with Artemisia pauciflora Web., there occur more alkaline deep solonetzes (the solonetz horizon begins 16 cm or deeper from the soil surface), or light-chestnut, deep, strongly solonetz soils, and sometimes solonetz-solonchaks. The solonetzization of the soils decreases when we move to communities of which the edificator is Ps. juncea; thus A. pauciflora—Ps. juncea communities are associated in the region with light-chestnut, moderately solonetz soils. The presence of Ps. juncea in the role of edificator in a Stipa community indicates an increase in the solonetz properties of light-chestnut soils, which are typical for Stipa communities. In cases where Ps. juncea is not the edificator, it does not act as an indicator of soil properties. Thus, depending on whether Ps. juncea is an edificator in communities formed by steppe or desert plants, it indicates, respectively, an increase or a reduction in the solonetz properties of zonal soils. Phytocenoses, in which the edificators are Anabasis salsa and Nanophyton erinaceum (Pall.) Leyss., are associated with more salinated soils in the region where our research was carried out: they are surface types or have shallow solonetz-solonchaks.

An indicator of a meadow type of soil is provided by Aneurolepidium ramosum (Trin.) Nevski, when it forms a community with Artemisia pauciflora and with Atriplex cana. Under these circumstances, A. ramosum indicates light-chestnut, moderately solonetz meadow soils, and, in our region, carbonate soils. Communities formed by various shrubs are associated with light-chestnut, strongly alkaline meadow soils. Apart from their meadow features, these soils differ from typical light-chestnut soils by their lighter mechanical structure. The most strongly alkaline types are the light-chestnut soils on which we find populations of Spiraea hypericifolia L. and Lonicera tatarica L. with a herbage layer formed by Stipa capillata, Elymus angustus Trin., Artemisia austriaca, A. sublessingiana, and A. marschalliana. The soils under growths of Spiraea alone and of Spiraea with Caragana balchaschensis are less alkaline; the carbonate horizon in these soils begins at a depth of 38-43-49 cm.

Summarizing the results of these investigations, we can without doubt assert that the vegetation cover allows an assessment not only of the type of soils dominant at the present time but also the direction of the soil-forming process.

## Summary

In the steppes of Central Kazakhstan, plant communities which have as edificators various species of feather-grass (Stipa lessingiana, S. sareptana, and S. rubens) and certain species of wormwood (Artemisia sublessingiana, A. marschalliana, and A. lercheana) in conjunction with the Stipa spp. noted, are associated with particular forms of light-chestnut soil. The colonization of petrophytes (including Ephedra sp.) is associated with a decrease in alkalinity. Communities of Atriplex cana and Artemisia pauciflora are indicators of alkaline soils (solonetzes). Meadow alluvial soils are indicated by Aneurolepidium ramosum communities.

## Literature Cited

Lavrenko, E. M. 1947. Geobotanical Zonation of the USSR. Moscow—Leningrad, Izd. Akad. Nauk SSSR.

Lavrenko, E. M. 1956. "Steppes and agricultural lands in steppe areas." In book: The Vegetation Cover of the USSR, Pt. 2. Moscow—Leningrad, Izd. Akad. Nauk SSSR.

Popov, M. G. 1940. "The vegetation cover of Kazakhstan." Tr. Kazakhsk. fil. Akad. Nauk SSSR, No. 18. Moscow—Leningrad, Izd. Akad. Nauk SSSR.

# PLANT INDICATORS OF SOILS OF THE WATERSHEDS
# OF THE SOUTHEASTERN DISTRICTS OF ROSTOV PROVINCE

## O. S. Gorozhankina

The significance of plants as indicators of natural conditions is generally recognized. In particular, individual plant species can be used as indicators in the compilation of maps of the regenerated vegetation cover of territories under development, where the indigenous plant communities are preserved only to a slight extent.

Indicators consist of plants which clearly react to changes in natural conditions but are stable with respect to different types of agricultural treatments and thus occur not only in indigenous cenoses but also in all (or at least in many) productive groupings of human origin. In other words, the species themselves are accurate not so much as associations but more as a reflection of the habitat.

Species satisfying these requirements were selected both on the basis of data in the literature and our own observations.

In particular, in diagnosing the variants of the steppe which constitutes the zonal vegetation of Rostov province, we used primarily representatives of perennial herbage, semishrubs, and low-growing steppe shrubs.

They were divided into the following three main ecological groups:

1. Comparatively mesophilic species, characteristic of the mixed herbage-sod-grassy steppe on chernozems. They include Amygdalus nana L., Inula germanica L., Limonium latifolium (Sm.) Ktzl., and others.

2. Species with a lower requirement of water (moderately xerophilic), also distributed in mixed herbage-grassy-sod steppe, but to some extent also entering the dry grassy-sod steppe on soils of the chestnut type. This group includes Medicago romanica Prod., Salvia stepposa D. Sch., Artemisia austriaca, and various other species.

3. Xerophilic species, located primarily in the dry and desert grassy-sod and wormwood-grassy-sod steppes: they include Artemisia lercheana Web., Kochia prostrata (L.) Schrad., Limonium sareptanum (Beck.) Gams., and Linosyris villosa D.C.

It is known that the dry and desert steppes are to a large extent complex, there being patches of more xerophilic solonetz vegetation and more mesophilic microdepression vegetation on the basic steppe background. This type of complexity is sometimes also observed in the chernozem grassy-sod steppe.

The aim of the present investigation was to obtain an objective assessment of the degree of reliability and sensitivity of the selected indicator plants and, in particular, the degree of their association with specific varieties of soil.

For this purpose, we used data from soil investigations carried out, with the participation of the author, by the Southern Institute for the Planning of the Water Economy (Rostov-on-the-Don).

The natural conditions of the territory studied (Sal-Manych interfluve and the slopes of Ergeni) are highly variable and change at short distances. In the western part, where the climate is less dry, southern chernozems and dark-chestnut soils develop on plakor areas, while further to the east, where there is an increase in the aridity of the climate, these soils give way to chestnut and light-chestnut soils.

To test the degree of sensitivity of the three ecological groups of plant indicators mentioned above, 845 profiles were made in the southern chernozems and also in dark-chestnut, and light-chestnut soils. In each soil variety, 60-70 profiles were included and these were made in plots differing in their economic state: on virgin land and on fallows of varying age.

The results showed that the indicator species of the first, relatively mesophilic group, are, in the region studied, associated with southern chernozems, while the second, moderately xerophilic, group is encountered both on chernozems and on dark-chestnut soils. Species of the third group, the most xerophilic, are located on all soils with the chestnut type of formative process, although their abundance shows a steady increase as moisture conditions deteriorate.

However, we did not observe an exact association of each group of indicator species with specific soil subtypes. Thus, Limonium latifolium, which is typical of chernozems, also ingresses on dark-chestnut soils, while Artemisia lercheana, a member of the moderately xerophilic group, is encountered not only on dark-chestnut but also on chestnut soils, although more rarely on the latter. On the other hand, desert species are observed not only on soils of the chestnut formation type, but also sometimes on southern chernozems.

All this would seem to indicate the inadequate sensitivity of the selected indicators to soil conditions. However, a comparison of the composition of indicator species on individual soil varieties within each subtype shows the opposite.

We thus see that the indicator species examined provide the possibility of quite accurate differentiation, not only of soil subtypes, but also smaller systematic units, represented by the solonetz and nonsolonetz varieties of each subtype. For example, solonetz southern chernozems can easily be differentiated both from nonsolonetz variants of the same subtype and from light-chestnut soils. They differ from the former by the presence of xerophilic desert species, and from the latter by the presence of a substantial number of plants of the mesophilic-steppe group. Still greater sensitivity is shown by the indicator species selected by us for chestnut soils.

The research results presented demonstrate that all the plants which we have identified can be considered as reliable and sensitive indicators of zonal soils in the southeastern districts of Rostov province.

Summary

Associations were demonstrated between Amygdalus nana, Inula germanica, and Limonium latifolium and southern chernozem soils, between Medicago romanica, Salvia tesquicola,* and Artemisia austriaca and southern chernozems and dark-chestnut soils, and between Artemisia lercheana, Kochia prostrata, Limonium sareptanum, and Lynosyris villosa and chestnut soils. The latter four species become more abundant with increasingly unfavorable humidity conditions.

*Text of paper gives S. stepposa—Tr.

# ECOLOGICAL AND GEOGRAPHICAL PREREQUISITES
# FOR GEOBOTANICAL INDICATION OF SALINIZATION
# OF SOIL-FORMING ROCKS

## N. G. Nesvetailova

The problem of using the vegetation cover as an indicator of salinated soils and soil-forming rocks is now attracting the attention of a very wide circle of research workers. The resolution of this problem has been dealt with in several specialized papers, which have discussed quite fully principal procedural problems of this approach (Vyshivkin, 1959; Nesvetailova, Rodman, 1959; Akzhigitova, 1960; Tagunova, 1961).

A considerable amount of factual material, which has been collected in this field of research as a result of work carried out by the All-Union Aerogeological Trust, leads us to an examination of some problems which have more general significance for the halidic indicator approach. In particular, we will mention the following:

1) the problem of the role of second-degree (in respect of abundance) components of a community in evaluating its indicator significance;
2) the problem of the accuracy and validity of the evidence provided by an indicator in relation to the width of its ecological range (local variability in salinization);
3) the problem of the consistency of the significance of an indicator under different geographical conditions in relation to the problem of the extrapolation of geobotanical data in haloidic indicator research (regional variability in salinization).

We have attempted to consider these problems using as examples associations of Artemisia lercheana Web., A. pauciflora Web., Anabasis salsa (C.A.M.) Benth., Halosnemum strobilaceum M.B., and Salicornia herbacea L.

With the aim of obtaining as reliable initial data as possible in working out the results of analyses of soil and subsoil aqueous extracts, we used only data which were completely reliable with respect to the species affiliation of the edificator. To avoid possible mistakes, we did not take into account analyses from areas where one might expect the area of distribution of the edificator to run together with those of closely related species difficult to distinguish from it.

This applies particularly to Artemisia lercheana Web., which can sometimes be confused with A. terrae-albae Krasch. and A. astrachanica P. Pol. in field work.

The following gradations were adopted for evaluating the degree of salinization (%):

| | |
|---|---|
| Negligible | up to 0.01 |
| Very slight (practically absent) | 0.01-0.25 |
| Slight | 0.26-0.5 |
| Moderate | 0.51-1.0 |
| Strong | 1.01-2.0 |
| Very strong | 2.01-5.0 |
| Exceptionally strong | 5.0 |

Basic to the qualitative evaluation of salinization was the concept of a salt geochemical phase, by which we mean sections of the soil-forming rocks which have uniform salinity. The phase representation includes all the ions of water-soluble salts, the content of which, expressed in meq %, is not less than 12.5 (with a total sum

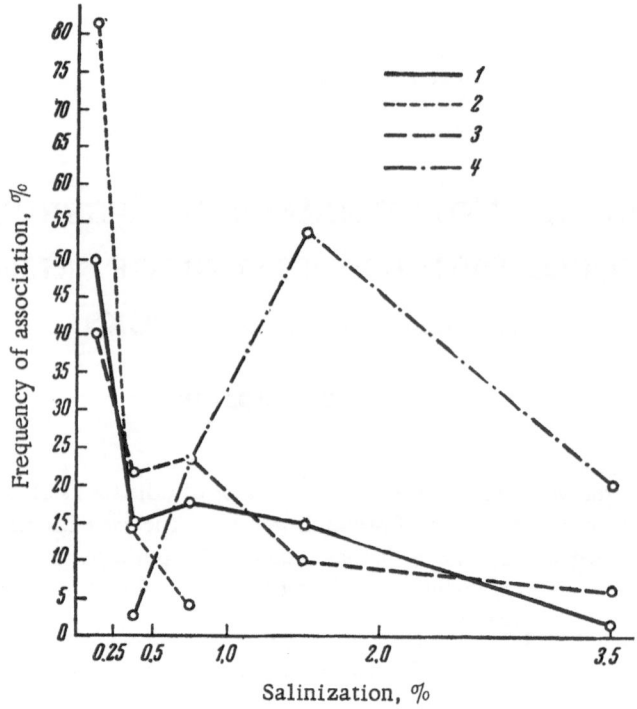

Fig. 1. Halidic indicator graphs of <u>Artemisia</u> <u>lercheana</u> Web., and its associations for the Caspian area: 1) <u>A. lercheana</u> curve; 2) same, for grass—<u>A. lercheana</u> association; 3) same, <u>A. lercheana</u> association; 4) same, saltwort—<u>A. lercheana</u> association.

of anions and cations equal to 100), these being arranged in order of decreasing quantity. The need for the elimination of minor and second-degree differences in the geochemical complex of a single type of soil-forming rock, differences which are usually not reflected by the vegetation (in particular the recombination of ions affecting in part the proportions of individual anions with cations), compelled us to employ not phases, but groups of phases for the assessment of the indicator properties of communities. Groups of phases are made up by bringing together phases with the same ion composition and the same dominant anion and cation.

The relationship of the associations examined with the salinization of subsoils is described in terms of the frequency of their occurrence under different salinization conditions and is illustrated by graphs, which we have named halidic indicator graphs, and in which the amount of the dry residue (as a %) is plotted on the abscissa, while on the ordinate axis we plot the number of actual phytocenoses of the particular association under which a certain type of subsoil salinization is established.

## The Connection of Edificator and Community as Indicators

The idea that the analysis of the indicator significance of individual edificator species, even if distinctive and widely distributed, cannot provide reliable data for the diagnosis of salinity without the use of condominants and some other components of the association, is not new. Confirmation of this viewpoint can be found in numerous research contributions devoted to problems of indicator research. The same conclusion is also reached from the analysis of factual material on the relationship of vegetation with the salinization of soils and sub-soils, as may be quite clearly demonstrated using <u>Artemisia lercheana</u> and its associations as an example.

<u>A. lercheana</u>, an edificator of several widely distributed associations in the Caspian area, has a very broad ecological range with respect to one of the most significant environmental factors in the area, the degree of salinization. The curve describing this range on the graph (Fig. 1) encompasses a field varying from prac-

158

tically nonsaline to very strongly salinated subsoils and has two raised areas: one, the more marked of the two, corresponds to nonsaline subsoils, while the second, which is very extended, incorporates a broad range from slight salinization to strong salinization. As may be seen from Fig. 1, this generalized curve can be divided into segments, corresponding to ecological niches occupied by individual associations of the particular edificator, or their variants, the differences between which are sometimes determined by the content and proportions of second-degree components. The curves describing each of these associations have a narrower range, are simple in type, and usually have a single clear peak, this constituting evidence of the greater precision of the indicator significance of the association as compared with the indicator. The associations mentioned can be arranged in some kind of ecological series, in which the curve for the grass—A. lercheana associations (including Festuca sulcata Hack., Stipa capillata L., S. sareptana Beck., and S. lessingiana Trin. et Rupr.) occupies the most left-hand part of the graph and is located in the area of one of the peaks of the general curve (the homologous nature of the curves for each of the associations of this group allows us to discuss them together, for the sake of convenience of presentation); the A. lercheana shows a clear shift to the right; and finally saltwort—A. lercheana associations occupy an extreme right-hand position, thus shifting in the graph primarily to the area of strong salinization.

The appearance of these individual curves are rather different. Thus, the curves for the saltwort—A. lercheana and grass—A. lercheana associations both have a single, clearly expressed peak, corresponding to the nonsaline and strongly saline subsoils, respectively. The curve of the A. lercheana association is very reminiscent of the general curve of the edificator which is characterized by two peaks.

The conclusion to be drawn from this is that in general, an indicator in halogenic indicator research should be an association of the edificator species rather than the edificator species itself.

This applies particularly to species having as wide an ecological range as A. lercheana; halogenic indicator research on the basis of species is apparently possible only with the most extreme halophytes, such as Salicornia herbacea, Halosnemum strobilaceum, and some other species having a very narrow ecological range.

The accuracy of the indicator evidence of individual associations can vary considerably depending on their potential ecological range. By no means all associations of the usually accepted amplitude can therefore serve as sufficiently reliable and precise indicators of a particular degree of soil and subsoil salinization. It is fully evident that the use for such purposes of associations of which the relationship with salinization is characterized by complex multipeak curves and a rather large span, leads to well-known practical difficulties and loses the more basic research approach, which we will again deal with below.

The Phenomenon of Local Variability of Salinization

By local variability of salinization, we understand a regular disparity of the results of analyses of subsoils under the same community under local conditions. This type of phenomenon occurs even in comparatively uniform conditions and within the limits of a single landscape type on a comparatively small territory. It can affect both the level of salinization and its qualitative make-up, and is observed in associations with a rather narrow ecological range as well as those with a broad range. An illustration of this in respect of the first group of communities is provided by any of the curves describing the specific A. lercheana association on the graphs, and as regards the second group, the point is illustrated by the material in a table showing the nature of the relationship of an Anabasis salsa—Eremopyrum orientale association with subsoil salinization in a comparatively small and quite uniform area in the Mangyshlak peninsula.

Recognized variation in salinization also occurs in multidominant associations of halophytes, e.g., Halosnemum strobilaceum, which are characterized by a very clear association with a limited group of external influences. Even within the limits of a single large sor, some differences are noted in the type of salinization of the sor deposits in different sections of an H. strobilaceum belt around its inner zone (see table).

As is obvious from the data shown, the phenomenon mentioned is widely distributed and is regular in character. An explanation for it should be sought in the plasticity of the plants, or their polymorphism.

Analyses of Aqueous Extracts of Subsoils

| Salinization | Number of analyses in associations | | |
|---|---|---|---|
| | Anabasis salsa — Eremorpyrum orientale | Halosnemum strobilaceum | |
| | Mangyshlak peninsula | Kaidak sor | |
| Slight.................... | 1 | — | — |
| Moderate................. | 7 | 1 | — |
| Strong.................... | 13 | 1 | 1 |
| Very strong............... | — | 2 | 7 |
| Exceptionally strong.......... | — | 4 | 1 |
| Qualitative composition: | | | |
| $Na^{\cdot}-Cl'$ .................. | 2 | 2 | 3 |
| $Na^{\cdot}-Cl'-SO_4''$............. | — | 3 | — |
| $Na^{\cdot}-Cl'-Ca^{\cdot\cdot}-SO_4''$........ | — | 1 | 1 |
| $Na^{\cdot}-Cl'-Mg^{\cdot\cdot}-SO_4''$ ......... | — | 1 | — |
| $Na^{\cdot}-SO_4''$ ................... | 4 | — | — |
| $Na^{\cdot}-SO_4''-Cl'$.............. | — | — | 2 |
| $Na^{\cdot}-SO_4''-Ca^{\cdot\cdot}$............ | 1 | — | — |
| $Na^{\cdot}-SO_4''-Ca^{\cdot\cdot}-Cl'$.......... | — | — | 3 |
| $Ca^{\cdot\cdot}-Cl'-Na^{\cdot}-SO_4''$.......... | — | 1 | — |
| $Ca^{\cdot\cdot}-SO_4''$ ............... | 11 | — | — |
| $Ca^{\cdot\cdot}-SO_4''-Na^{\cdot}-Cl'$.......... | 3 | — | — |
| Total ......... | 21 | 8 | 9 |

The variation in the qualitative aspects of salinization depends in the main on second-degree components of the geochemical phase index and may apparently be explained by the fact that individual indicators do not establish a relationship with all the components to the same degree, but are merely interrelated with some components, naturally the dominant ones.

Plant polymorphism can appear in a particular case by the occurrence of ecological and ecomorphological forms of the initial monolithic species, and these forms occupy habitats which differ somewhat in salinization. In those cases where changes of this type affect the edificator, we are in fact dealing with the subdivision of the initial association into a series of corresponding variants.

Apparently, this is what happens to associations of A. lercheana in the western and northwestern parts of the Caspian area.

The combination of analytical data on these ecologically varied association variants into a single group was clearly responsible for the complex shape of the curve characterizing the association of the degree of salinization with the community referred to, and for the presence of two peaks in the curve (Fig. 1).

We are apparently faced with a similar phenomenon in a group of A. salsa associations: in contrast to A. salsa (C.A.M.) Benth., which is widely distributed on sodium chloride and sodium sulfate salinated subsoils in the Caspian area, several authors (Momotov, 1953) have recently identified a special form, Anabasis salsa f. ramosissima, which is widely distributed on subsoils with calcium sulfate salinization.

The Phenomenon of Regional Variability of Salinization

The variability in salinity indices under one association shows an even greater increase if we examine the particular association on a territory which is sufficiently wide and complex from the physical and geo-

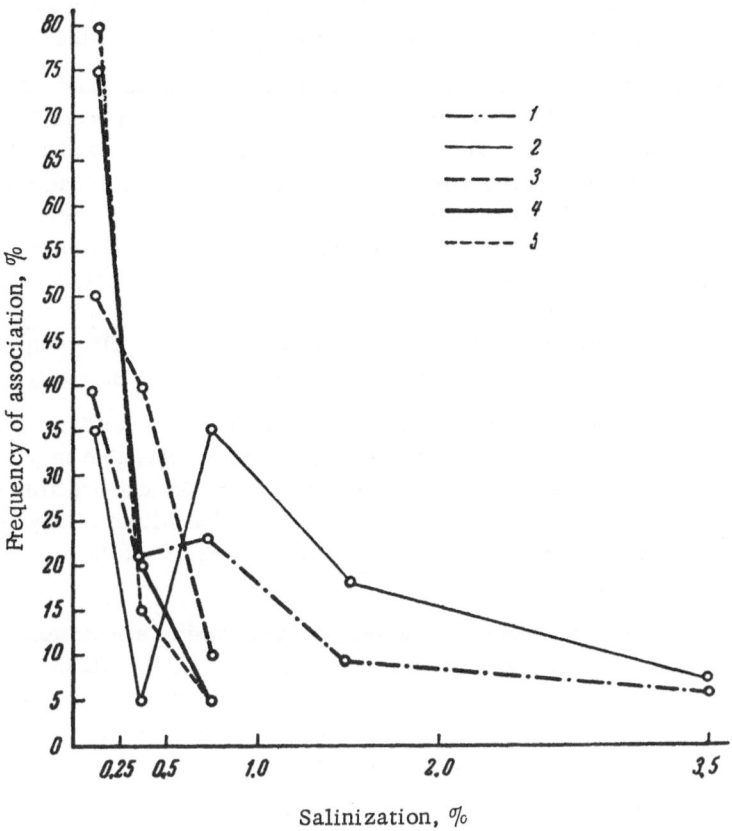

Fig. 2. Halidic indicator graphs of associations of Artemisia lerche-ana Web. for different regions of the Caspian area: 1) graph of A. lercheana association for the Caspian area as a whole; 2) same, in the northwest Caspian area (Black Earths, Sarpinsk depression); 3) same, in the northern Caspian area; 4) graph of grass—A. lercheana association in the northwest Caspian area; 5) same, in the northern Caspian area.

graphical aspect. This applies to both the degree of salinization and its qualitative aspects. As may be seen in Fig. 2, the nature of the relationship of A. lercheana associations with the degree of subsoil salinization undergoes some changes even in regions which are not far from one another, such as the Black Earths of the Sarpinsk depression and the. interfluve of the Volga and Ural Rivers, and the left bank of the Ural River.

The curves showing the ecological relationship of this association in the aforesaid regions differ both in their form and in the area covered on the graph: thus, the curve for the Volga-Ural interfluve has one peak and does not go beyond the moderate salinity limits, while the curve for the most western regions has two clear peak zones, one of which coincides with that of the previous curve, while the other corresponds to moderate salinization and stretches up to the very strong salinization area. The double-peak character of the curve for the Black Earths and the Sarpinsk depression reflects the presence of two different types of habitats occupied by this particular association: the massifs of firm and semifirm sands, and the watershed area.

An analogous pattern was also noted in the case of the Anabasis aphylla L.—Artemisia lercheana association. In the Volga-Ural interfluve, this association is found mainly in sor complexes and is related to strong salinization of subsoils; it is associated with moderate salinity on the right bank of the Ural River; and further to the southeast, in the region of the Sagiz and Émba Rivers, the association penetrates wide watershed areas and seems to be generally confined to subsoils of slight salinity. Behavior patterns of this type are characteristic for associations of the other edificator species studied and not only for A. lercheana associations. To illustrate the position regarding the regular changes in the qualitative form of salinity under a single association within different parts of its area, we may cite, as an example, the Anabasis salsa association.

In the Mangyshlak peninsula, this association shows a correlation with $Ca^{..}$ and $SO_4^{..}$, while in the Caspian depression it is mainly correlated with salts of the ions $Na^.$, $SO_4^.$, and $Cl'$ which occur in different quantitative proportions in various parts of the depression. Thus, as we move from the northern and eastern shores of the Caspian Sea, we note, corresponding to this movement to the periphery of the Caspian plain, an increase in the proportion of the $Na^.$ and $SO_4^.$ ions within the geochemical complex; in addition, the eastern profile is characterized by the presence of the $Ca^{..}$ and $SO_4^.$ ions and a gradual change in salinization from $Na^.-Cl'-SO_4^.$ (a direct change in the coastal sectors) toward a more sulfatic aspect, with an increase in the contribution of the $Na^.-Ca^{..}-SO_4^.-Cl'$ group of phases. The same also applies to Halosnemum strobilaceum M.B., an association of which shows a clear correlation with salinization characterized by the $Na^.-Cl''$ group of phases under some conditions, while in other conditions it is linked with $Na^.-Cl''-SO_4^.$ and even $Na^.-SO_4^.-Cl'$.

We thus see that curves constructed for a single indicator community under the conditions of Kazakhstan can change their form in different parts of the area, this being especially traceable in a southerly direction; further, a regular shift of the characteristic maximal value on the graph, in one direction or the other, is observed. The wide variability of salinity indices under a single association within a large region may thus be caused by combining analytical data for a series of ecogeographical variants of the association, as a consequence of only slight study of the association along the lines mentioned.

In practical halidic indicator work, special attention should be paid to associations whose relationship with salinization is characterized by complex multipeak curves, since it is just these associations which most frequently form ecogeographical variants. Associations which typically have clear single-peak curves usually remain rather constant with regard to their relationship with salinity conditions expressed on the graph, as may be seen in the case of grass—A. lercheana associations (Fig. 2).

On the basis of the data provided it can be confirmed that in halidic indicator investigations conducted in rather large territories some variability in the salinization under the same community is absolutely inevitable and depends on the ecological lability of the species and their ecogeographical variability, which leads to the occurrence of geographically modified ecogeographical variants of the particular association.

## Possibilities for the More Accurate Definition of the Indicator Significance of Communities

As is apparent from the aforesaid, a single association can have a rather broad ecological range with respect to salinity, this being in the main controlled by the plasticity of its components and their ecomorphological variability, leading to ecological and morphological differentiation within individual species. This suggests ways for increasing the accuracy and narrowness of the indicator significance of individual features, especially those which have a rather wide ecological range, which can in general lead to a tendency to a decrease in the scope of the indicator. The use for indicator purposes not of a definite edificator but of its individual associations also permits much greater precision and clarity in defining the indicator significance of a feature; further precision can be attained by identifying ecological and ecomorphological forms of the edificator of other components of the community and the communities corresponding to the variants taking into account their ecogeographical characteristics.

It should be borne in mind, however, that by no means all changes in the structure and composition of associations can serve as indicator features, since their appearance may be brought about by human, zoological, pyrogenic, and other effects, and is also dependent on the area of the components of the association. There are cases where a community apparently remains unchanged in all its features, but is associated with different subsoil salinization in different parts of its distribution area. This applies particularly to multidominant associations. One should make the reservation that very frequently our ideas as to the monolithic nature of edificators is due to their inadequate study along the lines mentioned or to a slight degree of morphological differentiation. It is quite possible that a detailed ecomorphological study, particularly of Anabasis salsa and Artemisia lercheana in different parts of their distribution areas, will reveal the existence of an extensive series of varieties or forms of these species. It would, however, be erroneous to assume that the above-mentioned ecological and ecogeographical variants of the associations must of necessity differ by definite, well-marked visual features, such as the appearance of special ecomorphological forms or the occurrence of new components within the composition of the association.

Changes in salinization within the limits range characteristic for the plant will to some extent be reflected in modifications in the vigor of the edificators of the association of other of its components. This means that, in determining the indicator significance of a community, special attention should be paid to the vigor of its components, since in many cases this can improve the precision of the information provided by the indicator community in each concrete case. In using this feature, however, it should be borne in mind that the growth condition in some groups of plants, such as annual or perennial semishrubs, can be affected to a significant extent by the meteorological and other conditions of the particular year.

The variability of salinity indices under a particular indicator, even under conditions of comparative topographic uniformity, is, as mentioned above, a phenomenon which is absolutely inevitable, since it is determined by the capacity of the plant to adapt to ecological factors operating within a rather wide range. This is precisely the reason why it seems to us incorrect and unrealistic to strive to always obtain uniform analytical data, absolutely in conformity for each indicator community. The procedure for developing indicator schemes should be primarily concerned with ways for generalizing analytical data with the purpose of finding an indicator, typical for the given community, which is duplicated in any salinization area of the same type. In this context, however, it is necessary to make sure that the scope of the particular indicator is in all cases understood in the same manner.

In the course of our work in halidic indicator research, we made an attempt to clarify these points, although a detailed discussion of them is not our purpose in the present communication. A particularly suitable approach in this context is the graphical method, which clearly demonstrates, by means of graphs, what is most typical for a particular association under given salinization conditions.

More detailed typological studies of a locality, with the aim of obtaining uniform analytical material, seems to us to be inadvisable, since a probable outcome would be the infinite subdivision of the territory occupied by the indicator and of the indicator itself, which will reduce its value, especially in small-scale work on large areas, and in the final analysis may not give the anticipated result.

The approach to an indicator, and the requirements regarding the accuracy of its indicational features, and the associated scope of an indicator community, should be determined exclusively by their usefulness and by the tasks of indicational research.

Thus, in detailed investigations intended to distinguish deposits which are similar in nature, very small, precise indicators should be used; in large-scale geologic indicator investigations of big territories, the increased differentiation of the indicator community creates substantial difficulties.

## Summary

In halidic indicator investigations, it is not the edificator species that is used as indicator, but the association of the species as a whole, or variants of this association, differentiated in accordance with the ecological and ecomorphological variants of the edificator.

Water extracts from soils underlying an association will vary to a certain degree under different physical-geographical conditions, as well as under relatively homogenous conditions, even within a small area. Variations in salinization of this kind are regular and are determined by ecogeographical variability and also by the replacement of ecological factors.

Extrapolation of halidic indicator patterns should be carried out on the basis of indicator zonation, taking into account the typology of habitats within each region.

## Literature Cited

Akzhigitova, N. L 1961. Plants as Indicators of Soil Salinization (under the conditions of Central Fergana). Author's Abstract of Candidate's Dissertation, Tashkent.

Vyshivkin, D. D. 1959. Research on Vegetation for Providing Information on Soil-Forming Rocks, Using the Mangyshlak Peninsula as an Example. Author's Abstract of Candidate's Dissertation, Moscow.

Momotov, L F. 1953. Soil-Vegetational Complexes of the Ust'-Urt. Tashkent.

Nesvetailova, N. G., and Rodman, L. S. 1959. "On some principles in the compilation of maps of subsoil salinization with the aid of geobotanical data." Nauchn. dokl. vysshei shkoly, seriya biol., No. 1.

Tagunova, L. N. 1961. The Development of the Vegetation Cover of the Northeastern Bank of the Caspian Sea (in relation to conditions of salinization in the soil-forming rocks). Author's Abstract of Candidate's Dissertation, Moscow.

# ON THE POSSIBILITY OF APPLYING
# THE GEOBOTANICAL METHOD IN THE SEARCH
# FOR SALINE-DOME STRUCTURES
# IN THE NORTH CASPIAN AREA

## A. M. Shvyryaeva

The distribution of vegetation is closely associated with the salt regime of the soil-forming rocks, as a result of which it may be suggested that the vegetation cover should to some extent reflect the anomalies in salinization which are associated with saline-dome structures. In this context, the problem arises of the geobotanical interpretation of the saline soil-geochemical indicator of mineral oil content.

The first attempts to study the possibility of using the vegetation cover for revealing saline anomalies, associated with saline-dome structures, were carried out by us in 1953 in the Caspian Kara-Kums, in a region where saline domes are widely distributed. The studies showed it is quite possible to use for this purpose maps of the salinization of soil-forming rocks, compiled from geobotanical data (Shvyryaeva and Starikova, 1955).

However, the effectiveness of applying the geobotanical method in searching for saline domes is to a large extent predetermined by the nature and degree of expression of surface geochemical phenomena associated with saline-dome structures.

As has been shown by research work, the degree of expression of saline anomalies is governed by the characteristics of the geological structure of the saline-dome structures, their depth of bedding, the form of migration of salts toward the surface, and also the general landscape conditions.

The most marked effect on overlying rocks is shown by outcrop and covered shallow-lying saline-dome structures.

As an example, we shall examine the Kamennyi covered-outcrop dome which is located in the southern part of the Northern Caspian area, and in which the saline nucleus lies at a depth of 4-5 m in the central part and 8-20 m on the periphery.

In spite of the fact that the dome is clearly evident in the relief, being raised 2-2.5 m over the surrounding area, it is characterized by very marked salinization with an overall dominance of sodium chloride in the ground salt complex. Thus, whereas in the surrounding area the ground shows marked $Mg^{..}-SO_4''-Cl''-Na^.$ salinization* with a total amount of salts in the order of 1-2 mg/liter, the ground located on top of the saline dome forms very strong $Cl''-Na^.$ salinization with a total amount of salt equal to 3-4 mg/liter.

---

*We adopted the following definitions of degree and type of salinization of soil-forming rocks from the results of analyses of aqueous extracts:

| Soils nonsaline, quantity of water-soluble salts less than | | | | 0.25% |
|---|---|---|---|---|
| " slightly salinated | " | " | | 0.26-0.5 |
| " moderately salinated | " | " | | 0.51-1.0 |
| " strongly salinated | " | " | | 1.01-2 |
| " very strongly salinated | " | " | more than | 2%. |

The type of salinization is given by the dominant ions, arranged in decreasing order of concentration. The dominant ions include those comprising not less than 25 meq %.

In accordance with the differences in salinization, the dome is also clearly indicated by vegetation; whereas on the surrounding plain we find Aeluropus litoralis (Gouan) Parl. solonchak meadows with A. litoralis and Tamarix ramosissima Ldb., communities of halophytes develop on the dome, including Suaeda confusa Iljin,, and in places Halosnemum strobilaceum M. B.

On the roof of erupted saline-dome structures, with the emergence on the surface (or near the surface) of a gypsum mantle, directly affecting the overlying deposits, saline anomalies are expressed by a marked degree of ground salinity with the predominance of sulfates in the salt complex. If the saline dome is expressed in the relief, the degree of salinization of the covering deposits, located on the arch of the saline-dome structure, can be below the surrounding salinity level as a consequence of leaching of readily soluble salts.

Thus, the Kusanbai dome, which is located on the northern coast of the Caspian Sea, has a gypsum mantle which is bedded at a depth of up to 12 m in its elevated part and up to 50 m in the marginal areas.

The soil-forming rocks which are located on the gypsum mantle acquire $Cl'-Na^{\cdot}-Ca^{\cdot\cdot}-SO_4''$ salinization, as a result of which the dome is readily distinguished from the general $Mg^{\cdot\cdot}-SO_4''-Cl''-Na^{\cdot}$ salinization, which is typical for the soils of the surrounding area.

A slightly raised part of the dome is somewhat desalinated and is characterized by strong salinization on a background of very strongly salinated soil-forming rocks.

In accordance with the salinity characteristics of the soil-forming rocks, the saline dome is clearly indicated by the vegetation. Whereas communities of Halosnemum strobilaceum and Aeluropus litoralis communities are distributed on the surrounding territory, a community of Anabasis salsa (C.A.M.) Benth. is developed on the elevated part of the dome.

The Bish-Chokho structure can serve as an example of an erupted saline dome with a gypsum mantle emerging to the surface in the central part of the north Caspian area.

This saline dome is clearly expressed in the relief, being raised 2-4 m above the surrounding territory. The close bedding of a gypsum mantle, in places emerging directly to the surface, is responsible for the high degree of $Ca^{\cdot\cdot}-SO_4''$ salinization of the area above the dome, as a consequence of which the saline dome can be clearly distinguished on the background of the surrounding nonsaline and slightly saline sands.

The specific nature of the salinization of the supradomal area is clearly reflected by the vegetation: Artemisia lercheana Web.—Eurotia ceratoides (L.) C.A.M. communities, incorporating Anabasis aphylla L., clearly differentiate the gypsum field of the dome from the background of widely distributed psammophyte communities.

Studies on a series of other erupted saline-dome structures, with the emergence of the saline stock or the gypsum mantle on the surface, showed that these have clearly expressed saline anomalies, well reflected by the vegetation associated with the arches of the domes or their compensational troughs.

The group of erupted saline domes, is, however, very restricted in its distribution. Within the limits of the Caspian depression, there is much wider development of covered saline-dome structures, these being located at varying depths under quaternary deposits.

The clearest saline anomalies on the surface are shown by covered, shallow-lying saline-dome structures. The reason for the occurrence of saline anomalies on these structures is the strongly saline ancient rocks (mainly argillaceous), which come close to the surface, and also the deep waters which migrate along the lines of disjunctive fissures.

These saline anomalies are reflected in strong salinization of the soil-forming rocks located on the saline-dome structures, and sometimes of the adjacent territory also.

Thus, the Babinskii saline dome, which is located near the eastern edge of the Volga delta on a marine plain only recently freed from sea water, is characterized by the very close (1-2 m) to the surface bedding of salinated clays, as a result of which there occurs concentration of salts and their accumulation on the surface, under nondrained conditions and the close bedding of groundwaters. As a result, the soil-forming rocks in the upper horizons are characterized by strong $Mg^{\cdot\cdot}-SO_4''-Na^{\cdot}-Cl''$ salinization with an increased content of Ca in the saline complex. Soil-forming rocks with a moderate level of $Mg^{\cdot\cdot}-SO_4''-Na^{\cdot}-Cl''$ salinization are distributed on the plain surrounding the dome.

The dome mentioned is demarcated only by the vegetation cover which provides evidence of the increased salinity of this sector as compared with the background salinity. Thus, whereas growths of Tamarix ramosissima Ldb. with Puccinellia dolicholepis Krecz. and Limonium gmelinii (Willd.) Ktze. are distributed on the surrounding marine plain, the vegetation cover on the area located above the dome is composed of a thin cover of L. gmelinii, with the inclusion of Frankenia hirsuta L., Limonium caspium (Willd.) Gams., and Puccinellia dolicholepis. Tamarix in this community occurs as isolated specimens.

We investigated the geochemical features of a group of closely bedded saline-dome structures in the region of the southern Émba (Kzyl-Kuduk, Alty-Kul', etc.).

These saline-dome structures are overlain with a surface of quaternary deposits, represented by arenaceous clays and sands not exceeding 70 m in thickness. Under the mantle of quaternary deposits, there is a complex of Turonian and Senonian rocks, or rocks of the Early Paleogene and Neogene. The saline nuclei in the domes lie at a depth of 200-1000 m. The domes are not evident in the relief.

A map of the salinization of the soil-forming rocks, compiled from geobotanical data, indicates that all the domes are reflected in the surface by more or less clearly expressed sites of maximum salinization. These anomalies in salinity are shown up by solonchaks and solonetzes with communities of halophytes—Halosnemum strobilaceum, Anabasis salsa, Atriplex cana C.A.M.—which can quite clearly be found on the overall background of Agropyron sibiricum—Artemisia lercheana communities which are dominant on the surrounding territory, made up of nonsaline or slightly saline sandy deposits (Shvyryaeva and Starikova, 1955). Saline anomalies associated with shallow-lying saline-dome structures and reflected by the vegetation were also observed by D. D. Vyshivkin (1955) in the Malii Uzen' region.

A strongly salinating effect on the covering deposits is shown by shallow-bedded saline dome structures even in cases where it is well expressed in the relief. Thus, the group of shallow-bedded saline domes which we studied on the region of the Ryn sands (Myn-Tyube, Koshalak, Sysyk-Tau, and Karauzek) are clearly evident in the relief and can easily be located by their vegetation.

The above-mentioned saline-dome structures are morphologically expressed by ridges, being composed of sands with a surface covering of loams. The relative elevation of the ridges is up to 10-15 m. Sors are located between the ridges. At the base of the ridges, and also on the periphery of the sors, Apsheron clays are found.

On the background of the surrounding nonsaline or weakly salinated sandy plain, the soil-forming rocks located on the domes are clearly differentiated by the strong and very strong level of $SO_4''-Cl''-Na^{\cdot}$ salinization on the hillocks and the exceptionally strong (up to 9 mg/liter) $Na^{\cdot}-Cl''$ salinization, with an increased magnesium and calcium content, in the sors.

In accordance with the strong degree of salinization of the soil-forming rocks on the territory of the saline-dome structures, Anabasis salsa communities are dominant, and as a result these domes are easily identified amongst the psammophytic vegetation which develops on the surrounding territory.

On the Myn-Tyube dome, communities of Atriplex cana C.A.M. and Limonium suffruticosum (L.) Ktze., incorporating Atraphaxis spinosa L., are indicators of Apsheron clays lying close to the surface, characterized by a very strong degree of $Mg^{\cdot\cdot}-Ca^{\cdot\cdot}-SO_4''-Cl''-Na^{\cdot}$ salinization.

On the slopes and tops of hillocks, formed by loams from the surface, the close bedding of Apsheron clays is clearly indicated by the inclusion of the above plant species in Anabasis salsa background communities.

On hillocks, made up of sandy loam deposits, the close bedding of Apsheron clays is indicated by communities of Artemisia lercheana, with the inclusion of Eurotia, Salsola arbuscula Pall., Limonium suffruticosum, Atriplex cana, and Anabasis aphylla.

Communities of Halosnemum strobilaceum and L. suffruticosum, with the inclusion of Salsola arbuscula, Atraphaxis spinosa, and Astragalus, constitute an excellent indicator of close-lying Apsheron clays on the Kosh-alak dome.

As a result of the fact that Astragalus and S. arbuscula have pale-colored stems, these communities are clearly demarcated on the background of the surrounding vegetation and clearly indicate outcrops of ancient salinated clays.

However, not all anomalies in salinity associated with closely bedded saline-dome structures are characterized by the development of halophyte communities.

Not infrequently, the geochemical features of closely bedded saline-dome structures, covered from the surface with sandy loam or sandy deposits are located by the inclusion of individual halophytic species in the background plant communities.

Thus, the eastern part of the Azgir elevation, which is well-expressed in the relief in the form of a 3-4 m ridge, is identified by the occurrence on the top of the ridge of such halophytic elements as Salsola laricina Pall., S. crassa M.B., Artemisia pauciflora Web., and Anabasis aphylla in the Artemisia lercheana community.

Lying at a depth of several meters, Apsheron clays show strong $SO_4'' - Ca'' - Na' - Cl''$ salinization.

We thus see that the group of shallow-bedded saline-dome structures always shows a salinating effect on overlying covering deposits.

Anomalies in salinization, associated with the effect of closely bedded, covered saline domes, are usually associated with the arches of saline-dome structures or their compensational troughs (if the latter are present). A geochemical feature of saline anomalies of this group of domes is the higher degree of salinity of the soil-forming rocks (as compared with the background level), with NaCl predominating in the salt complex and a high content of calcium.

According to the degree of expression of the saline anomalies, the latter are indicated either by halophytic communities or the inclusion of individual halophytes in the background plant communities.

Saline anomalies associated with the influence of deeply bedded covered saline-dome structures are less clearly expressed and, where conditions favorable for the migration of salts toward the surface are absent, cannot be identified by the vegetation.

However, in the presence of fissures leading to the migration of groundwaters to the surface, the soils located above the saline-dome structure undergo salinization. Depending on the degree of expression of saline anomalies, these domes are indicated by plant communities of varying degrees of halophily.

Thus, in the western part of the Ryn sands, surface geochemical features of covered, deeply bedded saline-dome structures, identified by geobotanical data, are differentiated by sectors with a substantial content of chlorides in the saline complex of their soil-forming rocks on the general background of $SO_4'' - Na'$ salinization. Such saline anomalies are differentiated by an abundance of Anabasis aphylla and saltworts in the background Artemisia lercheana plant communities.

However, in the case of their direct reflection in the relief, deeply bedded saline domes cause demineralization of the area and are marked by less halophytic communities.

The above shows that the plant cover reacts sensitively to salinization of soil-forming rocks and to varying extents reflects the characteristics of the surface geochemical phenomena of covered saline-dome structure.

This signifies that the geobotanical method is a valuable aid in prospecting for oil and natural gas.

## Summary

Saline anomalies were identified by means of geobotanical data. A comparison of the distribution of saline anomalies with a tectonic map of the region studied showed that many anomalies correspond fairly well with areas where saline-dome structures are developed. Some anomalies were displaced against the domes, or were not confirmed by gravimetric data. However, such anomalies were not numerous, so that this investigational trend seems to be promising.

## Literature Cited

Vyshivkin, D. D. 1955. "The procedure of compiling maps of ground salinization from geobotanical data." In book: Geobotanical Methods in Geological Research. Moscow.

Shvyryaeva, A. M., and Starikova, L. M. 1955. "The prospects of using geobotanical features for locating saline-dome structures." In book: Geobotanical Methods in Geological Research. Moscow.

# ON THE INDICATOR ROLE OF HALOPHYTIC
# VEGETATION UNDER THE CONDITIONS OF KHAKASSIA

## A. P. Samoilova

The steppes of Khakassia, together with the Minusinsk steppes, form a rather large steppe "island," located in the vast depression in the south of Krasnoyarsk territory, which is dissected by the Enisei River from north to south.

The peculiar halophytic vegetation of Khakassia is reflected in the ecological composition of the plants making up the vegetational cover of the salinated soils. While the edificators of halophytic phytocenoses in Central Asia are euhalophytes and xerohalophytes, in Khakassia the flora of salinated soils is more varied, as is general in the steppes of Siberia. The proportion of euhalophytes comprises 11.5% of the total halophytic flora composition, while xerohalophytes account for 5%; meadow-solonchak plants, or halomesophytes, play a substantial role in the make-up of the plant cover on salinated soils (32%). In addition to halophytic plants, the composition of Khakassian halophytic communities includes mesophytes and xerophytes; according to the present author's observations, the total number of ecological groups in the flora of salinated soils of Khakassia is close to ten. All this undoubtedly complicates the recognition of salinated soils by the vegetation cover.

In spite of the above-mentioned difficulties, the problem of utilizing halidic indicator research under Khakassian conditions does seem to us to be solvable. The indicator properties of halophytic vegetation in the region can be identified mainly by the two following approaches: 1) ecological series of lakeland halophytic associations, which indicate the level of lacustrine salinization, are examined for their capacity as indicators; 2) individual halophytic communities of plant species are used as indicators of soils (types and varieties of soils, and their type of salinization).

We shall now deal with the first type of indication. Observations on the lakeland halophytic vegetation of Khakassian steppes showed that in spite of the apparent variegation in the plant cover the distribution of phytocenoses shows a definite pattern, evident in that in each saline lake the main communities successively replace one another along the slope of the lacustrine depression and form ecological vegetation series. It was established that the character of ecological series of salinization is very different. One of the most mineralized lakes in Khakassia may well be the periodically desiccated Lake Kizyl-kul', or Abakan, which is located in the center of the Uibat steppe, and is well known on account of cooking salt having been extracted here for many years. The following series of associations has been described in the lake:

Zone I—growths of Salicornia herbacea L. on a typical solonchak;
Zone II—community of Kalidium foliatum (Pall.) Moq., soil also a typical solonchak;
Zone III—community of Elymus dasystachys Trin., on solonetzed soils;
Zone IV—growths of Lasiagrostis splendens (Trin.) Kunth. on solonetz, from which the shallow-turf feather-grass steppe begins.

As regards ecological series of vegetation around weakly salinated lakes, we will cite as an example the small lake of Altyn-kul', which lies within the Koibalsk steppes:

Zone I—growths of Phragmites communis Trin.;
Zone II—mixed herbage-sedge meadow (with Carex enervis C.A.M. and C. songorica Kar. et Kir.;
Zone III—Iris enstata—Atropis tenuifolia association;
Zone IV—community of Lasiagrostis splendens.

The composition of associations making up the lakeland ecological vegetation series can be varied even in lakes with similar types of salinization. However, for each group of lakes, differing by degree of mineralization (strongly, moderately, and slightly salinated lakes), there is a characteristic general ecological trend in the surrounding phytocenoses. In examining the possibility of indicator research, on the basis of the vegetation cover, into the degree of lacustrine salinization, one should bear in mind that in this region halidic indicators are not concrete ecological series (composed of well-defined associations), but types of ecological series, in which a decisive role is played by the ecological grouping of the edificators of the plant communities which make up the lakeland ecological series. The main types of ecological series for lakes with different degrees of mineralization may be illustrated in the following way:

Strongly salinated lakes

1. Associations of euhalophytes (Salicornia herbacea, Suaeda, Kalidium foliatum).
2. Associations of xerohalophytes.
3. Associations of haloxerophytes (usually communities of Lasiagrostis splendens).

Slightly salinated lakes

1. Associations of hygrophytes.
2. Associations of halomesophytes (salinated meadows).
3. Associations of haloxerophytes and xerophytes.

Moderately salinated lakes occupy an intermediate position between the types described in respect of the character of the surrounding vegetation.

The ecological series of associations around lakes with a high degree of mineralization reflect the parallel reduction in salinity and humidity of the soils with increasing distance from the lake; the most salinated soils and the communities of succulent euhalophytes corresponding to them occur in immediate proximity to the lake. On the lakes of moderate and slight salinity, the degree of soil salinization increases toward the center of the ecological series, while there is reduced salinization on the margins, and this too is reflected in the vegetation cover. Similar patterns were also found in studies of the lakeland halophytic vegetation in the Tuva steppes (A. P. Samoilova, 1951).

The second type of halogenic indicator research, noted by the author in Khakassia, is better known: this consists of the recognition, by means of the vegetation cover, of particular soil characteristics, with individual phytocenoses or plant species being used as indicators. As in the previous case, i.e., in the indication of mineralized lakes, the clearest relationship is seen in this case too between plant communities and the degree of soil salinization, that between communities and the qualitative make-up of salinity being much rarer.

The firmest relationship with a definite soil type is observed in associations formed by the euhalophytes Salicornia herbacea, Kalidium foliatum, and Suaeda corniculata (C.A.M.) Bge. These communities are noted only on typical solonchaks (damp, puffy, sometimes boggy types). It is interesting that associations of S. corniculata are found primarily around carbonate lakes of the soda type, which rarely occur in the Khakassian steppes, whereas Salicornia herbacea and Kalidium foliatum grow on chloride and chloride-sulfate solonchaks. Formations of Atropis tenuiflora (Turcz.) Griseb. constitute quite a good indicator of meadow soils with varying degrees of solonchakization. It is true that associations of A. tenuiflora are also encountered on meadow-bog-solonchak soils, but in this case the bogginess is revealed by a series of plants accompanying A. tenuiflora (carex enervis C.A.M., Bolboschoenus compactus (Hoffm.) Drob., and others).

Associations of Atropis tenuiflora + Aster tripolium L. are frequently found on meadow-bog-solonchak soils. A formation of Hordeum brevisubulatum (Trin.) Link. frequently becomes an indicator of meadow-solonchak, sometimes rather boggy, soils.

Indicator properties are somewhat less in the case of Saussurea papposa and Statice gmelinii Willd., communities which occur both on solonchaks and on solonchak soils. As regards the remaining halophytic phytocenoses of Khakassia (formed by Elymus dasystachys, Lasiagrostis splendens, Atropis tenuissima, Artemisia nitrosa Web., Plantago salsa Pall., and some other halophytic species), we virtually failed to find any relationship between them and specific soils. For example, associations of Atropis tenuissima (Litw.) V. Krecz. are found on dry salinated soils, but these soils may be very variable, e.g., undergoing solonchakization, strongly solonetzed, or solonchak-solonetz.

It is almost impossible to speak of any soil relationship in the case of communities of lyme-grass (Elymus dasystachys Trin., E. junceus (Fisch.) Nevski, and E. salsusuginosus (Turcz). Lasiagrostis splendens is observed on solonetz and solonchak chestnut soil, dark-colored solonchakized, sandy salinated, solonchak-solonetz soils, and others, besides. Communities incorporating L. splendens are exceptionally variable; L. splendens enters the vegetation cover both as an edificator and as a codominant, forming associations with typical halophytes or with components of the normal grassy-sod steppe. It can therefore be stated that it is not L. splendens communities but rather the species itself which acts as indicator of greater or lesser soil salinization under the conditions of Khakassia.

## Summary

In the steppes of Khakassia, studies of ecological series of associations acquire great importance in the determination of the successive changes of the types of soil salinization in lakeland depressions. Near strongly mineralized lakes, these series reflect a parallel decrease of salinity and soil moisture with increasing distance from the lake. Near slightly or moderately salinated lakes, the level of salinity tends to increase toward the central part of the ecological series and to decrease toward their extreme sections. Similar patterns have also been traced in the steppes of Tuva.

## Literature Cited

Samoilova, A. P. 1951. "The vegetation of the banks of some saline lakes in Tuva autonomous province." Tr. Tomsk. gos. univ., Vol. 116, biol. seriya.

# THE USE OF THE GEOBOTANICAL METHOD
# IN THE SEARCH FOR KIMBERLITE TUBES
# IN THE YAKUTIAN POLAR REGION

## I. I. Buks

In 1958, an expedition of the All-Union Aerogeological Trust began investigations designed to elucidate the possibility of using the geobotanical method in prospecting for indigenous deposits of diamonds (kimberlite tubes). These studies were continued in 1959.

The kimberlite tubes on which the observations were carried out were made up of horizontally bedded Cambrian carbonate rocks, representing five different rock strata. In all, we studied eight kimberlite tubes, of which one was located by itself in the upper course of the Motorchuna River, while the remaining seven were in the basin of the Merchimden River, a tributary of the Olenek River.

Of the types of vegetation developed in the region, the most common consists of Rhododendron—lichen, Vaccinium uliginosum—lichen, Alnus—lichen, and Ledum palustre—moss—lichen thin larch forest. Thin larch forest on kimberlite-containing rocks is characterized by a low-growing and very thin tree stand made up of Larix dahurica Turcz. The height of the tree stand fluctuates from 4-5 to 7-8 m, with a corresponding trunk diameter of 0.1-0.13 to 0.15-0.2 m. The number of adult larches on one hectare is 94, on the average. In thin larch forests, for which an undergrowth of alder and various willows (Salix xerophila Floder., S. kolymensis O. v. Semm.) is typical, the latter layer is usually thin. Also characteristic is a comparatively smooth surface, covered with a more or less uniform herbage-shrub and moss-lichen cover.

An exception is provided by a stratum of bituminous limestones and shales, on which, under certain relief conditions, there develops a substantially denser tree stand and undergrowth, and where the well-expressed medium- and large-sized hillock relief is responsible for the uneven distribution of the herbage-shrub and moss-lichen cover.

The vegetation on kimberlites, as compared with that on the surrounding rocks, develops more profusely. The height of the tree stand is 9-11 m on the average, with an average trunk diameter of 0.25 (individual specimens reach 15 m in height and 0.3-0.4 m in diameter). The average number of adult larches is 178 hectare.

The increased trunk diameter of larches growing on kimberlites is associated with their longer life in combination with an equal annual growth increment. Thus, the trunk diameter of the majority of larches, aged around 400 years and growing on limestones, reaches 0.15 m; their apices are usually shrivelled, while the pith is rotted,[*] evidence that they had reached their age limits (Starikov, 1957). At the same time, larches growing on tubes had a diameter of around 0.3 m and had a healthy pith although aged up to 550 years. This obviously indicates that the development conditions on tubes promote more favorable larch growth than conditions beyond the tubes. More abundant undergrowth is also observed on tubes than on the surrounding rocks.

The plant cover on the majority of tubes, although not on all, is characterized by the presence of an undergrowth of alder, which is characterized by greater height and cover than the undergrowth on the surround-

---

[*]All sawed larches with dry apices had a rotted pith.

ing rocks. In addition, kimberlite tubes are characterized by a clearly evident large- and medium-hummock microrelief, causing an uneven distribution of the grass-shrub and moss-lichen layers. The above-mentioned characteristics of the plant cover are, however, not typical for all kimberlite bodies.

In spite of these exceptions, more intensive vegetational development is quite typical where kimberlite bodies are located. The reason for this is, apparently, the chemical and physical properties of kimberlites, which differ from those of limestones. The ultrabasic igneous kimberlite rock is characterized by great mineralogical and, consequently, chemical variability, as compared with limestones. In particular, the composition of kimberlite includes apatite which contains phosphorus, mica, which contains potassium, and, in addition, a series of trace elements which also possibly have a favorable effect on the development of vegetation.

The characteristics of the vegetation cover of kimberlites as compared with those of the vegetation cover of the surrounding rocks is a factor in deciphering kimberlite bodies in aerial photographs. The increase in the height and cover of the tree and shrub layers, the denser grass and scrub layer, the differentiation of mosses and lichens according to the features of the microrelief—all constitute reasons for the darker tone of places where kimberlite bodies are located, on aerial photographs; kimberlite tubes are therefore represented in aerial photographs in the form of circular, isometric, dark spots. Dark strips extend from certain spots, indicating the zone of ablation from the tube.

In the case where a kimberlite body is not exceptional and is characterized by the intensive development of vegetation, it definitely has a dark tone on an aerial photograph but cannot necessarily be deciphered. The decipherability of the kimberlite tube depends in the first place on the vegetation cover on the rocks surrounding the tube. If the territory surrounding the tube has a thin larch forest with a rather thin tree stand and undergrowth and an evenly distributed grass-scrub and moss-lichen cover the tube can be clearly deciphered since the surrounding vegetation gives a paler tone on the aerial photograph. If there develops, on the rocks surrounding the tube, a relatively dense tree stand and undergrowth and there are substantial areas of moss on the ground cover, which together provide a dark tone, the tube can either not be deciphered at all or can be done so only very poorly.

Since most kimberlite bodies have a relatively dense shrub layer of alder, with a substantial quantity of mosses present in the ground cover and a good scrub canopy, good results may be attained through using colored pictures with SN-2 spectrozonal film.

It should be mentioned that the deciphering of kimberlite bodies on the basis of geobotanical features shown up in aerial photographs assumes that consideration is taken of many other geological and morphological factors which have a significant influence on the vegetation cover developed both on kimberlites and on the surrounding rocks. These factors comprise, in the main, the position of the kimberlite body in the relief, the resistance of the kimberlites making up the particular body to processes of erosion and denudation, lithological features of the surrounding rocks, the presence of tectonic disturbances, etc. All these factors affect the decipherability of tubes and, depending on circumstances, one factor or other is of leading importance.

It should not be assumed that the intensive development of vegetation is a feature exclusively of kimberlite bodies, although, with rare exceptions, this property is characteristic for them. Nonetheless, one may encounter sites in the surrounding landscape which, in shape and properties of the vegetation cover, are reminiscent of kimberlite tubes. Such zones are found more rarely in on-site studies than in aerial photographs, i.e., patches which are identical to kimberlite tubes in photographs are not always similar to them in situ.

The reasons for the appearance in photographs of spots of the "tube" type can be varied. They may represent a small site of an old fire, or a thermokarst depression, characterized by increased moisture and therefore vigorous willow growth, and so on. It is also possible, however, to find sites which are characterized, as are tubes, by a comparatively dense and tall tree stand and undergrowth, and a well-expressed microrelief, causing a specific type of ground-cover arrangement. The reasons for the formation of these spots may be varied but cannot always be explained satisfactorily, especially when only aerial photographs are used.

Summary

The analysis of aerial photographs of kimberlite bodies as well as observations on land have shown that vegetation can play a very important role in searching for these bodies. It was found that the vegetation cover is more intensely developed on the pipes than on the surrounding rocks (as expressed by greater height of trees and shrubs, denser grass, and scrub layers, and a better moss-lichen cover). There are, however, some inexplicable cases which do not conform to this general pattern.

Literature Cited

Starikov, G. F. 1957. Larch in Magadan Province. Problems of Geography of the Far East, coll. 3. Khabarovsk.

# THE USE OF BOTANICAL INDICATORS
# FOR THE DETERMINATION
# OF SOME CLIMATIC BOUNDARIES

## I. L. Krylova

The use of morphological features for indicator purposes is a field to which little research has been devoted. Normally, only the fact that a particular species is present is used in indicator work, and only in a few rare cases is reference made to the growth characters of plants and some of their morphological properties as indicational features (Nesvetailova, 1953).

In our investigation, we were concerned with the need to determine the position of the upper climatic boundary of forests on the slopes of the Crimean mountains. At present, the summit plateau and part of the slopes of the Crimean mountains are forest-less, while forests at their upper limits have been felled, so that the boundary has receded. As to where it reached previously, and whether forest on the Crimean mountain slopes had attained its natural climatic boundary, is unclear. The type of forest regeneration work in the upper parts of the slope depends on the solution of this problem.

Although attempting to determine approximately the position of the upper climatic limit of the forest, we decided to employ the growth characteristics of individual cases of pine undergrowth, which in some places had ascended beyond the forest boundary on the slopes.

For this purpose, specimens of pine undergrowth aged from 6 to 17 years were measured on slopes of varying exposure, and situated at heights of 1200 to 1450 m above sea level. For each of the small pines, we noted the age, determined by the number of whorls plus 4, the total height, the distance between whorls, i.e., the annual height increment of the trunk, and the number of branches in each whorl. In addition, we measured the length of needles and the density of needling, i.e., the number of needles per 1 cm shoot length. In all, about 700 specimens were measured.

Analysis of the results obtained revealed a substantial decrease in height of the undergrowth with increasing height on the mountain. This is particularly clear if we compare the average height (in cm) of the undergrowth at elevations of 1300-1360 and 1400-1450 m above sea level.

|              | 9-13 years       | 13-17 years      |
|--------------|------------------|------------------|
| 1300-1350 m  | $112.5 \pm 3.05$ cm | $170.6 \pm 3.54$ cm |
| 1400-1450 m  | $90.2 \pm 2.30$ cm  | $136.7 \pm 3.15$ cm |

The differences between average heights were significant in all cases. It is thus clear that the height of the pine undergrowth decreases sharply with increasing height up the mountain. The height of the undergrowth of corresponding age groups decreases by 20% per 100 m absolute height above sea level.

The depression of pines is particularly clear if we compare the growth of Pinus hamata D. Sosn. on the slopes of the Crimean mountains with its growth under favorable lowland conditions in Orel province. In the latter area the height of pines aged 15-20 years is 5-6 m (Vikhov, 1952), whereas Crimean pines of the same age attain a height of only $1\frac{1}{2}$ meters.

In addition to the decrease in height growth, higher elevations on the mountain also affect the increase in needling density. Up to a height of 1350 m, the density of needling, i.e., the number of needles per 1 cm shoot

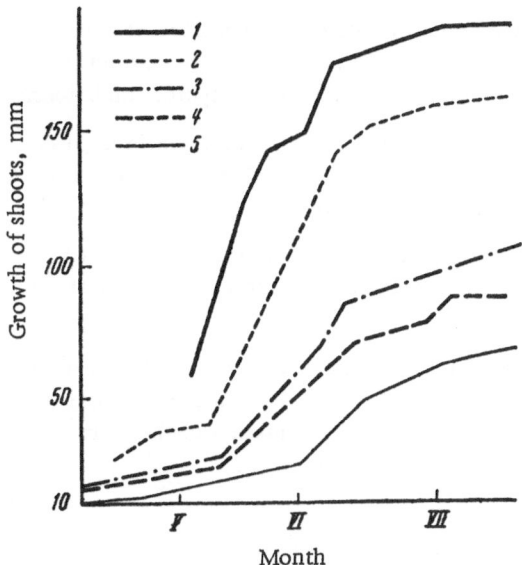

Growth of pine shoots, growing at different heights. Height above sea level: 1) 1150 m; 2) 1300 m; 3) 1380 m; 4) 1480 m; 5) 1520 m.

length, is $9.9 \pm 0.257$, while the corresponding figure above 1350 m is $10.6 \pm 0.210$. Analyses showed that the difference in needling density at different heights is statistically significant.

This result might well have been expected, since greater density of needling is an indication that the shoots at greater altitudes show less growth extension. Other features noted, such as the number of branches at a whorl and the length of needles, were not associated with height above sea level and do not have any indicator significance.

In order to establish where the boundaries of the zone of more or less satisfactory pine growth and that of the zone of acute depression are located, observations were made on the growth of pine shoots at 12 different habitats on the slopes and summit plateau, from 1150 m to 1520 m above sea level. At each habitat, specific shoots of the first order in three pines of different age groups were marked. Observations were made on 50 shoots at each habitat, so that in all about 600 shoots were brought under observation at all habitats. Each shoot was measured regularly every five days during the period of intensive growth and every ten days during the period of slow growth. The results of measurements at a number of habitats are shown in the figure.

It is apparent that the initial growth increment of shoots at high-level habitats is very retarded while the terminal increments almost coincide, but that the rate of growth incrementation and the final size attained differ very sharply. The boundary of severe deterioration of growth evidently occurs somewhere between 1300 and 1380 m above sea level.

It may be suggested that the upper climatic forest boundary runs in precisely this zone, i.e., it is the limit of more or less dense forests, above which there occur only the hardier specimens, forming a thin pine forest.

One basis for this conclusion is the great constancy of edaphic conditions during the entire course of the slope. The massif of the yaila is composed of a several-meter thick stratum of Jurassic limestones which constitutes the soil-forming rock throughout the entire slope. The data given were obtained on a slope with uniform exposure and with deep silt soils. Thus, changes in type of growth can only be attributed to changes in climatic conditions of which the most important is obviously temperature. Wind plays a much smaller role. According to observations carried out on pine growth in protected habitats and in habitats exposed to the wind effects, no influence of wind on the growth increments was noted—only the retention of growth increments was affected by the wind; in places not protected from the wind, the young growths frequently die by the spring of the following year.

The problem arises as to how close the relationship is between the height above sea level, and tree height close to the climatic boundary of the forest. Does a functional relationship exist between them, and is it possible, in the absence of an undergrowth above the present-day forest line, to predict the position of the upper climatic boundary of forests on the basis of the character of tree growth in the upper forest zone?

To obtain an answer to this question, observations were made in the "Denezhkin Kamen'" national forest in the Northern Urals.

Five profiles in different directions were made on the Elovskii Ural ridge, which runs in a southerly direction from a high-point of 869.4 m above sea level, in a belt between 700 and 850 m above sea level.

All the profiles were in a tall-herbage, thin spruce forest, and the particular area on the forest-planting plan was identified as having the following characteristics: composition 8E III IB; age class—V; volume—0.5; quality and yield index—Va.

During the entire course of the profile, edaphic conditions remained approximately constant: the slope was silty with montane meadow soils, and the grade of the slope was from 5 to 15°. This thus confirmed that the deterioration of growth along the profile was induced by changes in climatic, and not edaphic, conditions.

Each profile was divided into segments according to 10 m absolute height changes above sea level, using a height-measuring altimeter. Measurements were made of the height of 25 spruce trees (not selected), using a pole with divisions marked on it. In all, 1175 trees on the five profiles were measured.

The results of the height measurements were analyzed graphically and statistically, using analysis of variance methods. The statistical work was directed by the head of the department of forest evaluation at the Ural Forest Technical Institute, N. L. Leskov.

Analysis showed the coefficient of correlation between tree height and height above sea level is very high, from 0.68 to 0.90. This is evidence of a reliable relationship between these values and facilitated the use of this relationship for the construction of correlation equations. We used a correlation equation of the form

$$x_1 = x_1 + r \frac{\sigma_1^{c_1}}{\sigma_2^{c_2}} (x_2 - x_2),$$

where $x_1$ is the probable average tree height and $x_2$ is the height above sea level. Having interposed the corresponding values in this equation, we obtained five equations for each of the five profiles. We then made an equation with the mean arithmetical parameters, general for all the problems:

$$x_1 = 63.5 - 0{,}074x_2; \quad \sigma_{1,2} = \pm 1.2 \text{ m.}$$

By using this equation, with the height above sea level being interposed in place of $x_2$, it is possible to calculate the average height of trees at this level. In addition, one can calculate at what height above sea level the height of spruces reaches a definite given value. Thus if we, for example, take as the upper climatic boundary* the zone above sea level where spruces reach a height of 2 m, with all trees above this level being included in the zone of low-growing spreading trees and shrubs, then the formula suggested can be used to determine at what height the climatic boundary runs. Substituting the suggested values in the equation, we obtain $2 \text{ m} = 63.5 - 0.074 \ x^2$, i.e., $x = 831 \pm 1.2 \text{ m}$. This height is the climatic boundary of forest in the part of the ridge under study.

Naturally, the suggested equation is not universally applicable. In the first place, it is operative only for zones where the main factor affecting forest growth is a gradual deterioration of climate, and where the relationship between the height of trees and that above sea level is linear (for a height of 600-800 m in our example). Below this zone, where the thin forest ends and the dense forest commences, competitive relations begin to operate, as well as edaphac conditions, etc. Above 820 m, there is a curvilinear relationship which apparently depends on the special climatic regime close to the actual summit of the ridge. It is also obvious that the values of the equation coefficients will be modified for different rocks and different climatic conditions. There does, however, appear to be a basis for suggesting that the general character of the relationship remains the same.

It has thus been indubitably established that in the upper boundary of the thin-forest zone, where competitive factors do not play a great role and tree growth depends primarily on climatic effects, there exists a close functional relationship between the height of trees and the height above sea level (i.e., a characteristic of the climate). This permits the use of tree height as an indicator feature for determining the position of certain climatic boundaries.

---

*This boundary should, of course, be more correctly termed not the upper forest boundary, but the upper tree boundary, since it is the boundary not of the community, but of individual specimens of a definite size; however, in the Soviet literature, these two concepts are usually not differentiated.

Summary

The author used botanical indices for determining the upper climatic limit of forests on the Crimean mountains.  She located the line of contact between the zone of fairly good growth of Pinus hamata and the zone of acute growth depression.  A large number of young pine trees (from 6 to 17 years old) were measured on the hillsides, at altitudes ranging from 1150 to 1520 m above sea level.  The boundary line of acute growth depression was found to pass between 1300 and 1380 m above sea level.

In order to check these findings, we examined the correlation between tree height and their elevation above sea level in the North Urals.  The correlation coefficient proved to be very high (0.68-0.90).

Literature Cited

Vekhov, V. N. 1952.  "Biological features of some pine species under growing conditions in the western part of the Central Forest Steppe."  Candidate's Dissertation, Moscow.

Nesvetailova, N. G. 1953.  "On the vegetation of bituminous soils."  Byull. MOIP.  Novaya seriya, otd. biol., Vol. 58, No. 6.

# ECOLOGICAL COMPENSATION AND REPLACEABILITY
## AND THE EXTRAPOLATION
## OF PLANT INDICATORS

## B. V. Vinogradov

As a rule, plant indicators are determined in special key plots or standard objects and the botanical features identified in this way are extrapolated for analogous territories or objects.

At the present time, the development of plant indicator methods has reached a level such that after carrying out a definite amount of field work within each landscape it is possible to work out a more or less reliable system of botanical features for various natural conditions.

Among the main problems in the further development of the indicator method are the determination of the geographical rules of extrapolation of plant indicators, the study of the geographical variability of indicator functions, and the development of landscape studies and physical-geographical zonation, on the basis of which it should be possible to plan the preparation of indicator schemes and the establishment of their operational boundaries.

When an indicator moves from a key plot within which its indicator functions were established to a territory under investigation, we are immediately confronted with changes in the balance between vegetation and indicator factors, the following being the consequences: similar vegetation may be formed on different habitats, this being associated with the ecological compensation of factors; different vegetation may be formed on similar habitats or analogous indicator objects, this being associated with the ecological replaceability of vegetation.

A study of ecological compensation and replaceability should provide an answer to the basic problem in the theory and practice of plant indicators—their extrapolation. Determining the conditions under which the use of indicators is possible is no less important than preparing new indicator schemes.

Rübel (1936) refers to the "replaceability" of factors. A more precise view is that ecological factors such as light, heat, moisture and nutrient substances are not wholly replaced but may be only partially compensated (Shennikov, 1950). Unfavorable factors such as content of toxic compounds or factors operating indirectly such as mechanical structure may be replaceable to a great extent or completely. However, since in most cases it is difficult to make a clear distinction of these factors, the concept of compensation could well apply to all factor changes where the vegetation types remain similar, while the term replaceability could be adopted for vegetational changes in a single habitat. As a consequence of compensation of some ecological factors in different habitats, there develop similar, temporarily convergent associations (Shennikov, 1928). Although there may be a high degree of identity in respect of composition and structure, the developmental pathways of these associations diverge subsequently.

Three types of factor compensation may be distinguished: climatic, edapho-climatic, and edaphic. Each of these requires special study and detailed analysis. In this article we shall seek only to discuss systematically the forms of compensation, providing only brief examples in order to clarify the issue.

The types of compensation of climatic factors are quite widely recognized. Observations show a similar ecological effect from decreasing air humidity and increasing temperature, decreasing air dryness and increasing duration of dry winds, and vice versa. Inadequate duration of the growth period is compensated for by high summer temperatures, and precipitation by mists and air humidity. Determining the compensation of these factors has great significance in the extrapolation of indicators of climatic conditions. In addition, compensation of this type leads to ecological replaceability of vegetation under similar edaphic conditions. Thus, for example, the further one goes eastward, the more do northern species appear on steppe-zone soils. Dry Festuca sulcata—Stipa steppes are located on the dark-chestnut soils of Northern Kazakhstan, while on soils of this type in Northern Mongolia and Khingan less xerophytic montane steppes predominate, analogs of which in Northern Kazakhstan are confined to southern chernozems (Lavrenko, 1960).

Of most interest for elucidating the patterns of extrapolation of indicators is the edapho-climatic factor compensation type, which in turn can be subdivided into three classes: soil-climatic, hydro-climatic, and litho-climatic compensation.

Compensation of soil and climatic factors leads to the fact that vegetation similar in some particular features develops on different soils in different climatic zones.

Relatively poor soils in a hot climate may provide the same nutritional conditions as potentially richer soils in a cold climate. The Oxalidosa types of forests in the central taiga are associated with loams, though they may sometimes develop on carbonate rocks, shallow podzols or covered podzols, whereas in the south they are associated with the mixed forest located on the poorer sandy loams or sandy soils.

Moderate or inadequate soil moisture in the humid zones is compensated for by surplus soil moisture under arid conditions. Carex appropinquata Schum. acts as an indicator of moderate moisture in the forest zone and as an indicator of substantial moisture in the steppe zone (Ramenskii et al., 1956). On the other hand, the same sort of soil conditions are created on the dry and warm southern slopes, as in the plakor zones located more to the south. V. V. Alekhin's well-known "rule of precedence" (1936) is based on this.

Light, well-drained soils in an excessively moist climate create moisture conditions analogous to heavy soils in a moderately moist climate.

The hydro-climatic compensation of factors leads to a modification of the hydroindicational functions of vegetation in different climatic zones.

The water supply to deep-rooted phreatophyte and trichohydrophyte plants (Beideman, 1949, 1953, 1954) in the southern districts of arid zones derives from deeper groundwaters than in northern, or even more so, than subarid and neutral districts. According to the observations of E. A. Vostokova (1955), the Haloxylon aphyllum—Kalidium caspicum association in southern deserts is associated with groundwaters at depths of 10-20 m, the corresponding figures in northern deserts being 5-10 m.

In the case of hydrophytes, on the other hand, deep groundwaters and soil humidity in the moderately humid zone are compensated by shallow fresh groundwaters in arid conditions. Cynodon dactylon (L.) Pers. is distributed in the south of the forest zone on the soils where the groundwaters lie at a depth of 2 m, while under the conditions of the subtropical Wadi Araba desert they are located where the groundwaters are at a depth of up to 10 cm (Boyko, 1953).

Phreatophytes in the warm arid zones can use groundwaters of higher mineralization than in the moderately cool arid or subarid zones. An example of climatic-hydrochemical compensation is provided by Tamarix ramosissima Ldb., which in the southern deserts is associated with groundwaters with mineralization up to 7-15 g/liter and much more, while in the northern deserts the corresponding level is only up to 5-7 g/liter.

A reduction in the depth of permafrost is compensated for by a colder climate. As a result of climatic-geocryological compensation, woody species in Arctic regions are found where the permafrost is located at a deep level, while in dry, moderately cold regions trees are found where frozen ground is in the close proximity. According to the data of several authors, Larix sibirica Ldb. in the tundra grows where frozen ground is at a depth of 4-5 m, and in the taiga where such ground is at a depth of 1-2 m, while in the forest steppe frozen ground is found at a depth of 1 m under forest sites and is absent in forest-less localities.

As a result of climatic-geological factor compensation, similar plant groupings in different climatic conditions can be associated with various types of rocks.

With climatic-geochemical compensation, similar plant groupings develop both on normal soils with an unfavorable water and thermal regime in the steppe or tundra zone, and on soils with a high content of toxic metallic compounds in the forest zone.

Mixed herbage—F. sulcata—Stipa steppes with a poor mixed herbage are zonal in the dry steppe, while in the southern part of the forest zone of the Southern Urals they infiltrate onto serpentinites (Tyulina, 1929). Fragments of shrubby tundra also occur on serpentinites in the northern taiga and the forest tundra (Rune, 1953).

Climatic-lithological compensation is analogous to the above-mentioned climatic-edaphic compensation. Uniform conditions of ground moisture are established on unsaturated rocks in a moist climate and on water-accumulating rocks in a dry climate. A neutral or basic reaction of rocks in the forest zone is compensated by an acid reaction in the steppe zone. L. K. Pachoskii (1917) cites as an example Aurinia saxatilis which is associated with limestones in the forest steppe and with granites in the dry steppe. A series of plant groupings growing in humid zones on varied types of rocks, such as argillites, slates, or argillaceous limestones, are limited, in dry zones, to skeletal soils on granites, quartzes, and sands. This applies to birch and pine forests associated with outcrops of indigenous rocks in the steppe zone of Northern Kazakhstan.

Warm limestone and sandstone soils in the cold taiga zone produce a thermal and water regime analogous to that of normal soils in the moderately warm climate of the mixed-forest zone. As a result, Larix sibirica Ldb. in Central Siberia is frequently confined to limestones or sandstones, while in the European part of the USSR it is located on rocks of a varied mechanical structure.

The second reason for the zonal variability of the status of indicators is the zonal replaceability of vegetation associated with analogous habitats in different climatic zones. So far as soil and hydrological factors are zonal, they are not connected with apparent zonal replaceability of vegetation. Geological factors which are zonal are characterized by definite geographical variability in the ecological conditions created on rocks with similar lithological and geochemical conditions in different climatic zones (Viktorov, 1955). Kimberlites in moderately dry and warm climates have a negative effect on the development of vegetation and can be identified by patches of treeless patches which act as "levees" within a sea of trees, whereas in humid climates this effect is positive and the vegetation is more mesomorphic and dense than on the surrounding solid rocks. Zonal variability in the make-up of indicators is not examined in the present article, since it merely complicates indicator schemes without disturbing indicator relationships.

We thus see that as a result of the zonal compensation of climatic factors and the replaceability of vegetation, there are changes in the relationships of plants and communities and modifications in the make-up of indicators under similar natural conditions in different climatic zones. To exclude ecological compensation and replaceability of this type, indicator schemes should be extrapolated within the boundaries of territories with uniform edapho-climatic relationships, i.e., within the boundaries of geographical zones. Such indicators, which are extrapolated along geographical zones, may be termed zonal. Indicators of this group can be located along zones in a latitudinal direction tens of hundreds of times further than by the transverse extension of the zones in a southerly direction.

A single zonal limitation cannot, however, decide problems of indicator extrapolation, since the phenomena of compensation of edaphic factors and edaphic replaceability of vegetation are widely distributed within identical climatic zones.

As a result of compensation of different soil factors, highly different types of soil may reveal similar vegetational traits. This class of compensation includes the widely occurring physical-chemical compensation of soil factors. Soils which are of light mechanical structure and have a moderate humus content have the same quantity of available nutrient substances as heavier soils with an increased humus content. On light soils, plants can grow in the presence of a greater quantity of salts than on heavy soils. Salsola richteri Karel. in Western Turkmenia is distributed on sandy loam soils with a content of readily soluble salts up to 0.8-1.0%, while on gravel-silt-sand soils it can withstand salinization up to 1.7-2.0%.

Hydro-physical compensation of soil factors leads to the fact that heavy moist soils have the same quantity of available moisture as light soils which technically are drier.

Hydro-chemical compensation of edaphic factors is evident in that halophytes grow more successfully on moist, strongly salinated soils than on dry soils, although the latter may be more saline.

There also occurs chemical compensation, based on the relative ecological equivalence of certain substances.

Several groups of ecological factor compensation are also observed in the distribution of indicator of hydrogeological conditions.

Litho-hydrogeological compensation has also been identified. Close subsurface waters in soils of light mechanical structure create the same conditions of soil moisture as deeper waters in soils of heavier mechanical structure. In Western Turkmenia, the Tamarix ramosissima—Halosnemum strobilaceum association is located on sandy ground where the groundwaters are at a depth of 1-1.5 m, compared with 1.8-2.5 m in silty sands and sandy clays and up to 3-4 m in the heavier silty loams.

Litho-chemical compensation of hydrogeological factors includes cases where phreatophytes on light soils withstand high groundwater mineralization whereas on heavier soils they demand low water mineralization. It has been noted that Tamarix hispida Willd. on silty-sand solonchaks grows with groundwater mineralization of 30 g/liter or above, compared with 15 g/liter on loams.

Phenomena of hydrochemical compensation can be put in a special group. Plants, especially phreato- phytes, are able to draw fresh groundwaters from a greater depth, but somewhat mineralized waters from a lesser depth. In Western Turkmenia, one observed associations of Tamarix ramosissima Ldb. with maximum abundance on fresh groundwaters located at a depth of 3.5-4.5 m, compared with 1.8-2.4 m on slightly mineralized waters.

Another aspect which is highly important from the hydrologic indicator point of view is hydrogeological compensation such as the change in water provision through groundwaters and the capillary fringe of water sup- ply caused by interstices of increased moisture and perched water tables of condensation and infiltration origin. Changes of this type in sources of water supply and hydrogeological associations are observed in the trichohy- drophytes Haloxylon aphyllum (Minkw.) Iljin. and Salsola dendroides Pall.

Litho-geocryological factor compensation occurs in the zone where permafrost is located. Similar vege- tational groupings develop on ground of light mechanical structure with the permafrost level in the close prox- imity and on heavy ground with deeper bedding of permafrost.

Hydro-geochemical factor compensation is observed in those cases where the effect of dispersion halos of poisonous heavy-metal compounds has a result which is ecologically similar to the effect of an unfavorable soil moisture regime. Identical xeromorphism is shown in Katanga by steppes of Andropogon filiformis on normal dark-colored soils and by groupings including metallophilic flora on moister soils enriched with CO and Cu compounds (Duvigneaud, 1960).

Soil- or eluvial-geological factor compensation is evident in the fact that the deep bedding of certain rocks creates conditions which are similar to those produced by the surface bedding of others. Easily weathered rocks close to the surface can have the same vegetational cover as solid, difficultly weathered, deeply bedded rocks with deep soils and a comparatively large accumulation of eluvium.

One observes factor compensation, not only promoting the expression of particular geological conditions in the vegetation cover but also of factors inhibiting their expression. The absence of indicators is observed where there is exceptionally deep bedding of indigenous rocks under a thick layer of covering deposits but also where there is close bedding of rocks but a high permafrost level. Identical biogeochemical accumulation of elements is noted on rocks which are poor in mineral ores and on ore-containing bodies covered with flowing groundwaters (Rankama, 1954).

In addition to the compensation of factors, the ecological replaceability of vegetation must be taken into account in the compilation and amplification of indicational schemes. This replaceability leads to the factor that different vegetation can be located on analogous soils, groundwaters, and rocks. The replaceability of vegetation is observed especially frequently in studies of indicators of geological formations (and other factors operating indirectly). The occurrence of vegetational replaceability only complicates indicator schemes, but does not result in errors in indicator research, provided there has been sufficient study.

We thus see that as a result of edaphic compensation and the replaceability of vegetation, changes occur in the relationship of plants and the type of indicators of the same natural conditions in different landscapes of climatically uniform territories. In order to exclude the phenomena of edaphic compensation and indicator replaceability, it is necessary to extrapolate within the boundaries of territories characterized by constant relationships of vegetation with edaphic factors. Territories of this type with a consistent relationship of vegetation, relief, soils, moisture, and geological structure are treated in the literature as geographical landscapes and their components, locality, landmarks, and phases. The extrapolation of indicators within geographical zones is thus most rationally effected from a single landscape or its component to another, analogous landscape. Those indicator features which are extrapolated from one locality to another, or from one landmark to another, may be termed local or landscape features. Morphologically similar landscapes have been designated landscape-analogs. Many indicators can be located in landscapes which are tens or hundreds of kilometers away, but are analogs, with greater success than in landscapes which are situated in close proximity, but are genetically and morphologically different.

The properties of the development of the vegetation cover in a particular territory contribute to the fact that in different regions there are changes in the floristic composition of the indicator vegetation, changes in the synecological (phytocenotic) areas of the plants, and differences in the developmental stage of the plant cover.

Geographic replaceability of vegetation, linked with floristic differences between individual provinces or districts, is observed. Thus, chestnut soils in the Ukraine are occupied by Stipa ucrainica—Artemisia taurica associations, in Western Kazakhstan by Stipa sareptana—Artemisia lercheana, in Eastern Kazakhstan by S. sareptana—A. sublessingiana, in Tyan'-Shan by S. glareosa—A. serotina, and in Tuva) by S. Decipiens—Cleistogenes squarrosus.

Historical replaceability is associated with the fact that several phytocenoses still do not occupy their ecological area. Haloxylon aphyllum (Minkw.) Iljin. is located in the upper Quaternary littoral plain of Western Turkmenia and in the Lower Uzboi valley where groundwaters are bedded at a depth of 4-7 m, while according to E. A. Vostokova (1953) they are linked with groundwaters of moderate depth (10-20 m) in the Tertiary central deserts of Central Asia. Replaceability of this type is encountered on the border of the progressional area of Larix sibirica Ldb. and other species.

Syngenetical replaceability of vegetation is associated with the fact that in some parts of the ecological area of a species there are strong competitors, the composition of which changes in different regions. P. D. Yaroshenko (1950) gives this as an example: Pteridium aquilina L. is displaced by groupings of Sambucus in the Talysh mountains, while in Imeretian mountains, on the other hand, it is displaced by groupings of Kolkhid oak-beech forest.

As a result of the different forms of genetic and geographic replaceability, changes occur in the relationships of plants in different geographical regions and in the make-up of indicators of particular natural conditions. To exclude these forms of indicator replaceability, the area of extrapolation of indicators should be restricted to genetically identical territories, which in physical geography are named geographical regions, provinces, zones, and districts. The indicator features located within these boundaries may be termed regional.

To surmount regional limits, use is made, on the one hand, of the above-mentioned system of landscape analogs, which Gams (1918) has felicitously termed isocenoses. Besides the above-mentioned isocenoses of the chestnut-soil zone, we might cite Acacia tortilis—Aristida pungens in the Algerian Sahara, and Ammodendron conollyi—Aristida karelini in Central Asia. Our thin mixed-grass steppes composed of Stipa capillata L. and Festuca sulcata Hack. and the short-grass prairies of Bouteloua gracilis and Bulbilis dactyloides constitute distant analogs.

Thus, the basis of indicator extrapolation should be the landscape principles of differentiation: zonal, regional, and morphological. In accordance with this, indicators can be divided into three groups on the basis of patterns, methods and distances of extrapolation.

1. Zonal indicators retain identical indicator significance within the boundaries of geographical zones, subzones, and belts. Their extrapolation is limited by the boundaries of the zones. They lose or change their indicator significance when moving from one zone to another.

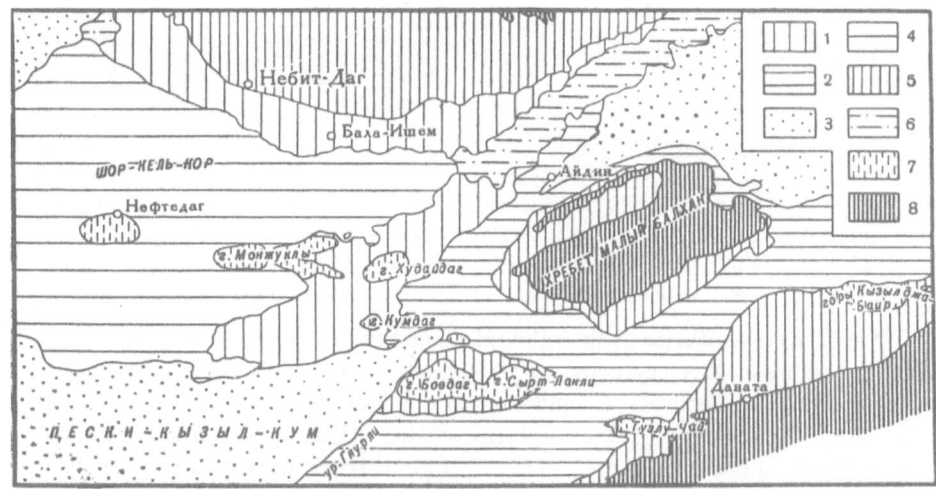

Fig. 1. Systematic map of a group of landscapes in the Pribalkhan district of Western Turkmenia (compiled by B. V. Vinogradov and V. P. Miroshnichenko): 1) stony-sand and stony-clay deserts; 2) clay deserts; 3) sandy deserts; 4) solonchak deserts; 5) foothill deserts; 6) river valleys; 7) outcrops of indigenous rocks; 8) montane landscapes.

2. Regional indicators retain identical indicator significance within the boundaries of geographical regions, provinces, districts, and zones. Their extrapolation is limited by the boundaries of the regions. Their indicator significance is lost or modified when moving from one region to another.

3. Landscape or local indicators retain identical indicator significance within the boundaries of definite structural components of the geographical mantle, i.e., landscapes, sites, landmarks, and phases. Their indicator significance is lost or changed when moving from one landscape or its component to another.

Local indicators are of most practical interest, since indicator investigations are carried out within a single geographical zone or in a single region. Indicator features are worked out in key plots in each landscape, landmark, site, and phase, and then extrapolated to territories with similar physical and geographical conditions. The landscape method permits the control of the extension of indicators by means of additional features (relief, geological structure, hydrography, and position in an ecological series). Many groupings change their indicator significance in different landscapes, sites, and landmarks, but retain it on analogous territories.

Landscape control is based on the fact that factors may be compensated with respect to a single feature (vegetation), but compensation is excluded with respect to a complex of features (landscape).

As a result of determining their relationships with landscape, many groupings of small value as indicators significantly increase their usefulness in landscape combinations and ecological series. And in general, our experience in using hydrologic indicators in Western Kazakhstan (Vinogradov, 1958) and the investigations of Z. V. Karamysheva (1960) on the geological relationships and indicators of soils in Northern Kazakhstan showed that entire ecological vegetation series or landscape combinations, rather than individual plants or even distinct groupings, constitute reliable indicators of particular conditions. The same groupings in different ecological series or varied elements of a single series indicates different natural conditions. If we use these series and combinations in the role of indicators, we exclude the phenomenon of replaceability to a large extent.

Landscape maps were employed by us for the extrapolation of indicators beginning with work in Turkmenia in 1951-1953. Soil and geobotanical features of recent tectonic movements, determined in key plots differentiated in each landscape, are extrapolated on the landscape map. The major groups of landscapes are shown on the latter (Fig. 1). There is a special set of indicators of tectonic structure for each group of landscapes.

In 1954-1955, soil indicators were extrapolated according to the system of geographic landmarks in the Northern Caucasus. The indicators were studied in key plots in distinct types of landmark, and then extrapolated to landmark-analogs, which correspond to the units of the landscape map (Fig. 2).

185

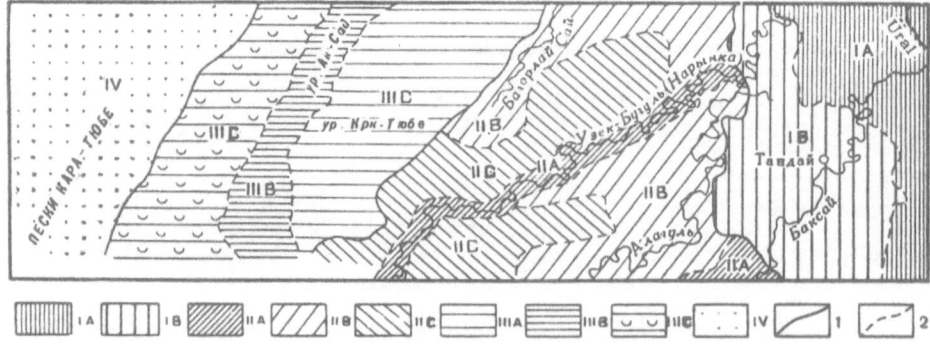

Fig 2. Map of a group of sites in the alluvial plain on the lower course of the Ural River (compiled by B. V. Vinogradov, N. G. Konopleva, V. V. Kuznetsov, V. K. Markovskii, and T. A. Popova). I) Landscape of the Ural River valley: IA) group of sites in the low floodplain; IB) same, high floodplain. II) Landscape of extra-flood-plain inundations of the Ural River: IIA) group of inundation sites with long-functional riverbeds; IIB) same, with riverbeds functional for a brief period; IIC) same, broad inundation areas without riverbeds. III) Landscape of ancient alluvial plain: IIIA) group of lowland sites with isolated Agropyron desertorum—wormwood depressions; IIIB) same with quackgrass depressions and limans; IIIC) same, with solonchak depressions. IV) Landscape of sandy deserts and firm lowland sands: 1) landscape boundaries; 2) boundaries of groups of sites.

The greatest development of the landscape method was attained in the course of finding indicators for hydrogeological deciphering in Northern Kazakhstan and the North Caspian area in 1955-1958. This provided the basis for preparing large-scale maps of sites and the differentiation of hydrologic landmarks (Vinogradov, 1958).

Large-scale landscape maps can thus be used for the extrapolation of indicators of geological structure, site maps can be used for the extrapolation of indicators of groundwaters, and landmark maps for the extrapolation of soil indicators.

The following conclusions may be drawn on the basis of the material presented:

1. Different types of compensation of ecological factors and vegetation replaceability lead on extrapolation to a breakdown of the indicator functions of vegetation and to changes in indicator schemes.

2. Different indicator features have a different extrapolation range, and thus zonal-, regional-, and landscape-restricted indicators are distinguished.

3. The most effective method of excluding compensation and replaceability is the landscape control of indicator extrapolation, which consists in the extrapolation of indicators according to analogous landscapes, sites, landmarks, ecological series, and so on.

Summary

Different types of compensation of ecological factors affect the extrapolation of indicators, disturbing the indicative functions of the plants and altering indicator schemes. Different indicator characters have different ranges of extrapolation. Indicators limited with respect to zone, region, and landscape may be distinguished. The most effective method of excluding compensation and replaceability is landscape control, which consists of the extrapolation of indicators according to analogous landscapes.

Literature Cited

Alekhin, V. V. 1936. "The vegetation of the USSR in the major zones." In book: G. Val'ter and V. V. Alekhin. Principles of Botanical Geography. Moscow—Leningrad, Biomedgiz.

Beideman, I. N. 1949. "The role of the vegetation cover in the water-salt regime of soils." Pochvovedenie, No. 7.

Beideman, I. N. 1953. "Eco-biological principles in the interchange of the vegetation cover (using the depression of the Eastern Zakavkaz'e as an example)." Bot. zhur., Vol. 38, No. 4.

Beideman, I. N. 1954. "The development of vegetation and soils in the depression of the Eastern Zakavkaz'e." In collection: Problems in the Improvement of the Fodder Basis in the Steppe, Semi-Desert, and Desert Zones of the USSR. Moscow.

Viktorov, S. V. 1955. The Use of the Geobotanical Method in Geological and Hydrogeological Research. Moscow, Izd. Akad. Nauk SSSR.

Vinogradov, B. V. 1950. "On the relationship of vegetation with groundwaters in steppe landscapes of Northern Kazakhstan and the use of the vegetation as an indicator in the hydrogeological deciphering of aerial photographs." Izv. Akad. Nauk SSSR, seriya geogr., No. 1.

Vostokova, E. A. 1955. "The application of the geobotanical method in hydrogeological research in deserts and semi-deserts." In collection: The Application of the Geobotanical Method in Geological Research. (Tr. VAGT, No. 1.) Moscow, Gosgeolizdat.

Vostokova, E. A. 1953. "Vegetation as an indicator of geological and hydrogeological conditions in deserts and semi-deserts in connection with their reclamation."

Karamysheva, Z. V. 1960. "Development of steppe vegetation in the hummocky plain of Central Kazakhstan." Bot. zhur., Vol. 45, No. 8.

Lavrenko, E. M. 1960. Geobotany. Soviet Geography. Moscow, Geografgiz.

Pachoskii, I. K. 1917. Description of the Vegetation of Kherson Province, Vol. II. Kherson.

Ramenskii, L. G., Tsatsenkin, I. A., Chizhikov, O. N., and Antipin, N. A. 1956. The Ecological Evaluation of Fodder Lands by the Vegetation Cover. Moscow, Sel'khozgiz.

Rankama, K. 1954. On the use of trace elements in solving some problems of applied geology (Russian translation from English). In book: Geochemical Methods of Prospecting for Ore Deposits. Moscow, IL.

Tyulina, L. I. 1929. "On the evolution of the vegetation cover of the foothills of the Southern Urals." Trudy Zlatoustovskogo obshch. kraevedeniya i Gos. Il'menskogo zapovednika, No. 1.

Shennikov, A. P. 1928. On Convergence amongst Plant Associations. Essays on Phytosociology and Phytogeography. Moscow, Izd. "Novaya derevenya."

Shennikov, A. P. 1950. Plant Ecology. Moscow, Izd. "Sov. nauka."

Yaroshenko, P. D. 1950. Scientific Principles Concerning the Vegetation Cover. Moscow, Geografgiz.

Boyko, H. 1953. "Ecological solution of some hydrological and hydroengineering problems." Proc. Ankara Sympos. Arid Zone Hydrol., UNESCO.

Duvigneaud, P., and de Denacyer-Smet. 1960. "Action de certains métaux lourds du sol (cuivre, cobalt, manganèse, uranium) sur la végétation dans le Haut-Katanga. Rapp. sol et végétation dans le Haut-Katanga. Rapp. sol et végétation." Prem. Colloq. Soc. bot. de France. Paris.

Gams, H. 1918. Prinzipenfragen der Vegetationsforschung. Zürich.

Rübel, E. 1936. "The replaceability of ecological factors and the law of the minimum." Ecology, Vol. 16, No. 3.

Rune, O. 1953. "Plant life on serpentines and related rocks in the north of Sweden." Acta phytogeogr. Sue., Vol. 31. Uppsala.

# EXPERIENCE IN THE INDICATOR INTERPRETATION
# OF GEOBOTANICAL MAPS
# IN THE NORTH CASPIAN REGION

## A. M. Shvyryaeva and G. A. Mikhailova

Any geobotanical map, in conformity with present ideas on content, will of course contain elements reflecting the indicator significance of the vegetation.

The first condition of indicator interpretation is a correspondance of the scale of the indicator map and that of the indicator object. Thus, on a small-scale map reflecting zonal patterns, we can attempt to indicate the boundaries of zones and subzones and other phenomena of the same scale, but cannot expect to locate on it geobotanical features of the distribution of specific lithological types or surface geochemical phenomena of saline-dome structures. Generally, therefore, a geobotanical map should be of the same scale or larger than the scale of the indicator map derived from it.

In the compilation of indicator maps, we applied the following forms of interpretation:

1. Direct interpretation, in which the geobotanical map can be directly converted into an indicator map on the basis of sufficiently extensive published data or special research.

2. Multistage interpretation, in which the indicator map obtained by direct interpretation is in turn converted into another indicator map.

3. Complex interpretation, based on the synthesis of several types of indicator map into a single complex map.

Quite frequently, the geobotanical map itself is used in complex interpretation. It is obvious that the range of analyzable interlandscape relationships is narrower in the first case than in the second and third.

In the present report, we shall deal with experience in the indicator interpretation of a geobotanical map for the purpose of clarifying certain problems associated with the geological mapping of the lower Ural River region, carried out by a VAGT expedition.

The geologic indicator investigations were carried out by the method of standard plots. On these plots, the geological structure of which had earlier been described by geologists, careful geobotanical and geomorphological descriptions were made, boreholes were prepared, and samples of soil, soil-forming rocks, and groundwaters were selected for lithological and chemical analysis. By comparing the geobotanical descriptions with different elements of the landscape, we determined the indicator significance on the plant communities and their complexes and compiled a geoindicational scheme. In this connection, it should be borne in mind that the geologic indicator scheme of one region cannot be used without verification in a region with different physical and chemical conditions. Simultaneously with the setting-up of the standard plots, we carried out geobotanical mapping work involving the wide use of aerial-photography data.

In the process of geobotanical mapping, we established in detail the characteristics of the vegetation cover and the patterns of its distribution. This was achieved not by detailed mapping, which encumbers the map with numerous small contours, but by combining contour areas so as to show complexes of plant communities. The necessity for indicating complexes and combinations of plant communities is due to the fact that the

Geologic Indicator Scheme for the Lower Ural River Region

| Plant indicator communities | Soils | Lithology of the soil-forming rocks | Salinity of soil-forming rocks | | Indicator significance | | | | | |
|---|---|---|---|---|---|---|---|---|---|---|
| | | | | | Depth of bedding and degree of mineralization of the first water-carrying horizon‡ | | Age of soil-forming rocks | Genesis of soil-forming rocks | Geomorphological conditions of the habitat | |
| | | | degree* | type† | depth of bedding, m | mineralization | | | | |
| Communities of Petrosimonia triandra (Pall.) Simonkai | Puffy solonchaks | Sands | Moderate | $Na^·–Cl'–SO_4–Mg^·$ | 2-3 | Saline | New Caspian (modern) | Marine | Almost undifferentiated marine plain | |
| Communities of Halosnemum strobilaceum M.B. (near sors) | Solonchaks | Clays | Very strong | $Na^·$ $Cl'$ | 0.5-2 | Brine | New Caspian (modern) | Sor | Periphery of sor troughs in a ridged or hillocky marine plain | |
| Communities of H. strobilaceum incorporating saltworts and swede (litorai) | Encrusted puffy solonchaks | Sands, rarely clays | Strong–very strong | $Na^·–Cl'–SO_4–Mg^·$ | 2.5-5 | Brine | New Caspian (modern) | Marine | On elevated sites of a slightly differentiated marine plain | |
| Communities of Atriplex cana C.A.M. | Solonchakized solonetzes | Heavy loams, rarely clayey sands | Same | $Na^·–Cl'–SO_4–Ca^{··}$ | 2-4 | Brine | New Caspian (early) | Liman | Troughs and sor-adjoining depressions in a ridged or hummocky marine plain | |
| Communities of Artemisia pauciflora Web. | Solonetzes | Heavy loams, clay, or clayey sand | Strong | $Na^·–Cl'$ | 3.5 and deeper | Saline | Same (modern) | Liman-alluvial | Liman-like depressions in a ridged delta plain | |
| Communities of Aeluropus litoralis (Gouan) Parl. | Delta solonetzed | Sands | Weak–moderate | $Na^·–Cl'–SO_4$ | 2-4 | Brine | New Caspian (early) | Delta | On low plakor sites of a slightly undulating plain | |
| Communities of A. litoralis incorporating freshwater-loving species | Slightly solonetzed | Sands | Very weak–weak | – | 2-4 | Slightly salinated | New Caspian (modern) | Delta and alluvial | Same and on terraced ridges | |
| Communities of Aeluropus litoralis incorporating halophytic species | Solonchakized solonetzes | Sands | Strong | $Na^·–Cl'–SO_4$ | 2-4 | Saline | New Caspian (early) | Delta | In minor depressions or on low plakor sites of a delta plain | |
| Communities of Glycyrrhiza glabra L. | Meadow | Sands | Very slight | – | 2.5-4 | Fresh or slightly salinated perched water table | New Caspian (modern) | Alluvial | Dry riverbeds of temporary water currents in river valley | |
| Communities of Kalidium caspicum L. with saltworts | Delta solonchakized | Clays, loams | Strong | $Na^·–Cl'–SO_4–Mg^{··}$ | 3-5 | From saline to brine | Same | Delta | Elevated sites of slightly differentiated delta plain | |
| Communities of Anabasis salsa (C.A.M.) Benth. (in ancient delta) | Solonchakized solonetzes | Loams, more rarely clays, sandy loams in places | Strong–very strong | $Na^·–Cl'–SO_4–Ca^{··}$ | 5-7 | Saline | Verkhnekhvalynskii | Marine | Flat spurs (hillocks) of ancient delta plain | |
| Communities of A. salsa (in marine plain) | Same | Loams or clay, occasionally clay | Same | $Na^·–Cl'–SO_4$ | 7-15 (20) | Saline | Same | Marine | Flat spurs (hillocks) of ridged marine plain | |
| Communities of Artemisia lercheana Web. | Brown (fallow) meadow-steppe or brown saliniferous | Sandy loams, occasionally sandy loams | Very weak–weak | – | 3.5-5 | Slightly salinated (perched water table) | New Caspian (modern) | Liman | Liman-like depressions on a flat-hillocky ancient delta plain | |
| Communities of A. lercheana with the inclusion of Agropyron desertorum and Stipa sareptana | Same | Same | Very weak | – | 3-5 | Fresh or slightly salinated perched water table | Same | Deluvial | Microsinkholes and hollows in ridged ancient delta and marine plains | |
| Communities of Artemisia arenaria + Elymus giganteus Vahl. | Primitive sandy | Sand | Same | – | 1.5-3 | Same | Verkhnekhvalynskii | Marine | Deflation basins on hummocky sandy plain | |

*Degree of salinization of soil-forming rocks: very slight—up to 250 mg/100 g; slight—251-500 mg/100 g; moderate—501-1000 mg/100 g; strong—1000-2000 mg/100 g; very strong—above 2000 mg/100 g.

†Type is determined by the results of analyses of aqueous extracts. Ions present in the amount of not less than 15 meq % (dominant ions), arranged in decreasing order of content.

‡Degree of mineralization of groundwaters: fresh and slightly brackish—up to 3 g/liter; salinated—3-5 g/liter; saline—5-50 g/liter; brine—above 50 g/liter.

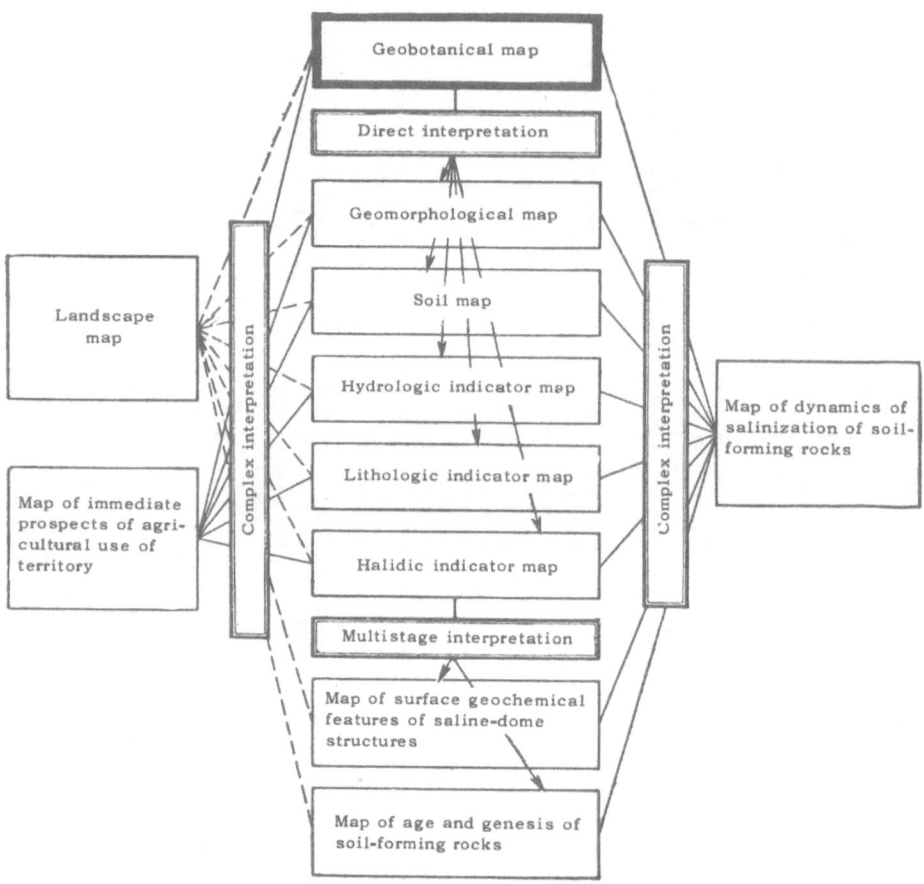

Fig. 1. Scheme of succession of indicator interpretations of a geobotanical map.

correctness of interpretation of a geobotanical map and the accuracy of specialized maps prepared on its basis depend on the specificity of the area of the geobotanical contours which characterize to varying degrees the properties of the natural situation.

In geobotanical mapping, special attention should be paid to the appearance of plant communities or individual species which are not typical for the particular natural conditions. They are frequently associated with features of the geological structure, such as recent tectonic movements, faults, saline-dome tectonics, closely bedded ore bodies, and other similar factors which influence the hydrogeological conditions, the geochemical regime of the soil-forming rocks, etc. Communities and plants which are nontypical for the landscape studied are frequently distributed in a narrow local area and are often dropped in normal mapping. Such communities or scatterings of individual plants may be noted on the map by provisional symbols, if the scale does not permit their precise distribution to be marked.

As a result of the studies carried out, we compiled a geobotanical map which was interpreted into a series of special maps by means of an indicator scheme. By direct interpretation, we obtained a lithologic indicator map (the lithological structure of the soil-forming rocks), a hydrologic indicator map (the depth of bedding and the level of mineralization of the first water-carrying horizon), a halidic indicator map (salinization of the soil-forming rocks), and soil and geomorphological maps. The following were compiled by the multistage interpretation method: a map of the age and genesis of the soil-forming rocks and a map of surface geochemical features of saline-dome structures; and by complex interpretation we compiled a landscape map, a map of the immediate prospects for the agricultural use of the territory and a map of the dynamics of salinization of the soil-forming rocks. The order of indicator interpretation is shown in the scheme (Fig. 1).

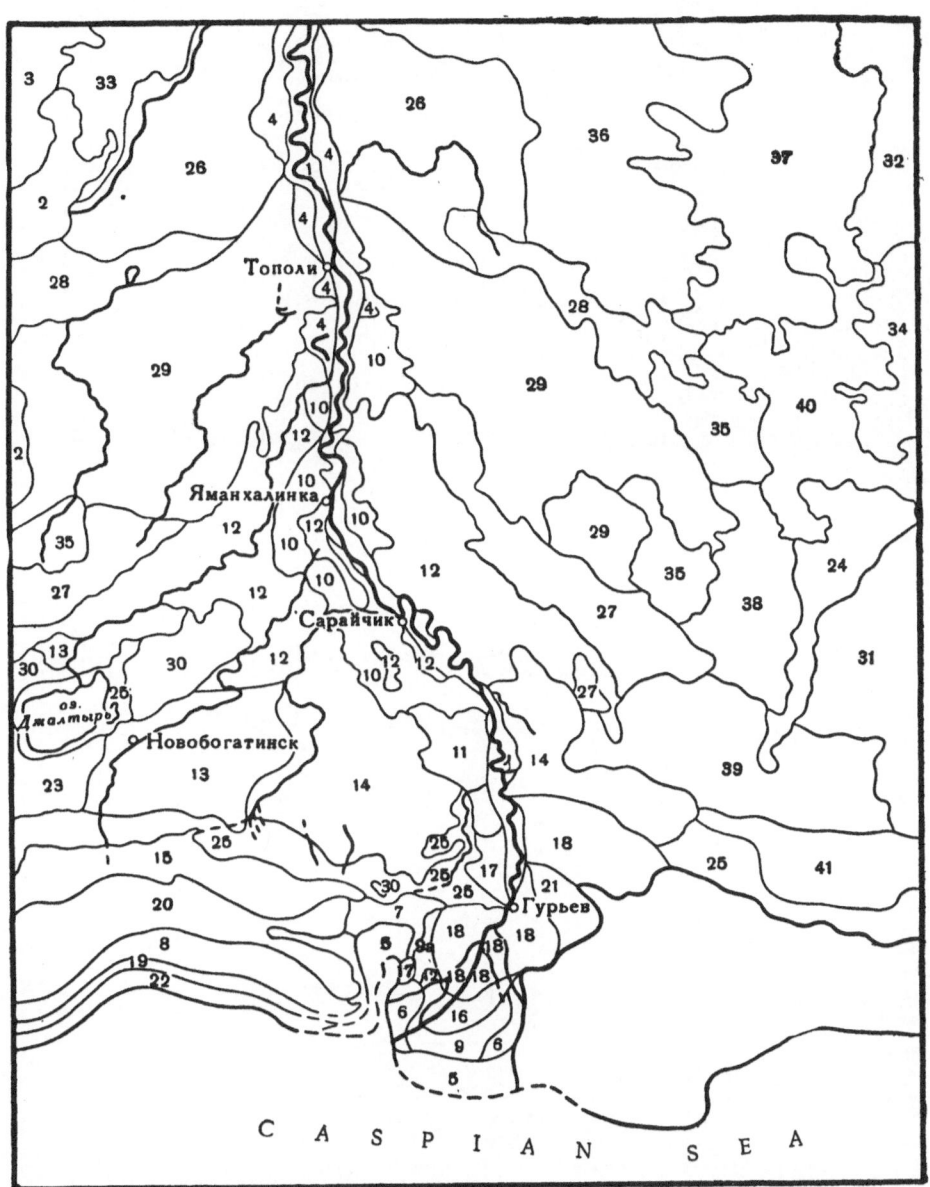

Fig. 2.  Geobotanical map scheme of the lower Ural River region (for key see p. 196).

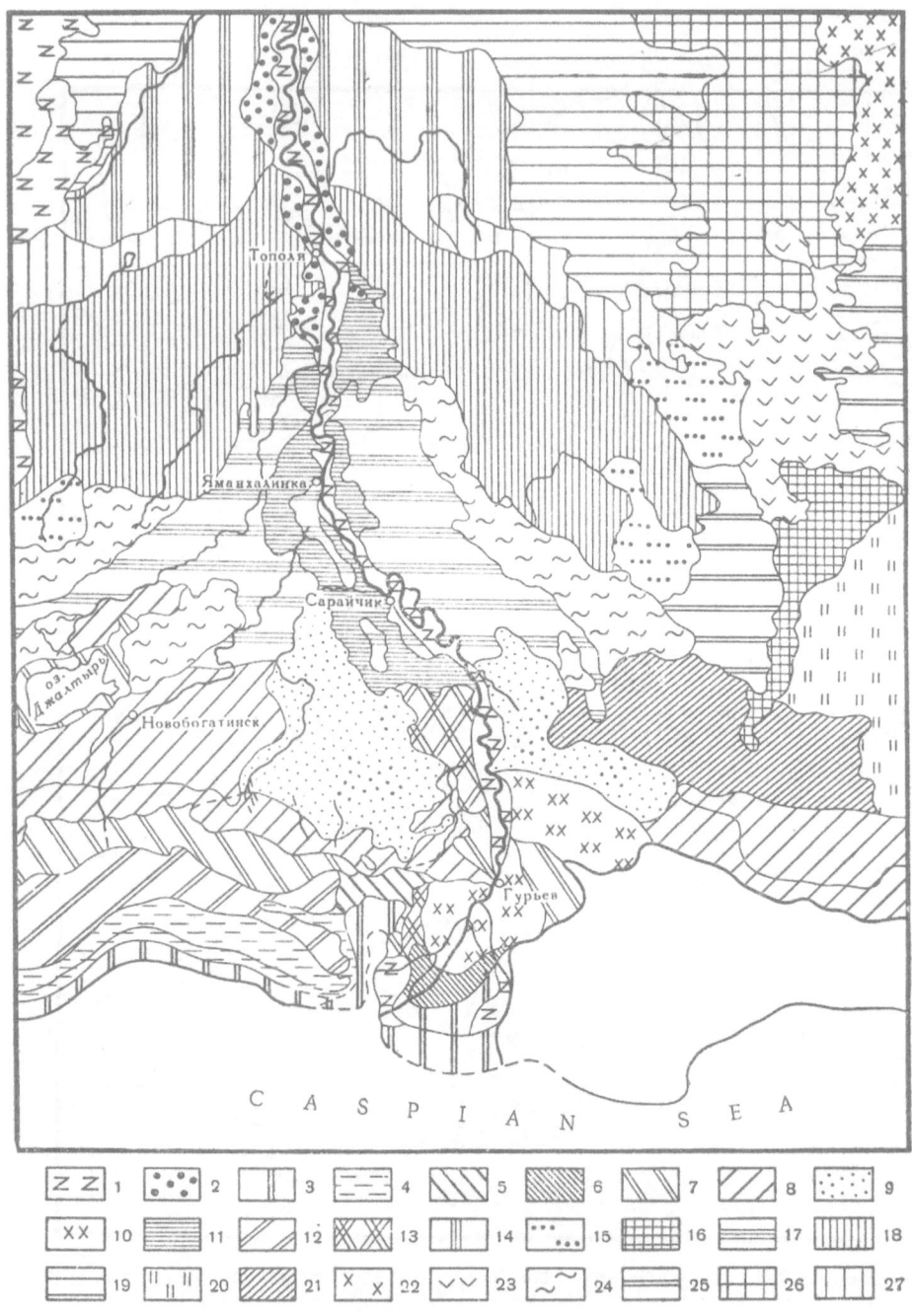

Fig. 3. Map scheme of salinization of soil-forming rocks of the lower Ural River region (for key see p. 197).

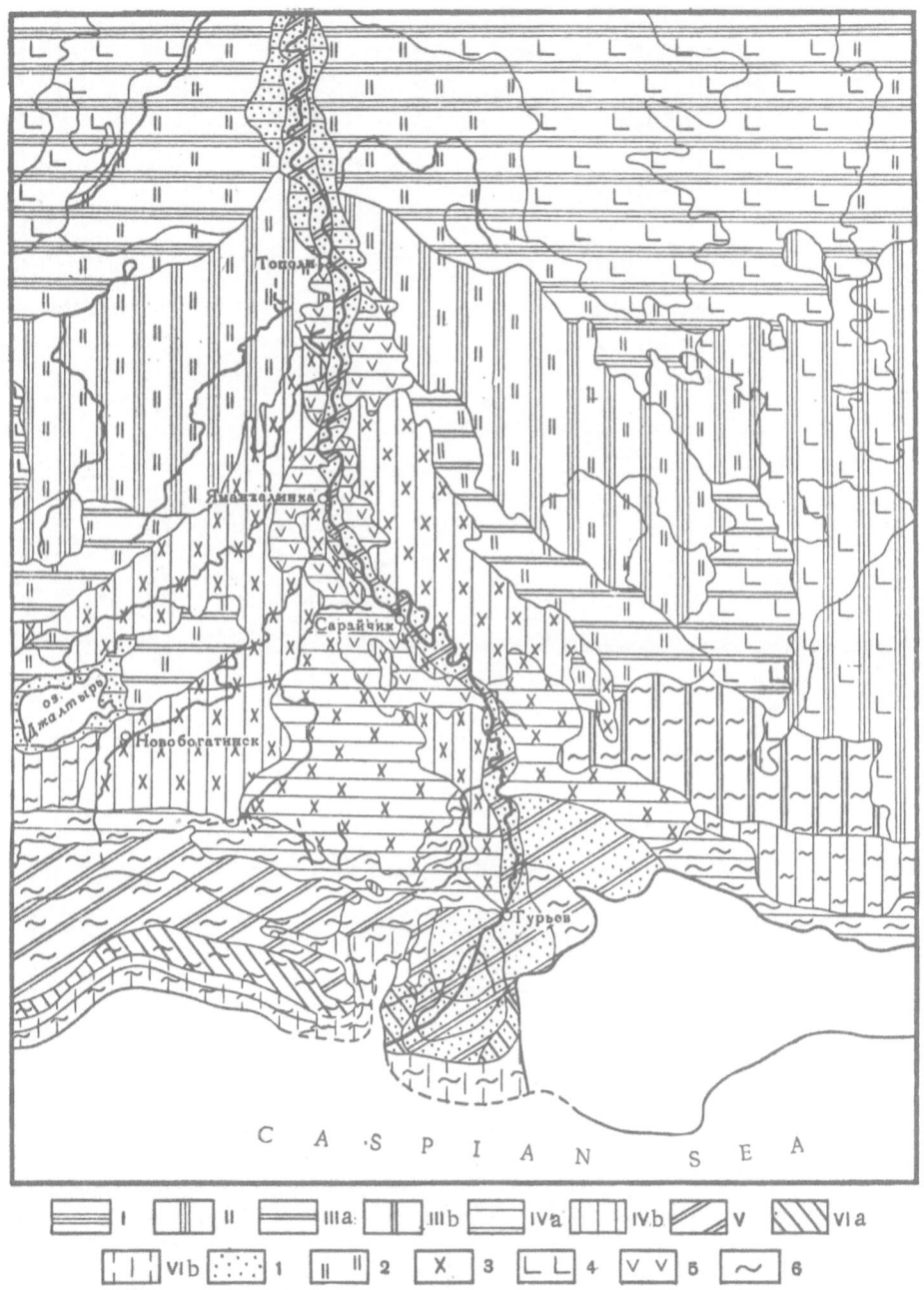

Fig. 4. Map scheme of age and genesis of soil-forming rocks of the lower Ural River region. Age of soil-forming rocks: I) regions of the predominant distribution of soil-forming rocks of Verkhnekhvalynskii age (maximal transgression): Verkhnekhvalynskii—on plakor sites, modern—in depressions; II) Verkhnekhvalynskii rocks in the first stage of depression of basin (to 10-12 m)—marine; Verkhnekhvalynskii rocks in the second stage of depression of basin (to 16 m) — delta; III) Verkhnekhvalynskii in the second stage of depression of basin (to 16 m): a) Verkhnekhvalynskii rocks on plakor sites now in depressions; b) Verkhnekhvalynskii rocks; IV) regions of predominant distribution of soil-forming rocks of New Caspian age (maximal basin transgression 22 m); a) New Caspian rocks; b) New Caspian rocks on plain, Verkhnekhvalynskii rocks on outliers; V) New Caspian stages of basin recession — 25.4 m; VIa) New Caspian stages of basin recession — 28 m; b) New Caspian present-day. Genesis of soil-forming rocks: 1) alluvial deposits; 2) marine deposits on elevations, liman-alluvial and delta deposits — in dry riverbeds and depressions; 3) delta deposits in depressions, marine deposits on hillocks; 4) marine deposits on elevations, sor deposits in sors, and liman-alluvial deposits in depressions; 5) delta and alluvial deposits; 6) marine deposits.

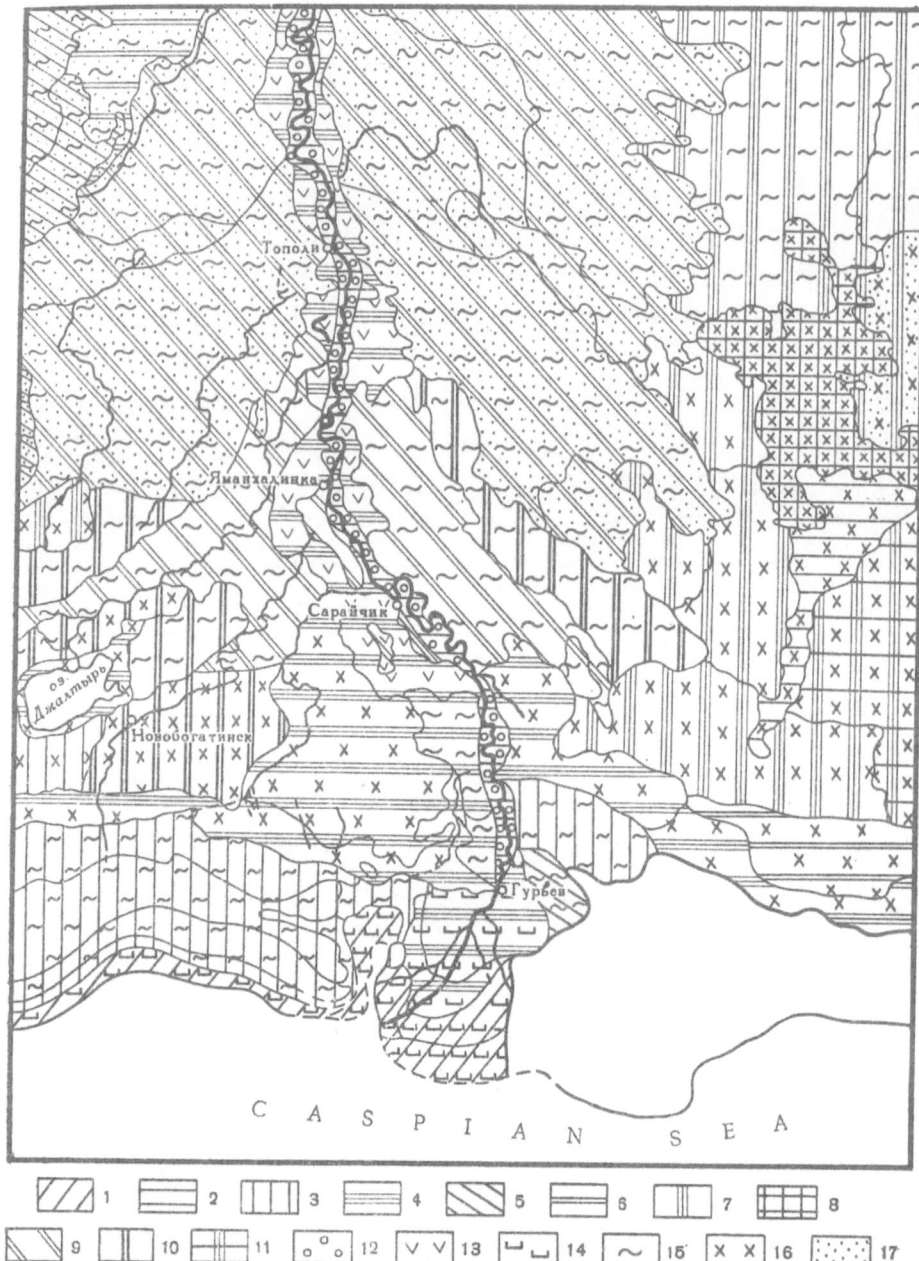

Fig. 5. Map scheme of the depth of bedding and the level of mineralization of the first water-carrying horizon of the lower Ural River region. Depths of bedding of first water-carrying horizon: 1) from 0 to 1.5 (2.0) m; 2) from 0 to 4.0 (5.0) m; 3) from 1.2 (2.0) to 3.0 (3.5) m; 4) from 2.5 to 5.0 m; 5) from 3.5 to 5.0 m; 6) from 0.5 to 5.0 m; 7) from 0.5 to 7.0 (15.0) m; 8) from 0.5 to 10.0 (15.0) m; 9) from 1.5 to 7.0 (8.0) m; 10) from 3 to 6.0 (8.0) m; 11) from 0.5 to 20 m. Mineralization of groundwaters: 12) fresh; 13) fresh—slightly brackish; 14) salinated, in places saline; 15) saline; 16) brine; 17) regions where fresh and slightly salinated perched water tables are located.

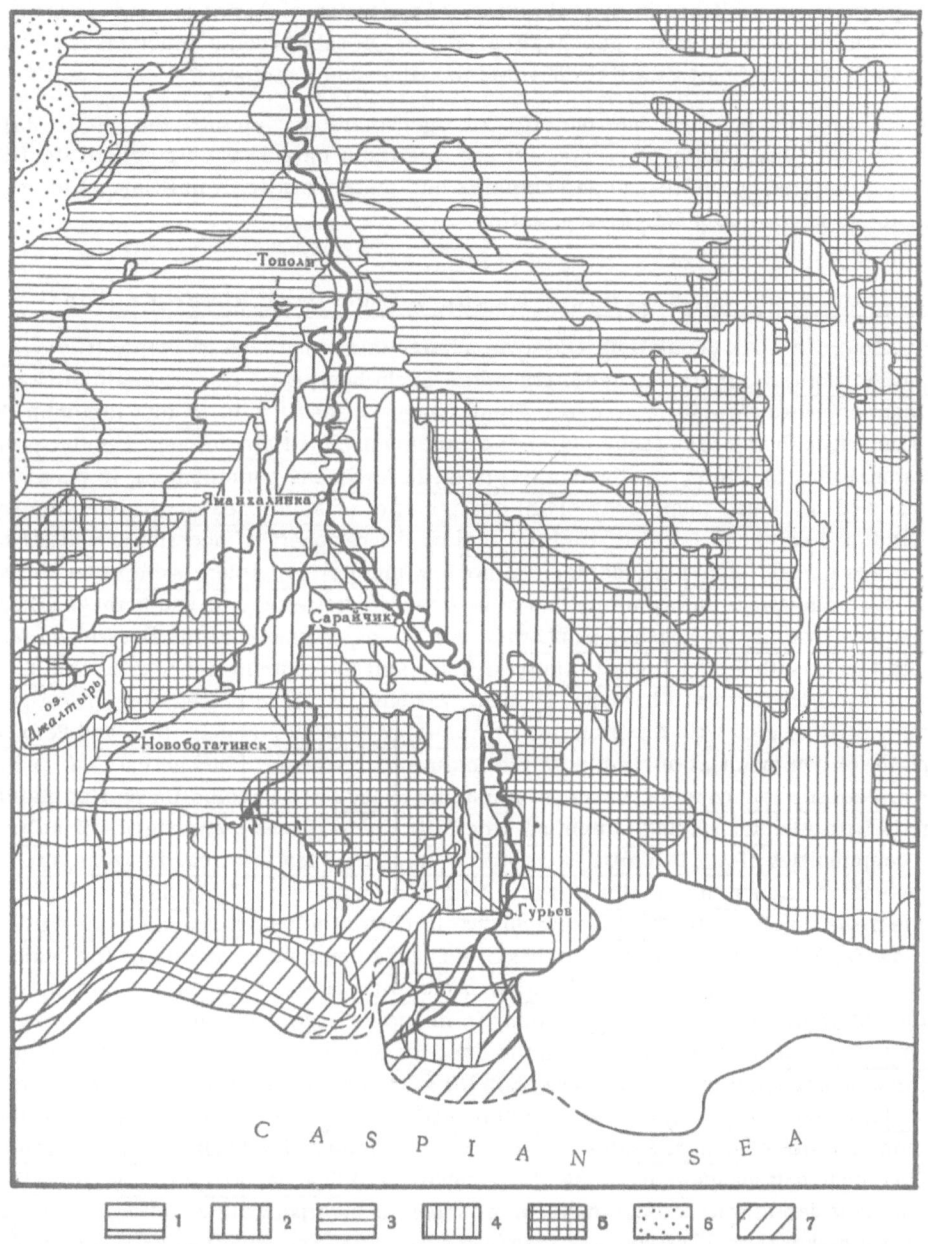

Fig. 6. Map scheme of immediate prospects for the agricultural use of the lower Ural River region (compiled by S. V. Viktorov). 1) Regions suitable for early reclamation under farm crops (partly carried out already); 2) regions of possible liman irrigation and organization of local reservoirs. Possible forest plantings in dry, particularly deep riverbeds; 3) regions of liman hay crops with possible undersowing of salt-resistant forage plants in the largest limans; 4) poor inconvenient pastures (with predominance of solonchaks); 5) wormwood-saltwort pastures; 6) sandy pastures; 7) regions of possible reed growths.

Extracts from the indicator scheme, the geobotanical map (Fig. 2), and also the map schemes of halidic indicator research (Fig. 3), hydrologic indicator research (Fig. 4), age and genesis of the soil-forming rocks (Fig. 5), and the immediate prospects of agricultural use of the territory (Fig. 6) are given below.

The landscape map was published in the collection, "Application of Aerial Methods in Landscape Investigations" (1961), while the map of surface geochemical features of saline-dome structures is given in the present collection in A. M. Shvyryaeva's article devoted to saline-dome tectonics.

Indicator maps compiled by direct interpretation of a geobotanical map show the relationship of the vegetation cover with any individual landscape component and as a result have limited practical significance. However, indicator maps compiled by multistage and complex interpretation reveal multilateral intralandscape relationships and therefore have wide practical application.

Key to the Geobotanical Map Scheme of the Lower Ural River Region

1) Poplar—willow—Tamarix populations and hydrophytic—mixed herbage groupings (Glycyrrhiza uralensis, Sonchus, Xanthium, Sophora alopecuroides, reed, groundsel, buckwheat, etc.) on floodplain of Ural River; 2) communities of psammophytes (Artemisia arenaria, Elymus angustus, bezlistnik, Bromus tectorum, Euphorbia seguieriana, etc.) on ridged-hummocky sandy plain, and communities of Halosnemum strobilaceum in the periphery of occasional sors; 3) Agropyron sibiricum—Artemisia lercheana communities, in places with Kochia prostrata on flat parts of the ridged-hummocky plain; communities of A. lercheana with psammophytes on semi-firm hummocky sands and communities of H. strobilaceum on the periphery of occasional shallow sors; 4) communities of Glycyrrhiza glabra, Alhagi pseudalhagi, Limonium gmelinii, Atriplex, Acroptilon picris, sometimes with Tamarix on ridged elevations and communities of hydrophilic grasses and mixed herbage (Hierochloë, Agropyron, Fagopyrum, Glycyrrhiza, Sonchus, Euphorbia, etc.) in dry channels of the flat-ridged plain; 5) communities of reed, in places with Typha angustifolia on a flat plain; 6) communities of reed and T. angustifolia with the inclusion of hydrophytic mixed herbage (Sonchus, Atriplex, etc.), in places with willow groves on a flat plain; 7) communities of reed with the inclusion of Aeluropus litoralis, Puccinellia, and Petrosimonia triandra on flat plain; 8) communities of reed, with the inclusion of Senecia, Pluchea caspica, and Atriplex, and communities of Petrosimonia triandra, in places including Suaeda and Puccinellia on flat plain; 9) communities of Tamarix with Atriplex sp. and Limonium gmelinii, and in places communities of Aeluropus litoralis on flat plain; 10) communities of Ae. litoralis, Acroptilon picris, Atriplex tatarica, Dodartia, Eremopyrum orientale and saltworts, and Limonium gmelinii; places with communities of Glycyrrhiza on flat plain; 11) solonchak meadows made up of Ae. litoralis communities, in places including Limonium gmelinii, Atriplex, Acroptilon picris, bassia, Salsola sp., and with patches of a Suaeda community on almost flat plain; 12) solonchak meadows made up of communities of Ae. litoralis, Acroptilon picris, Agropyron ramosum, and Limonium gmelinii on almost flat plain; communities of hydrophytic mixed herbage made up of Glycyrrhiza, Alhagi pseudalhagi, Agropyron repens, and other species in the dry beds of temporary watercourses; communities of Anabasis salsa on occasional small hummocks (outliers); 13) solonchak meadows comprising communities of Ae. litoralis, Suaeda, Artemisia sp., Agropyron ramosum, Atriplex tatarica, and other species in flat, liman-like depressions; communities of Anabasis salsa on hillocks and communities of Halosnemum strobilaceum on low plakor sites; 14) solonchak meadows made up of communities of Suaeda, Atriplex, Ae. litoralis, and Artemisia sp. in wide shallow depressions and communities of saltworts incorporating Kalidium caspici and Halostachys caspica on low plakor sites; 15) solonchak meadows made up of communities of Suaeda and Ae. litoralis incorporating Kalidium caspici and Halostachys caspici on flat plain, sometimes with patches of depressed reed and Atriplex in microdepressions; 16) communities of Suaeda incorporating K. caspici and Atriplex sp., in places with Tamarix, on almost flat plain; 17) communities of Suaeda, in places including Halosnemum strobilaceum and Salicornia herbacea, on flat plain; 18) communities of Suaeda and H. strobilaceum, in places with K. caspici, Atriplex, and Nitraria schoberi on an almost flat plain, and communities of Atriplex in the dry beds of temporary watercourses; 19) thin growths of Tamarix with Petrosimonia triandra on flat plain; 20) communities of P. triandra and Puccinellia in places including Tamarix, Halostachys caspica, and Ae. litoralis on flat plain; 21) communities of Petrosimonia angustifolia and Salsola crassa, with K. caspici and Anabasis aphylla, on almost flat plain; 22) growths of Salicornia herbacea with rare groups of reed, in places including Aster on

flat plain; 23) communities of Halosnemum strobilaceum on slightly elevated flatland sites, communities of Anabasis salsa on occasional hillocks and communities of Ae. litoralis in liman-like depressions; 24) communities of Halosnemum strobilaceum, Statice sp., in places Anabasis salsa, on low inter-sor watersheds of flat-ridged plain; 25) communities of H. strobilaceum with a profusion of halophytes, with Kalidium caspici and Halostachys caspica on flatland and communities of Ae. litoralis with Halosnemum strobilaceum and saltworts on microdepressions; 26) communities of Anabasis salsa on plakor sites of a flat-ridged plain, with troughs and liman-like depressions occupied by communities of Artemisia lercheana including a small admixture of Agropyron desertorum and Anabasis aphylla; in beds of dry watercourses, communities of Agropyron repens, A. ramosum, Artemisia lercheana, and other species, or communities of Hierochloë, Glycyrrhiza, Alhagi pseudalhagi, Artemisia, and other species; 27) communities of Anabasis salsa on plakor sites of flat-ridged plain; communities of Artemisia lercheana with Agropyron ramosum or A. desertorum, or sometimes Artemisia pauciflora communities, in liman-like flat depressions; 28) communities of Anabasis salsa on plakor sites of a flat-ridged plain, with communities of Artemisia pauciflora incorporating saltworts and Agropyron ramosum, or communities of Artemisia lercheana on liman-like depressions and dry watercourses; 29) communities of Anabasis salsa on plakor sites of a flat-ridged plain, with troughs and liman-like depressions occupied by communities of Artemisia lercheana (in places with Agropyron desertorum or Limonium gmelinii, Acroptilon picris, Artemisia sp., etc.), sometimes by communities of A. pauciflora; communities of Agropyron repens, A. ramosum, Artemisia lercheana, and other species; 30) communities of Anabasis salsa on plakor sites of an almost flat plain; communities of Artemisia pauciflora (rarely A. lercheana) in shallow flat depressions; 31) communities of Anabasis salsa on plakor sites of an undulating plain; communities of Artemisia lercheana with Anabasis aphylla on microsinkholes and interhillock depressions, with communities of Halosnemum strobilaceum on the periphery of sors; 32) communities of Anabasis salsa on plakor sites of microhummocky plain; communities of Artemisia lercheana, frequently with Agropyron desertorum, in troughs and microsinkholes, and communities of H. strobilaceum on periphery of sors. In places, solonchak meadows from communities of Agropyron repens, A. ramosum, Ae. litoralis, with other species in depressions; 33) communities of Anabasis salsa on plakor sites of flat-ridged plain; in troughs, communities of Artemisia lercheana, in places with Agropyron desertorum, Stipa sareptana, or Anabasis aphylla, and communities of H. strobilaceum on periphery of sors; 34) communities of Anabasis salsa on plakor sites of a sloping-ridged plain; communities of Artemisia lercheana with Agropyron desertorum and Anabasis aphylla in shallow interhillock depressions; communities of Artemisia lercheana or A. pauciflora, or communities of Atriplex cana with Artemisia pauciflora in interhillock depressions and depressions near sors; communities of H. strobilaceum on the periphery of sors; 35) communities of Anabasis salsa on plakor sites of a flat-ridged plain, with patches of Artemisia lercheana communities in microsinkholes; communities of Artemisia pauciflora or Artemisia sp. in liman-like depressions, and communities of H. strobilaceum on periphery of sors; 36) communities of Anabasis salsa, and communities of Artemisia pauciflora or lercheana on plakor sites of flat, slightly undulating plain, with grass—mixed herbage communities in troughs and interhillock depressions, and communities of H. strobilaceum on the periphery of the occasional sors; 37) communities of Anabasis salsa on plakor sites of sloping-ridged plain with communities of Artemisia lercheana in microsinkholes; communities of H. strobilaceum and Atriplex cana with Artemisia pauciflora on the periphery of sors; 38) communities of A. salsa on plakor sites of a ridged plain with communities of Artemisia lercheana and Anabasis aphylla in microdepressions and communities of H. strobilaceum on the periphery of sors; 39) communities of Anabasis salsa on plakor sites of slightly undulating plain, with solitary specimens of H. strobilaceum or Kalidium caspici on the lower parts of slopes; communities of Artemisia lercheana or A. pauciflora with Atriplex cana in flat shallow depressions; communities of H. strobilaceum on the periphery of occasional sors; 40) communities of A. salsa on plakor sectors of ridged plain with communities of Artemisia lercheana with Agropyron desertorum and Stipa sareptana in microsinkholes; communities of H. strobilaceum and Atriplex cana with Artemisia pauciflora on the periphery of the numerous large sors; 41) communities of A. salsa on plakor sites of small-hummock plain and communities of H. strobilaceum and solonchak meadows on depressions.

## Level and Type of Salinity of Soil-Forming Rocks in the Lower Ural River Region

1) Very slight (−); 2) very slight—slight (−); 3) moderate ($Na^{\cdot}-Cl'-SO_4''-Mg^{\cdot\cdot}$); 4) moderate—strong ($Na^{\cdot}-Cl'-SO_4''-Mg^{\cdot\cdot}$); 5) moderate—strong ($Na^{\cdot}-Cl'-SO_4''$; $Na^{\cdot}-Cl'-SO_4''-Mg^{\cdot\cdot}$); 6) moderate—strong

$(Na^{\cdot}-Cl'-SO_4''-Mg^{\cdot\cdot};\ Na^{\cdot}-Cl'-SO_4'')$; 7) strong—very strong $(Na^{\cdot}-Cl'-SO_4''-Mg^{\cdot\cdot})$; 8) strong—very strong $(Na^{\cdot}-Cl'-SO_4'';\ Na^{\cdot}-Cl'-SO_4''-Mg^{\cdot\cdot})$; 9) strong—very strong $(Na^{\cdot}-Cl'-SO_4''-Mg^{\cdot\cdot};\ Na^{\cdot}-Cl'-SO_4'')$; 10) moderate—strong $(Cl'-Na^{\cdot}-Mg^{\cdot\cdot}-SO_4'';\ Na^{\cdot}-Cl'-SO_4'')$; 11) very slight—slight; moderate (in places) $(Na^{\cdot}-Cl'-SO_4'')$; 12) moderate—strong; strong $(Na^{\cdot}-Cl'-SO_4''-Mg^{\cdot\cdot};\ Na^{\cdot}-Cl'-SO_4'')$; 13) strong; very slight—slight $(Na^{\cdot}-Cl'-SO_4'')$; 14) strong—very strong; very slight—slight $(Na^{\cdot}-Cl'-SO_4''-Ca^{\cdot\cdot})$; 15) strong—very strong; very strong $(Na^{\cdot}-Cl'-SO_4''-Ca^{\cdot\cdot};\ Cl'-Na^{\cdot})$; 16) very strong; strong—very strong $(Cl'-Na^{\cdot};\ Na^{\cdot}-Cl'-SO_4'')$; 17) slight—moderate; strong—very strong; very slight—slight $(Na^{\cdot\cdot}-Cl'-SO_4'';\ Na^{\cdot}-Cl'-SO_4''-Ca^{\cdot\cdot})$; 18) strong; very slight—slight $(Na^{\cdot}-Cl'-SO_4''-Ca^{\cdot\cdot};\ Na^{\cdot}-Cl')$; 19) strong—very strong; very slight—slight; very strong $(Na^{\cdot}-Cl'-SO_4'';\ Cl'-Na^{\cdot})$; 20) strong—very strong; moderate; very strong $(Na^{\cdot}-Cl'-SO_4'';\ Na^{\cdot}-Cl';\ Cl'-Na^{\cdot})$; 21) strong—very strong; strong; very slight—slight $(Na^{\cdot}-Cl'-SO_4'';\ Na^{\cdot}-Cl'-SO_4''-Ca^{\cdot\cdot})$; 22) strong—very strong, strong; very slight—slight $(Na^{\cdot}-Cl'-SO_4''-Ca^{\cdot\cdot};\ Na^{\cdot}-Cl'-SO_4'')$; 23) very strong; very strong; strong—very strong $(Na^{\cdot}-Cl'-SO_4'';\ Cl'-Na^{\cdot};\ Na^{\cdot}-Cl'-SO_4''-Ca^{\cdot\cdot})$; 24) strong—very strong; strong; slight; very slight—slight $(Na^{\cdot}-Cl'-SO_4''-Ca^{\cdot\cdot};\ Na^{\cdot}-Cl';\ Na^{\cdot}-Cl')$; 25) strong—very strong; very strong; moderate; strong—very strong $(Na^{\cdot}-Cl'-SO_4'';\ Cl'-Na^{\cdot};\ Na^{\cdot}-Cl';\ Na^{\cdot}-Cl'-SO_4''-Ca^{\cdot\cdot})$; 26) strong—very strong; very strong; strong; very slight—slight $(Na^{\cdot}-Cl'-SO_4'';\ Cl'-Na^{\cdot};\ Na^{\cdot}-Cl'-SO_4''-Ca^{\cdot\cdot})$; 27) very strong—strong; very strong; strong; very slight—slight $(Na^{\cdot}-Cl'-SO_4''-Ca^{\cdot\cdot};\ Cl'-Na^{\cdot};\ Na^{\cdot}-Cl'-SO_4'')$.

## Summary

Geobotanical maps of the normal type, when accompanied by data on the significance of plant communities as indicators of ecological conditions, can be interpreted into different indicator maps; the authors distinguish three types of interpretation: 1) direct interpretation, i.e., the direct conversion of a geobotanical map into a corresponding indicator map (lithologic indicator, hydrologic indicator, etc.); 2) multistage interpretation, when a single indicator map, obtained by means of interpretation, is converted into another map (e.g., a halidic indicator map being transformed into a map of geochemical regions); and 3) complex interpretation, when several indicator maps are combined into one (as when a map of indicators of different types of habitats is derived from hydrologic indicator, lithologic indicator, and other maps).

# EXPERIENCE IN THE COMPILATION OF
# INDICATOR REFERENCE BOOKS

## E. A. Vostokova, A. V. Shavyrina, N. N. Preobrazhenskaya,
## and L. N. Tagunova

The possibility and prospects of using the geobotanical method in geology and hydrogeology (Lavrenko, 1954; Viktorov, 1955) has now been clearly established.

Attempts to present indicator data in a popular manner were made many years ago. The first was apparently the key compiled by B. A. Keller to the soils and parent rocks on the basis of the vegetation for the Zaisank district (Keller, 1912). This key was, however, intended only for the assessment of certain rather conditional gradations in the level of nutrient elements and the moisture content in soils.

Of great importance from the procedural point of view was a book by L. G. Ramenskii, L. A. Tsatsenkin, O. N. Chizhikov, and N. D. Antipin (1956), which dealt with theoretical principles and described the procedure for the ecological assessment of habitats by the vegetation cover. This book is of particular assistance to the geobotanist, rather than the geologist; in addition, plants are examined in it as indicators of certain rather tentative levels of moisture, salinity, and so on (frequently there are no strictly quantitative details), and this renders difficult the practical use of the data provided.

L. V. Larin's book (1953), which deals with the assessment of soils and agricultural lands on the basis of vegetation, and the "Atlas of Plant Indicators of Habitat Conditions," compiled by the botanist Krüdener with the assistance of several highway engineers (Krüdener et al., 1941), are much more suitable aids to the large-scale use of the geobotanical method in the practice of geological and hydrogeological research.

In the course of geobotanical work during the geological expeditions of the former Ministry of Geology and Conservation of Mineral Resources, geobotanists prepared materials for the comprehensive instruction of geologists in indicator principles, right from the very inception of the geologic indicator approach.

This material was very varied. The simplest material consisted of reference herbaria, i.e., herbaria of the main plant indicators with a brief explanatory text (these were usually prepared and used under field conditions). Also compiled were albums with photographs of plant indicators and examples of geologic and hydrologic indicator deciphering, which provided a more complete and more scientifically based view of indicator research, but which were more complicated and very laborious to compile. The staff of the geobotanical laboratory of VSEGINGEO is therefore now working on the production of indicator reference books.

At the basis for the compilation of these reference books is the hypothesis that there is a close interrelationship between the structure of the ground, the depth of bedding, and the degree of mineralization of subsurface waters, on the one hand, and the character of the vegetation on the other. This hypothesis is of course fundamental to the theory of indicator geobotany (Viktorov, 1955).

Plant communities which have a high degree of concurrence with indicator objects are regarded as indicators in the reference books.

Reference has frequently been made in geobotanical indicator work to the geographic localization of phytogenic indicators. Before proceeding to the compilation of the reference books, we therefore carried out a regional study of the territory of the USSR to determine the regions in which similar geologic indicator patterns are retained to a significant extent.*

Two regions were recognized in the work of producing the reference books: 1) the desert region, comprising the southern desert subzone; 2) the forest region, comprising the western area of the southern taiga and broad-leaved forest zones.

The procedure for compiling the reference books consisted of two stages: first, the determination of the indicator significance of plant communities within the region; and second, the actual compilation of the reference book.

To determine the indicator significance of plant communities, i.e., to discover indicator communities in a particular region, wide use was made of published data and, in addition, factual material was collected in key plots.† In the desert region, studies were made on indicator patterns in 12 key plots, located in different parts of the region, consideration being paid to ensuring that all the varied conditions of the desert zone were covered. In the forest region, factual data were collected in 8 key plots, also located in different parts of the region. The study of the indicator role of the vegetation in the key plots was frequently accompanied by indicator mapping.

On the basis of the field collections as well as published and background material, a generalized indicator table was compiled which showed the indicator significance of the plant communities in the particular region. This generalized indicator table is the basic reference book. The only indicators utilized are those plant communities with high levels of indicator significance and validity. The following indicator objects were selected: for the reference book on the desert region—the mechanical structure of the ground (top 10-m layer), the depth of bedding, and the level of mineralization of subsurface waters (lying at a depth of 20-25 m from the surface); and for the reference book on the forest region—the mechanical structure of the ground to a depth of 5-10 m, the depth of bedding of groundwaters or a perched water table to a depth of 5-10 m, and the thickness of peat deposits.

After having identified the main plant communities, since knowledge of these plants is necessary for recognizing plant indicator communities. For this purpose, floristic lists were compiled for each group of associations, by which means the dominance and constancy of the species could be determined. The dominant species of the communities, and in some cases the constant species also, were selected for inclusion in the album part of the handbook.‡

The "Reference Book on Plant Indicators of Groundwaters and Terrains in the Southern Deserts of the USSR" is constructed in the following manner. To begin with, the following six main desert types are identified and briefly described on the basis of geomorphological, terrain, geobotanical, and, to some extent, geological features: sandy, sandy loam—clayey, gravelly, gypsum—stony, loess, solonchak; river valleys and oases are also treated.

Each type is described according to a single plan and is illustrated with aerial photographs of the typical landscape, landscape photographs, and a profile representation of vegetational distribution in the particular type of desert. After each description of a desert type, a brief key is given to landscape plants and hydrologic indicators, illustrated with figures of those aspects of the plant of value for diagnostic purposes.

A substantial section of the reference book consists of an album of plant drawings. For each species, a drawing is given of the whole plant or part of the plant, a photograph of the plant in a cenosis, and a schematic profile illustrating the indicator significance of the plant.

---

* The nature and principles of this zonation are not dealt with here.
† Key plots were selected in localities which were most typical as regards geological and geomorphological conditions and where the vegetation cover had been least disturbed. The area of a key plot varied from 100 to 3000 km².
‡ In the desert region, dominants in indicator communities were particularly evident. For simplicity of presentation, frequent use is made of the term "plant indicator"; it should be taken to mean plants which were dominant in indicator phytocenoses.

At the end of the "Reference Book" are found auxiliary tables, indexes of Russian, local, and Latin plant names, a subject index, and also explanations of certain specialized terms.

The reference book compiled for the forest region was constructed somewhat differently; it lacks a description of types of locality, but includes specifications of terrain soils and groundwaters in relation to vegetation, and has therefore been termed a "Reference Guide to the Terrain Soils and Subsurface Waters on the Basis of Vegetation, for the Western Area of the Forest Zone."

The reference book is intended for determining the mechanical structure of the top (5-10 m) layer of soils of Quaternary deposits, the presence of shallow-bedded (to 5 m) subsurface waters, and for the economic evaluation of territories and engineering-geology site reconnaissance.

It consists of the following sections:

I. A brief physical and geographical description of the region.

II. A guide to the terrain soils and depths of bedding of subsurface waters on the basis of vegetation (it was constructed on the basis of the lithochemical principle; descriptions of vegetation closely associated with relief conditions are given as the main indicational features).

III. An album of plants which are dominant or constant for specific plant indicator communities. Each plant drawing is accompanied by a brief explanatory text.

Reference tables and indexes of Russian and Latin plant names are given at the end of the book.

All the sections of the reference guide are illustrated with photographs of plant indicator communities and individual plants, facilitating their recognition under natural conditions, and schematic profiles illustrating their indicator significance.

In the introductions to the reference handbooks, explanations are given of their structure as well as methods for using them, particular attention being paid to the process of identification in the lithochemical tables.

Summary

One of the most effective forms of introducing indicator geobotany in the practice of agriculture, geology, and other fields of scientific and economic activity is the preparation of reference books of plant indicators. The authors have prepared two indicator reference books—one for the southern deserts of the USSR (the Kara-Kum, the Kyzyl-Kum, and the Ustyurt) and one for the western part of the forest zone. In the first manual, special attention was paid to the edificator species of diverse communities, and in the second, to the indicator value of individual landscape units, possessing both geobotanical and geomorphological characteristics.

Literature Cited

Viktorov, S. V. 1955. The Use of the Geobotanical Method in Geological and Hydrogeological Research. Moscow, Izd. Akad. Nauk SSSR.

Keller, B. A. 1912. "Botanical and geographical studies in the Zaisansk district of Semipalatinsk province." Trudy Obshch. estestvoispyt. Kazansk. univ., Vol. 44, No. 5.

Lavrenko, E. M. 1954. "The basic trends in geobotanical research in the USSR in relation to the needs of the national economy." In book: Problems of Botany, II. Moscow—Leningrad, Izd. Akad. Nauk SSSR.

Larin, I. V. "Experience in the identification, from the vegetation cover, of soils, parent rocks, relief, agricultural lands, and other elements of the landscape in the central part of Ural province." Trudy obshch. izuch. Kazakhstana, VII, No. 1. Kzyl-Orda.

Larin, I. V. 1953. Determination of Soils and Agricultural Lands from the Vegetation Cover. Moscow, Sel'khozgiz.

Ramenskii, L. G., Tsatsenkin, I. A., Chizhikov, O. N., and Antipin, N. D. 1956. Ecological Assessment of Forage Lands from the Vegetation Cover. Moscow, Sel'khozgiz.

Krüdener, A., Becker, A., Escher, W., Mussgang, R., and Zacharias, I. 1941. Atlas Standortkennzeichender Pflanzen. Berlin.

# ON THE POSSIBILITY OF USING THE GEOBOTANICAL
# METHOD FOR MAPPING QUATERNARY DEPOSITS

## N. N. Preobrazhenskaya

Prospecting for quaternary deposits with the aid of the geobotanical indicator method is widely used in the geological expeditions program of VAGT during research work in arid regions (Voronkova, 1955), it having proved possible in several cases to differentiate lithologically similar strata of differing origin on the basis of geobotanical data (Shvyryaeva, 1955).

There has been less research work along these lines in the forest zone, although some experience has been gained (Ososkov, 1899; Biské, 1949; Mraz, 1958; and others). Particular interest in this respect attaches to a paper by K. Mraz (1958), in which the author describes the principles for the use of forest communities for mapping Quaternary deposits and weathering crusts. It is mentioned in the paper that indicator mapping in Czechoslovakia is based on the amalgamation of the phytocenotic method with the method of comparative ecology or with the study of ecotypes (Pogrebnyak, 1944; Vorob'ev, 1953). The author mentions the following successive stages in the carrying out of indicator mapping: a) investigation of the ecology and indicator suitability of the species; b) investigation and identification of plant communities and their classification; c) studies on plant communities and determination of vegetational types as indicators of Quaternary deposits and weathering crusts; and d) mapping of Quaternary deposits with the help of the vegetation.

As indicators, the author uses plant communities (taking into consideration the ecology of the species which make up the communities).

Tables included in the text of Mraz' paper and constructed on the ecological-phytocenotic principle are of interest. These tables demonstrate the relationship between plant communities and various types of Quaternary deposits. For the identified indicator communities, lists of plants are given in the tables, and data are provided on the ecology of the plants.

The sequence of work in the mapping of Quaternary deposits from the vegetation, proposed by K. Mraz, coincides to some extent with our procedure for compiling geobotanical indicator maps.

We carried out a program on the mapping of Quaternary deposits on the basis of geobotanical data in the broad-leaved forest zone with the aim of elucidating the possibilities of the indicator method in this region.

The work was carried out along the following lines.

Prior to departure for field work, a preliminary indicator table was compiled on the basis of published data, thus establishing possible plant communities which are indicators of lithological varieties of Quaternary deposits, at the same time collecting data on the ecology of the plants making up these communities.

In the field, a careful study was made of the distribution of plant communities in relation to particular environmental conditions (using the method of standard vegetational description, ecological profiling, and so on), and reliable indicator communities of different Quaternary deposits were identified. As a result of this work, we identified many new individual plant indicators as well as indicator communities, the indicational significance of which was in agreement with data in the literature.

On the basis of this type of detailed analysis of the vegetation cover, geologic indicator tables, were compiled which reflected the role in indicator research of the vegetation in relation to particular

TABLE 1. Scheme of the Relationship of Plant Communities with Different Types of Quaternary Deposits in the Broad-Leaved Forest Zone (fragment)

| Plant communities | Indicated lithological varieties of deposits | | | |
|---|---|---|---|---|
| | fluvioglacial | loess | eluvial-deluvial | alluvial |
| I. Pine forests | Sands, sometimes with eolic conversion | — | — | Sands |
| II. Dry meadow with dominance of Festuca ovina, Nardus stricta, Hieracium, Thymus | — | Sands | Sands | — |
| III. Pine forests with inclusion of broad-leaved species; beech-pine, oak-pine, hornbeam, beech, and oak forests with inclusion of pine | Sands and sandy loams, closely underlain by limestone, loam, carbonate sandstone | — | Sands and sandy loams, closely underlain by limestone or loam | — |
| IV. Willow groves, sometimes with Prunus spinosa; grass—mixed herbage meadows with smooth bromegrass, Anthoxanthum odoratum, and Agropyron repens | — | — | — | Sands, sometimes interstratification of sands and loams |
| V. Beech forests with sedge, Oxalis, Galeobdolon—Asperula and some other communities; dry meadows of mixed herbage and grasses with Agrostis vulgaris, Achillea millefolium, etc. | — | Loams, of substantial thickness, sometimes loams closely underlain by limestones, marl | — | — |
| VI. Hornbeam forests with sedge, mixed herbage and fern communities, hornbeam-beech forests with Asperula—mixed herbage communities; complex oakwoods | — | — | Loams, sometimes closely underlain by limestones, marl | — |
| VII. Teucrium—mixed herbage meadows, and meadows of Salvia verticillata with mixed herbage | — | — | Thin eluvium on limestone, marl | — |
| VIII. Floodplain grassy oakwood | — | — | — | Argillaceous sands, interstratification of sands and loams |
| IX. Oak forests, sometimes including spruce, beech, and hornbeam, and Molinia—Deschampsia, sedge—mixed herbage communities, etc. | — | — | — | Loams, sometimes with interlayers of shingle (ancient alluvial deposits) |

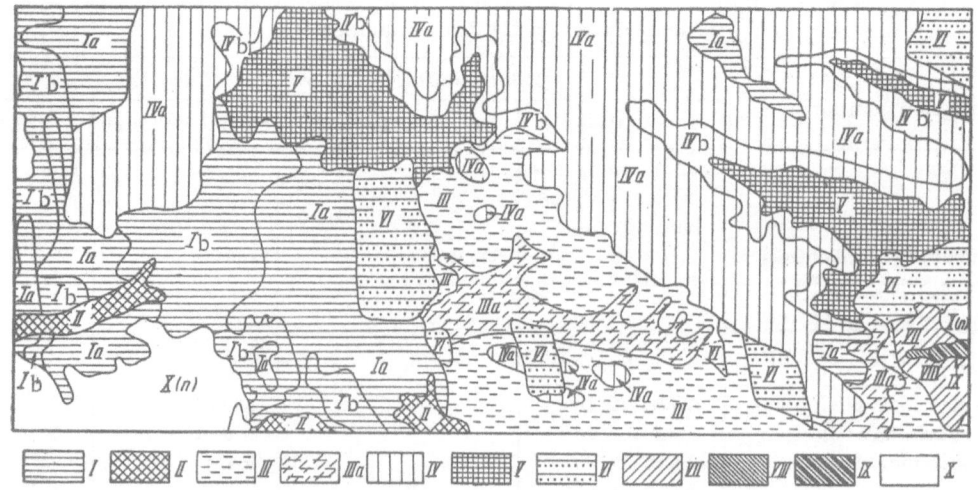

Fig. 1.  Scheme of the geobotanical indicator map(see Table 2).

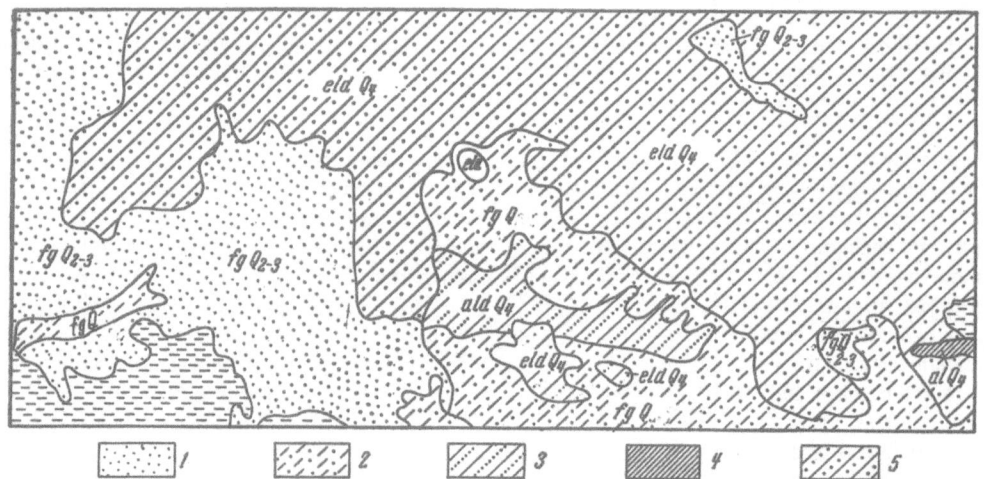

Fig. 2.  Map scheme of Quaternary deposits, compiled on the basis of geobotanical indicator investigations: 1) fluvioglacial small-grained sands ($fgQ_{2-3}$); 2) fluvio-glacial sands, closely underlain by loams or clays (fgQ); 3) alluvial-deluvial deposits; sandy loams, underlain by loams and clays ($aldQ_4$); 4) alluvial loams ($alQ_4$); 5) eluvial-deluvial sandy loams and sands on peaks and slopes of ridges and hillocks, closely underlain by limestones $eldQ_4$).

types of deposits.  A small part of a schematic generalized type of geologic indicator table is presented in  Table 1.

Simultaneously with the detailed study of vegetation, geobotanical mapping was carried out, as a result of which geobotanical indicator maps were  compiled, these being subsequently interpreted, in accordance with the geologic indicator table, into a map of Quaternary deposits.  For this purpose, the key to the geobotanical map includes, besides the designation of the communities, the conditions indicated by the communities, i.e., the key represents a special variant of the geobotanical table of plant communities.  In the course of mapping, use was made of different taxonomic units of the communities (associations, groups of associations).

Figures 1 and 2 represent the schematic geobotanical indicator map and the map of Quaternary deposits of the same site, compiled on the basis of the geobotanical indicator map (Table 2).

204

TABLE 2. Scheme of Geobotanical Indicator Research in the Mapping of Quaternary Deposits (see Fig. 2)

| Index | Plant community | Principal plants | Relief | Conditions indicated by the vegetation | | |
|---|---|---|---|---|---|---|
| | | | | lithological varieties of Quaternary deposits | Soils | depth of bedding of ground-waters, m |
| Ia | Hieracium—Festuca pine wood | Pine, Festuca ovina, Hieracium pilosella, Thymus | Upper and middle parts of slopes of ridges and hillocks | Fluvioglacial sands | Podzolic | More than 5 |
| Ib | Pine groves | — | — | Sands | — | — |
| II | Oak-pine forest with admixture of hornbeam, birch, and with strongly developed undergrowth | Pine, oak, hazel, buckthorn Nardus stricta, Vaccinium myrtillus, strawberry, bracken, Asarum, green mosses | Lower parts of slopes of ridges | Fluvioglacial sands, underlain by clays | Podzolic, gley, sandy | 1-3 |
| III | Mixed herbage-fescue oakwood with admixture or pine and strongly developed undergrowth | Oak, pine, Festuca rubra var. heterophylla, Majanthemum, Paris, Briza, etc. | Middle and upper parts of slopes and ridges | Fluvioglacial sands, underlain by loams | — | More than 3 |
| IIIa | Hornbeam-oak forest with admixture of pine and strongly developed undergrowth | Oak, hornbeam, pine, hazel, buckthorn, Vaccinium myrtillus, Majanthemum, etc. | Lower parts of slopes of ridges, bottom of troughs | Alluvial-deluvial sandy loams, underlain by loams and clays | Sod; slightly podzolic | Up to 3 |
| IVa | Sedge-mixed herbage beech wood with isolated specimens of pine | Beech, pine, hairy sedge, Asperula, pechenochnitsa, etc. | Watershed plateaus and upper parts of slopes | Eluvial-deluvial sandy loams and sands, closely underlain by limestones | Slightly and moderately podzolic sandy loam | More than 10 |
| IVb | Asperula beech wood with admixture of pine | Beech, pine, Oxalis, Majanthemum, Asperula, etc. | Slopes and bottoms of gullies | Eluvial-deluvial sandy loams on limestones and carbonate sandstone | Moderately and slightly podzolic gleying | 3-5 |
| V | Mixed herbage hornbeam forest with admixture of pine | Hornbeam, pine, Asperula, Vaccinium myrtillus, Athyrium, Asarum, etc. | Slopes and watersheds | Eluvial-deluvial sands and sandy loams of limestones | Humified-carbonate, slightly podzolic | More than 3 |
| VI | Glades, overgrown with brush made up of oak, hornbeam and, more rarely, beech | — | — | Same | — | — |
| VII | Hieracium—fescue meadow | Festuca ovina, Hieracium pilosella, Anthoxanthum odoratum | Upper parts of slopes of ridges and hillocks | Eluvial-deluvial sands and sandy loams | Podzolized | More than 10 |
| VIII | Anthoxanthum odoratum—avens meadow | Anthoxanthum odoratum, avens, clover | Slopes and bottom of trough | Alluvial-deluvial loams | Meadow | Up to 2 |
| IX | Mixed herbage—Deschampsia meadow | Deschampsia | Bottom of trough | Alluvial loams | Meadow-bog | 0.5-1.0 |
| X | Plowed land | — | — | — | — | — |

A comparison of the maps shows that geobotanical data permit the differentiation of the contours not only of distinct lithological types of deposits but also genetic types. It is in this context clear that each lithological type of deposit is indicated by a definite plant association or a specific group. Thus, fluvioglacial sands are distinguished by the distribution of different types of pine forest, whereas sands, closely underlain with limestones or loams, are distinguished by the distribution of hornbeam-oak and oak forests with a substantial proportion of pine, etc. Under these circumstances, the presence of pine in the tree stand is evidence of a sandy—sandy loam mechanical structure of surface deposits, while dominance of broad-leaved species (oak, hornbeam) is evidence of the proximity of loam or limestone. It should be noted that thin alluvial-deluvial deposits in shallow valleys and ravines do not influence the distribution of forest vegetation, which in this case is an indication of some other deposits.

We thus see that it is possible to map Quaternary deposits from the vegetation of the forest zone; the prospects are especially good in view of the application of aerial methods. The combination of aerial observations and the deciphering of aerial photographs with geobotanical indicational mapping should lead to a marked increase in the speed of geological survey work.

## Summary

The author presents a scheme for indicator investigations in the mapping of Quaternary deposits, which differs somewhat from the scheme of the Czech investigator Mraz. The first stage of the work consists of the construction of a preliminary indicator scheme (based on published and unpublished data). The previously established indicators are verified in the field by the method of standard plots and ecological profiles. Indicator mapping is widely used to obtain greater precision of the indicator scheme. This permits more rapid work than Mraz' system, in which mapping begins only after completion of the whole cycle of field work leading to the establishment of the indicator scheme.

## Literature Cited

Biské, G. S. 1949. "Experience in applying aerovisual observations in surveying Quaternary deposits in Karelia." Izv. Karelo-Finsk. filiala Akad. Nauk SSSR, No. 4.

Vorob'ev, D. V. 1953. Forest Types in the European Part of the USSR. Kiev, Izd. Akad. Nauk UkrSSR.

Voronkov, L. F. 1955. "Experience in using the geobotanical method in compiling a lithological map of ancient alluvial deposits." Tr. VAGT, No. 1, seriya geobot. Moscow, Gosgeoltekhizdat.

Ososkov, P. A. 1889. "Distribution of Lower Cretaceous iron-containing rocks in the region of the Zasursk forests." Materials toward an Understanding of the Geological Structure of the Russian Empire, No. 1. Izd. MOIP.

Pogrebnyak, P. S. 1944. Principles of Forest Typology. Kiev, Izd. Akad. Nauk UkrSSR.

Shvyryaeva, A. M. 1955. "Experience in using geobotanical features for differentiating lithologically similar strata of differing origin." Tr. VAGT, No. 1, seriya geobot. Moscow, Gosgeoltekhizdat.

Mraz, K. 1958 (1959). Vegetaйni typy jako pomůcka při mapovani kvarternich pokryvnych útvaru a zvětralionovych plaštů. Antropozoicum, 8.

# HYDROLOGIC INDICATOR PROPERTIES
# OF THE VEGETATION IN ZONES OF INADEQUATE MOISTURE
# AND THEIR REPRESENTATION IN AERIAL PHOTOGRAPHS

## T. A. Popova

The vegetation is an important component of the landscape and constitutes an important indicator of shallow-lying subsurface waters. Different plant species are associated with a specific depth of bedding and degree and type of mineralization of groundwaters; within the boundaries of a landscape, these associations are usually very constant.

The use of vegetation as a feature in searching for subsurface waters, using aerial survey data, depends on the possibility of deciphering vegetation. The main features of the vegetation which are reflected in aerial photographs and may have indicator significance are as follows: 1) species composition; 2) degree of mesomorphism; 3) degree and character of soil cover; 4) distribution of the species over the surface; 5) rhythm of development; 6) contour patterns of communities; 7) type of combinations of fragments and complete phytocenoses; and 8) relationship of plant communities with other elements of the landscape.

### Species Composition

The closest relationship with groundwaters is observed in phreatophyte plants, the roots of which absorb water from the capillary fringe or the zone of groundwater saturation. Determination of the species composition and information on the biological features of the plants permits an assessment to be made of the water regime of the habitat and the subsurface waters. In aerial photographs, trees and bushes can be deciphered by the type of crown projection, their tone, their intrinsic features, and the shading. In dry regions, these large living forms are sparsely distributed and are associated with sites where there are closely bedded groundwaters and thus their deciphering is of considerable interest.

Thus, Populus diversifolia Schrenk. grows best where there are fresh groundwaters bedded at a depth of 3-5 m (Akhmedsafin, 1947). P. diversifolia forms tugai forests along rivers and freshwater lakes in the zone supplied with river and lake waters, while in the proximity of outcrops of water sources, it occurs in the form of individual specimens. In large-scale aerial photographs,* P. diversifolia can be recognized by the large, rather arched projection of its crowns, which have a decorative, almost round shape and a protuberant-cellular appearance. The shading of the crowns is characterized by a broad, semicircular shape.

Haloxylon aphyllum (Minkw.) Iljin, has a tree-like appearance and reaches a height of 6-8 m in the presence of fresh or slightly mineralized groundwaters of the sulfatic type at a depth of 4-8 m. In aerial photographs, the crown projections of this plant have a peculiar shape and a grey tone and a dotted appearance. Dark shadows duplicate the form of the crowns, in which branching commences close to the ground.

Tamarisks are associated with fresh and slightly mineralized groundwaters bedded at a depth averaging 4-6 m. The vigorously branching dense tamarisk bushes display their large dimensions in

---

*We are referring in all cases to black-and-white photographs using panchromatic-type film.

large-scale aerial photographs and show a uniform circular shape and a slightly crenulate margin. The tone of images of the bushes is very dark, almost black.

Reeds in sands can be deciphered by their dark web-like appearance on a light background, this being caused by the characteristic growth of reed with its long, creeping rhizomes.

## Mesomorphism

The degree of mesomorphism of the species making up a community reflects their ecological characteristics and, consequently, the water regime of the root-inhabiting horizon of the soil.

Xerophytes in the steppe and desert zones are associated with plakor conditions and obtain an adequate supply of moisture from atmospheric precipitation. Groundwaters on these sites are bedded at a comparatively deep level and are characterized by high mineralization. The overwhelming majority of xerophytes have a xeromorphic structure: well-developed palisade tissues and covering structure, reduced leaves, and a low content of chlorophyll. All these characteristics contribute to the fact that xerophytes have pale-grey and grey tones in aerial photographs.

Mesophytes in districts with a dry climate are associated with sites having, in addition to atmospheric precipitation, a supplementary moisture supply drawn from ground or flood waters. Mesophytes typically have a mesomorphic structure; considerable foliar development, large soft leaves with a well-developed spongy parenchyma and a high chlorophyll content, and a deep green color (Ivanov, 1931). Their images in aerial photographs have a dark or very dark tone.

Hygrophytes, with a hygromorphic structure, are confined to places with surplus moisture, and in their external features are similar to mesophytes, their images in aerial photographs also showing dark tones.

## Projective Covering

The majority of the southern soils (serozems, light-chestnut soils, and solonetzes) give a very pale tone when in the dry condition. Accordingly, the higher the projective plant covering, the darker is the tone of the photographic image. Meadow soils have a darker tone than steppe and desert soils, but they also are paler, when in a dry condition, than the mesophilic vegetation growing on them. The same pattern is thus retained: the denser the vegetation, the greater is the projective covering and the darker is the tone on the photograph. In southern regions, where the development of vegetation is limited by deciphering of moisture, increased humidity promotes the formation of particularly dense communities with a high projective covering and composed of the most mesophilic species. In extreme conditions, with a very marked water deficiency, very thin communities, made up of typical xerophytes, are developed. As a result of a combination of factors, therefore, the most humid localities are represented by a dark tone, while the dry habitats have a pale tone.

Different patterns are observed in hygro-hydrophilic vegetation, where the background image to the vegetation is provided not by the soil but by the water surface. In cases where the photograph is taken with the sun at a high position or when the bottom of the reservoir has a dark color, the water surface is represented by a very dark or black tone, while the dense vegetation has a lighter tone. Compact growths of reeds in limans are represented in such cases by a paler tone than the open water surface.

## Distribution of Species over the Surface

Each plant species typically has a characteristic morphological structure and thus uneven distribution of the species in a community determines the type of image. The diffuse distribution of specimens in dense monocenoses or cenoses of morphologically similar species results in the formation of a uniform tone. Monocenoses of large plants give a small uniform figure. In complex phytocenoses, any type of group distribution, mosaicism or spacial sinusoid pattern leads to a particular form of image.

208

## Rhythm of Development

Different vegetational types differ from one another by their rhythm of seasonal development. For example, steppe communities are characterized by early-spring development and summer dormancy, while southern-desert communities show the development of early-spring sinusia of ephemerals and ephemeroids and have a brief growth period. Meadows, on the other hand, develop uniformly and for a long period. In dry regions, communities of phreatophytes show particularly prolonged vegetative growth. While the entire plakor vegetation fades, phreatophyte communities remain green and fresh. The darkest shades in the late summer and autumn aerial photographs correspond to communities of phreatophytes.

## Form of Contours

The shape of contours in developed communities is determined by the specific habitat. In dry-climate regions, the shape of contours for the majority of communities duplicates the shape of the habitat in relation to moisture conditions. Thus, the straight-line distribution of tamarisk bushes in deserts is frequently associated with the lines of tectonic disturbances, along which groundwaters can seep (Miroshnichenko, 1954; Viktorov, Vostokova, Voronkova, 1955; Vinogradov, 1955). In semi-desert regions, where local infiltration lenses of fresh groundwaters lie on a background of mineralized waters, the form of the lenses is duplicated by communities of perennial mesophytes and, especially, phreatophytes. In foothills and small-hummock plains, outcrops of water sources are confined to the lower parts of ravines, and we find developing around the sources communities of mesophytes, whose contours are rather triangular in shape. The form of contours of tamarisk tugaic derivatives along dry riverbeds marks the location of groundstreams. In sandy massifs, the contour form of mesophilic communities indicates the zones of seepage of fresh groundwaters.

## Type of Combinations of Fragments and Communities

A particular type of combination of fragments of plant communities in complexes corresponds to definite habitat and thus types of complexes can be examined as hydrologic indicators. The classification of types of complexes suggested by E. I. Ivanova and V. M. Fridland (1954) is of particular interest in this context.

According to this classification, four classes of soil complexes are differentiated in the dry steppes and semi-deserts. They are genetically associated with the development of vegetation and each is characterized by a specific type of water regime and depth of water table. These complexes can be deciphered in aerial photographs by the type of image. M. S. Simakova (1959) noted that a striated pattern is typical for meadow complexes. We attained similar observations on the territory of the ancient delta of the Ural River. Meadow-steppe and steppe complexes produce a patch picture.

## Relationship of Plant Communities with Other Elements of the Landscape

Regular relationships of vegetation with the relief, soils, and zoöcenoses, permit the use of elements of the landscape as direct deciphering features for the vegetation.

Knowledge of biotic relationships justifies the use of zoöcenotic features for deciphering vegetation. Thus, in the area of the Sarpinsk lakes, the widespread distribution of gopher warrens is associated with Artemisia pauciflora Web. growths. This species is not connected with groundwaters, and grows on solonetzes, under which there lie groundwaters of high mineralization having no contact with the surface through upward capillary movement. A. pauciflora thus provides indirect evidence of the absence of groundwaters, in the proximity. The presence of gopher warrens is of substantial assistance in deciphering A. pauciflora growths. Heaps of excavated soil are usually surrounded by a fringe of rather dense vegetation, and therefore appear in aerial photographs as pale dots, surrounded by a dark ring.

Soils can provide important information in several circumstances. Quackgrass growths of Agropyron repens (L.) P.B. are widely distributed in sinkholes on the right bank of the Volga. These growths are represented by different associations; those of the glycophytic series indicate the presence of lenses of fresh waters, while those of the halophytic series indicate the presence of lenses of groundwaters with somewhat higher mineralization. Deciphering quackgrass growths of the different series is effected on the basis of the tone, and also from the specific finely reticulated picture, which is produced as a result of disintegration of the surface of the solonetz soil.

The relief can provide decisive information. Within the boundaries of the ancient delta of the Ural River, quackgrass growths of A. repens indicate the presence of various groundwaters. It is obvious on land that these associations are also different, but the distinctions cannot be discerned in aerial photographs. For purposes of deciphering, we must bring in indirect features, and, in particular, the relief. Quackgrass growths with an admixture of Artemisia lercheana indicate mineralized waters, while pure quackgrass growths with a high projective covering indicate fresh waters. In aerial photographs, these quackgrass growths cannot be distinguished by photograph tones, but can be differentiated by their position on the relief; the former are confined to enclosed sinkholes, while the latter are associated with watercourse-like hollows, connecting large flat depressions, the local drainage systems.

## Summary

Characteristics essential to the deciphering of subsurface waters by their vegetation cover include the following: 1) species composition; 2) degree of mesomorphism; 3) degree and character of the soil cover; 4) distribution of the species over the surface; 5) rhythm of development; 6) contour patterns of communities; 7) type of combinations of fragments and complete phytocenoses; and 8) relationship of plant communities with other elements of the landscape. To illustrate these aspects, the author examines various cases of hydrogeological indicator research. Particular attention is paid to the hydrologic indicators Haloxylon aphyllum, Populus diversifolia, Agropyron repens, and Tamarix spp.

## Literature Cited

Viktorov, S. V., Vostokova, E. A., and Voronkova, L. F. 1955. "The use of geobotanical characteristics for locating tectonic disturbances." Tr. VAGT, No. 1, seriya geobot.

Vinogradov, B. V. 1955. "Examples of the relationship of vegetation and soils with recent tectonics." Bot. zhur., Vol. 40, No. 6.

Ivanov, L. A. 1931. Plant Physiology. Moscow—Leningrad, Sel'khozgiz.

Ivanova, E. N., and Fridland, V. M. 1954. "Soil complexes of dry steppes and their evolution." In book: Problems in the Improvement of the Fodder Basis in the Steppe, Semi-Desert, and Desert Zones of the USSR. Moscow, Izd. Akad. Nauk SSSR.

Miroshnichenko, V. P. 1954. "Experience in developing and applying aerial methods for studying recent and present-day tectonic movements within the boundaries of foothill plains of the accumulative-eolic type." Tr. Labor. aérometodov Akad. Nauk SSSR, III, Moscow, Izd. Akad. Nauk SSSR.

Simakova, M. S. 1959. "Procedures in mapping soils in the Caspian depressions on the basis of aerial photographic data." In book: Soil and Geographic Investigations and the Use of Aerial Photography in Mapping Soils. Moscow, Izd. Akad. Nauk SSSR.